Microfluidic Bio-Sensors and Their Applications

Microfluidic Bio-Sensors and Their Applications

Editor

Krishna Kant

Basel • Beijing • Wuhan • Barcelona • Belgrade • Novi Sad • Cluj • Manchester

Editor
Krishna Kant
Universidade de Vigo
Vigo, Spain

Editorial Office
MDPI
St. Alban-Anlage 66
4052 Basel, Switzerland

This is a reprint of articles from the Special Issue published online in the open access journal *Biosensors* (ISSN 2079-6374) (available at: https://www.mdpi.com/journal/biosensors/special_issues/microfluidic_bio_sensors).

For citation purposes, cite each article independently as indicated on the article page online and as indicated below:

Lastname, A.A.; Lastname, B.B. Article Title. *Journal Name* **Year**, *Volume Number*, Page Range.

ISBN 978-3-0365-8860-5 (Hbk)
ISBN 978-3-0365-8861-2 (PDF)
doi.org/10.3390/books978-3-0365-8861-2

Cover image courtesy of Krishna Kant

Contents

About the Editor

Krishna Kant

Krishna Kant is a Marie-Curie Researcher and a A/Professor at the Center for Research in Nanomaterials and Biomedicine, University of Vigo, Spain. He is an expert in the field of Microfluidics and Biosensing for the Design and Manufacture of Micro-/Nanobiomedical devices. He received his Master's in Technology (M.Tech.) in Nanobiotechnology from Amity University in 2009, and his Ph.D. in Nanotechnology from Flinders University, Australia, in 2014. His research interests include microfluidics, cell separation, and point-of-care testing devices. He has published over 40 peer-reviewed articles. His research has gained increasing recognition through important awards and honors such as the MSCA individual fellowship, the Lady Davis fellowship in postdoctoral research, and the ARC fellowship during his Ph.D. He has also received awards from various institutions in Hackathons and the EMBO visiting fellowship.

Editorial

Microfluidic Bio-Sensors and Their Applications

Krishna Kant [1,2]

[1] Biomedical Research Center (CINBIO), University of Vigo, 36310 Vigo, Spain; krishna.kant@uvigo.gal
[2] Centre for Interdisciplinary Research and Innovation (CIDRI), University of Pertoleum and Energy Studies, Dehradun 248007, India

Citation: Kant, K. Microfluidic Bio-Sensors and Their Applications. *Biosensors* **2023**, *13*, 843. https://doi.org/10.3390/bios13090843

Received: 17 August 2023
Accepted: 23 August 2023
Published: 24 August 2023

Biosensors are a promising tool for a wide variety of target analyte detection and enable point-of-care diagnostics with reduced volume and space. The integration of sensing elements into microfluidic systems has attained great interest in research. Due to the advanced manufacturing and availability of sensor systems, they have become very cost-effective to produce. Improvements in the field of designing and fabricating microfluidic chips have been significant in recent years. The use of microfluidic systems with integrated sensors have been developed by various new cost-effective approaches like 3D-printed microfluidics, polydimethylsiloxane (PDMS) soft lithography, and laser cutting and micro-milling. These microfluidic chips have been used with various technologies like electrochemical, optical, fluorescence, etc., for their sensing application. These techniques have also forced researchers to alter these microfluidic sensor systems according to changing experimental requirements in a simple and cost-effective manner. Recently, the enhancement of wearable microfluidic- and smartphone-based sensors has become a developing research field for promising widespread point-of-care testing. However, microfluidic chips and developed applications remain in their initial development phases, and they have not yet been largely commercialized. Microfluidic and biosensor advancements are growing hand in hand, both in research and product development, and thus their miniaturization for use in commercial products is also taking place. Microfluidic technology and integrated biosensing represent future goals in this area. Investigating the integration of artificial intelligence and electronics sensor with biological elements to better understand and mimic natural systems. This will eventually enable the development of new products for mankind and society. This Special Issue covers the dedicated innovations across a variety of topics in this area, from sensing to manufacturing and implementation methods to novel microfluidic-based sensors for biological purposes. Researchers have reported on the latest developments in microfluidic-based sensing in glucose, graphene application and applications for cancer cells. The multiplexed sensors and other types of integrated sensor including electrochemical, optical, magnetic, and other transduction types with microfluidics chips are highlighted for the interest for early career researchers.

This Special Issue contains ten excellent papers (eight research articles and two review papers). The research articles focus on the microfluidic-assisted synthesis of metal organic frameworks (MOFs) for sustained drug delivery [1]. It explores the high surface area and addition of functional surfaces to make them ideal drug delivery vehicles. The first article concerns glucose biosensor, presenting the quantification, excellent linearity, temperature calibration, and real-time detection using a resistor and capacitor, combined with a PDMS microfluidic channel [2]. Another article presents an RFID-based glucose microwave biosensor and achieves a non-contact measurement of the concentration of glucose [3]. These studies present innovative applications of microfluidic and biosensors for glucose. Similarly, a graphene oxide-based electrode is presented, promising low-cost and easy fabrication of sensors sensitive for selective label-free DNA biosensing [4].

One of the research articles presents integrated microfluidic sensors for rapid analyte detection on antibody-decorated beads for two acute kidney injury biomarkers [5]. A novel silicon-based lab-on-chip sensor to perform low-density and high-resolution multi-assay analysis for DNA amplification and hybridization is also presented to complement this Special Issue [6]. In the study 'Surfactant-Based Microdroplets and Monitoring Bacterial Gas Metabolism with Droplet-Based Microfluidics', the authors confirmed that gas cross reaction took place between droplets formed by fluorinated oil and the PFPE–PEG–PFPE surfactant [7]. This allowed the researchers to choose the substrate and material accordingly to avoid any cross reactivity. A further application of microfluidic chips has been demonstrated in cellular study, where an optically induced dielectrophoresis technique was used to alter the cell microenvironment and functionalities [8]. The published review articles focus upon the field of cancer detection. Two different approaches (paper-based and microfluidic channel-based) are discussed, including their advancements in cancer biomarker detection and in capturing CTC cells [9,10].

With the collection of papers in this Special Issue, we have taken a step towards the advancement of new microfluidic devices for various microfluidic-based biosensing applications.

Funding: Krishna Kant acknowledges the European Union's Horizon 2020 research and innovation program under the Marie Sklodowska-Curie grant agreement no. (894227).

Acknowledgments: The authors are grateful for the valuable opportunity to serve as Guest Editors of this Special Issue, "Microfluidic Biosensors and Their Applications". We are thankful to all the authors for their contributions to this Special Issue. We also thank the editorial assistance and publishing staff of *Biosensors* for their support.

Conflicts of Interest: The authors declare no conflict of interest.

References

1.   Bendre, A.; Hegde, V.; Ajeya, K.V.; Thagare Manjunatha, S.; Somasekhara, D.; Nadumane, V.K.; Kant, K.; Jung, H.Y.; Hung, W.S.; Kurkuri, M.D. Microfluidic-Assisted Synthesis of Metal—Organic Framework—Alginate Micro-Particles for Sustained Drug Delivery. *Biosensors* **2023**, *13*, 737. [CrossRef] [PubMed]

2.   Ma, Y.; Qiang, T.; Gao, M.; Liang, J.; Jiang, Y. Quantitative, Temperature-Calibrated and Real-Time Glucose Biosensor Based on Symmetrical-Meandering-Type Resistor and Intertwined Capacitor Structure. *Biosensors* **2021**, *11*, 484. [CrossRef] [PubMed]

3.   Gao, M.; Qiang, T.; Ma, Y.; Liang, J.; Jiang, Y. RFID-Based Microwave Biosensor for Non-Contact Detection of Glucose Solution. *Biosensors* **2021**, *11*, 480. [CrossRef] [PubMed]

4.   Asefifeyzabadi, N.; Holland, T.E.; Sivakumar, P.; Talapatra, S.; Senanayake, I.M.; Goodson, B.M.; Shamsi, M.H. Sequence-Independent Dna Adsorption on Few-Layered Oxygen-Functionalized Graphene Electrodes: An Electrochemical Study for Biosensing Application. *Biosensors* **2021**, *11*, 273. [CrossRef] [PubMed]

5.   Thiriet, P.E.; Medagoda, D.; Porro, G.; Guiducci, C. Rapid Multianalyte Microfluidic Homogeneous Immunoassay on Electrokinetically Driven Beads. *Biosensors* **2020**, *10*, 212. [CrossRef] [PubMed]

6.   Iemmolo, R.; La Cognata, V.; Morello, G.; Guarnaccia, M.; Arbitrio, M.; Alessi, E.; Cavallaro, S. Development of a Pharmacogenetic Lab-on-Chip Assay Based on the In-Check Technology to Screen for Genetic Variations Associated to Adverse Drug Reactions to Common Chemotherapeutic Agents. *Biosensors* **2020**, *10*, 202. [CrossRef] [PubMed]

7.   Ki, S.; Kang, D.K. Gas Crosstalk between PFPE–PEG–PFPE Triblock Copolymer Surfactant-Based Microdroplets and Monitoring Bacterial Gas Metabolism with Droplet-Based Microfluidics. *Biosensors* **2020**, *10*, 172. [CrossRef] [PubMed]

8.   Chu, P.Y.; Hsieh, C.H.; Lin, C.R.; Wu, M.H. The Effect of Optically Induced Dielectrophoresis (ODEP)-Based Cell Manipulation in a Microfluidic System on the Properties of Biological Cells. *Biosensors* **2020**, *10*, 65. [CrossRef] [PubMed]

9.   Dukle, A.; Nathanael, A.J.; Panchapakesan, B.; Oh, T.H. Role of Paper-Based Sensors in Fight against Cancer for the Developing World. *Biosensors* **2022**, *12*, 737. [CrossRef] [PubMed]

10.   Bhat, M.P.; Thendral, V.; Uthappa, U.T.; Lee, K.H.; Kigga, M.; Altalhi, T.; Kurkuri, M.D.; Kant, K. Recent Advances in Microfluidic Platform for Physical and Immunological Detection and Capture of Circulating Tumor Cells. *Biosensors* **2022**, *12*, 220. [CrossRef] [PubMed]

biosensors

Article

Microfluidic-Assisted Synthesis of Metal—Organic Framework—Alginate Micro-Particles for Sustained Drug Delivery

Akhilesh Bendre [1], Vinayak Hegde [1], Kanalli V. Ajeya [2], Subrahmanya Thagare Manjunatha [3], Derangula Somasekhara [4], Varalakshmi K. Nadumane [4], Krishna Kant [5,*], Ho-Young Jung [2,*], Wei-Song Hung [3,*] and Mahaveer D. Kurkuri [1,*]

[1] Centre for Research in Functional Materials (CRFM), JAIN (Deemed-to-be University), Jain Global Campus, Bengaluru 562112, Karnataka, India; akhilesh.bendre@jainuniversity.ac.in (A.B.); vinayakhegde2020@gmail.com (V.H.)
[2] Department of Environment and Energy Engineering, Chonnam National University, 77 Yongbong-ro, Buk-gu, Gwangju 61186, Republic of Korea; ajeyhegde94@gmail.com
[3] Advanced Membrane Materials Research Center, Graduate Institute of Applied Science and Technology, National Taiwan University of Science and Technology, Taipei 10607, Taiwan; mssubrahmanya@gmail.com
[4] Department of Biotechnology, JAIN (Deemed-to-be-University), School of Sciences, JC Road, 34, 1st Cross Road, Sudharna Nagar, Bengaluru 560027, Karnataka, India; d.somasekhara@jainuniversity.ac.in (D.S.); kn.varalakshmi@jainuniversity.ac.in (V.K.N.)
[5] Biomedical Research Center (CINBIO), University of Vigo, 36310 Vigo, Spain
[*] Correspondence: krishna.kant@uvigo.es (K.K.); jungho@chonnam.ac.kr (H.-Y.J.); wshung@mail.ntust.edu.tw (W.-S.H.); mahaveer.kurkuri@jainuniversity.ac.in (M.D.K.)

Abstract: Drug delivery systems (DDS) are continuously being explored since humans are facing more numerous complicated diseases than ever before. These systems can preserve the drug's functionality and improve its efficacy until the drug is delivered to a specific site within the body. One of the least used materials for this purpose are metal—organic frameworks (MOFs). MOFs possess many properties, including their high surface area and the possibility for the addition of functional surface moieties, that make them ideal drug delivery vehicles. Such properties can be further improved by combining different materials (such as metals or ligands) and utilizing various synthesis techniques. In this work, the microfluidic technique is used to synthesize Zeolitic Imidazole Framework-67 (ZIF-67) containing cobalt ions as well as its bimetallic variant with cobalt and zinc as ZnZIF-67 to be subsequently loaded with diclofenac sodium and incorporated into sodium alginate beads for sustained drug delivery. This study shows the utilization of a microfluidic approach to synthesize MOF variants. Furthermore, these MOFs were incorporated into a biopolymer (sodium alginate) to produce a reliable DDS which can perform sustained drug releases for up to 6 days (for 90% of the full amount released), whereas MOFs without the biopolymer showed sudden release within the first day.

Keywords: adsorption; MOFs; microfluidic chip; drug delivery; sodium alginate

Citation: Bendre, A.; Hegde, V.; Ajeya, K.V.; Thagare Manjunatha, S.; Somasekhara, D.; Nadumane, V.K.; Kant, K.; Jung, H.-Y.; Hung, W.-S.; Kurkuri, M.D. Microfluidic-Assisted Synthesis of Metal—Organic Framework —Alginate Micro-Particles for Sustained Drug Delivery. *Biosensors* **2023**, *13*, 737. https://doi.org/10.3390/bios13070737

Received: 4 May 2023
Revised: 23 June 2023
Accepted: 12 July 2023
Published: 17 July 2023

1. Introduction

The administration and delivery of therapeutic and diagnostics agents to an afflicted site in the body is one of the major challenges in the treatment of various diseases. Conventional methods of drug administration involve various direct administration strategies, such as oral, ocular, transdermal, and intravenous methods of delivering drugs without any provision of a "carrier". These methods are sometimes ineffective as they suffer various limitations, such as loss of drug function and efficacy, decreased selectivity and transport to the target site, and undesirable effects on other untargeted tissues. Drug delivery systems (DDS) can overcome all these limitations. Drug delivery systems (drug carriers) maintain the physiochemical properties of the drug under the biological environment of the body and can provide a controlled sustained release of drug molecules over a prolonged time [1].

Some of these carriers can also provide secondary effects, such as optical diagnostics and anti-microbial properties. Based on their origin, the drug delivery systems can be categorized as organic, inorganic, and hybrid systems [2–4]. Among the inorganic materials used, one such drug delivery system uses high-surface-area materials, such as metal—organic frameworks (MOFs), for drug delivery.

MOFs are high-surface-area materials (usually >1000 m^2/g) that can adsorb various materials onto their surface. Due to their extensive physical and chemical properties, MOFs are ideal candidates for various applications, including gas adsorption, energy storage, catalysis, wastewater treatment, and biomedical applications [5–7]. Their porous structure and their capacity for surface modification make MOFs potential drug carriers [8–12]. Among various known MOFs, Zeolitic Imidazole Framework-67 (ZIF-67) is one of the least employed frameworks in drug delivery [13–15]. ZIF-67 is known to have a very high surface area as well as a simple room-temperature synthetic route. Batch synthesis methods, such as room-temperature direct mixing, solvothermal, and hydrothermal methods, are generally used in the preparation of such MOFs, but recently, microfluidic platforms have been employed to synthesize nanoparticles, including materials such as MOFs [16–18]. ZIF-67 contains cobalt (Co^{2+}) ions, which can have a cytotoxic effect on the cells, but the addition of other metals such as zinc (Zn^{2+}), which is biologically relevant, can improve the physical stability of ZIF-67 without compromising the desired qualities. The microfluidic approach is a more controlled method of nanoparticle synthesis when compared to bulk methods [19]. It provides proper control and regulation of the mixing strategies, flow parameters, and flow rates used to control the physiochemical characteristics of the synthesized nanoparticles. Properties such as crystal size, growth rate, surface functions, and synthesis characteristics such as continuous production, scalability, reproducibility, and uniformity can be controlled and varied according to the function of the nanoparticles and the cost–time constraints of the synthesis process [20–22]. Further modification, such as encasement in an organic polymer matrix, helps further preserve the carrier and hence the drug while providing a platform for further modification [23–27]. Sodium alginate is a prime candidate for such application as it is non-toxic, anionic, natural, abundant, and highly biodegradable and biocompatible and has a hydrophilic nature and high porosity and mechanical strength.

The present article aims to provide a comprehensive understanding of the microfluidic synthesis of ZIF-67 as well as its bimetallic derivative Zn-ZIF-67, comparing it to the bulk synthesized ZIF-67 in terms of its structure and drug adsorption/loading properties. Further, sodium alginate beads were synthesized, drug-loaded MOFs were incorporated inside the beads, and the drug release profiles were studied. There have been a very limited number of studies regarding the microfluidic synthesis of ZIF-67. We present a novel synthesis of its bimetallic variant for the first time and its combination with sodium alginate for drug delivery applications. This work aims to increase the understanding of such materials and their potential as functional drug delivery agents.

2. Materials and Methods

2.1. Materials

The polydimethylsiloxane (PDMS) (SYLGARD 184 silicone elastomer and curing agent) used for device fabrication was purchased from Dow Corning Pvt. Ltd., Bangalore, Karnataka, India. The primary metal salt, Cobalt (II) Nitrate hexahydrate ($Co(NO_3)_2.6H_2O$), and linker, 2-Methylimidazole (2-MIM), were purchased from AVRA Pvt. Ltd., Bangalore, Karnataka, India. The secondary metal salt, Zinc Nitrate hexahydrate ($Zn(NO_3)_2.6H_2O$), metal salt (linker), Calcium chloride ($CaCl_2$), and biopolymer, sodium alginate, were purchased from NICE Chemicals Pvt. Ltd., Bangalore, Karnataka, India. The drug diclofenac sodium was purchased as tablets from CIPLA, and the tablets were crushed into a powdered form. The MTT assay kit (EZcount MTT Cell Assay Kit) was purchased from HiMedia Laboratories Pvt. Ltd., Bangalore, Karnataka, India. Dulbecco's Modified Eagle Medium (DMEM), Dimethyl Sulfoxide (DMSO), and phosphate-buffered saline (PBS) buffer were purchased from Sigma Aldrich (India) Pvt. Ltd., Bangalore, Karnataka, India.

MCF-7 breast cancer cells were acquired from the Department of Biotechnology, JAIN (Deemed-to-be-University), Karnataka, India.

2.2. Microfluidic Device Fabrication

The PDMS microfluidic synthesis chip (as shown in Figure 1) was fabricated using the standard photolithographic technique with a channel width and depth of 150 and 100 µm, respectively, with the design based on a microfluidic device for fluoride detection previously reported by our group [28]. There are two inlets and one outlet with 1 mm polytetrafluoroethylene (PTFE) tubings inserted into them. The tubings are connected to their respective syringes (or let into a glass beaker for the outlet tubing). The 'S'-shaped part of the chip promotes the vigorous mixing of reactants; the cylindrical region (diameter = 2 mm) present after the mixing of the reactants promotes further mixing and might cause larger, heavier-sized particles (>1–2 µm) to settle down due to loss of fluid velocity and also promote the hydrodynamic focussing of homogenous particles towards the outlet.

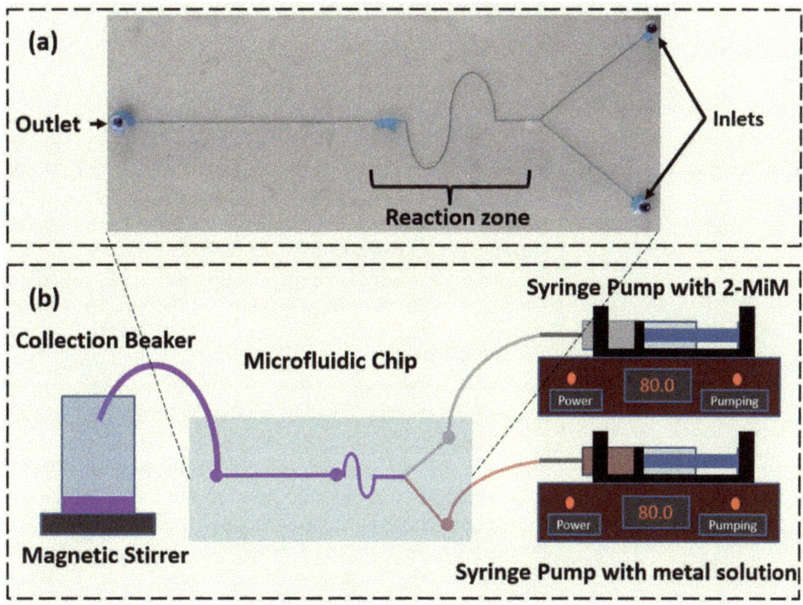

Figure 1. The (**a**) microfluidic chip with the inlets, channels, and reaction zone shown (dyed with methylene blue dye) and (**b**) setup for the synthesis of MOF materials.

2.3. Synthesis Methods

2.3.1. Synthesis of ZIF-67 and Zn-ZIF-67

Conventional synthesis of ZIF-67 was performed by dissolving 897 mg of $Co(NO_3)_2.6H_2O$ and 1982 mg of 2-MIM in 30 mL of methanol each via sonication. Both solutions were mixed in a beaker and stirred continuously for 24 h. The precipitate was collected, washed with methanol, and dried in a hot air oven overnight. A Y-channel with a serpentine mixing region was used for the microfluidic synthesis of ZIF-67 and Zn-ZIF-67 (MZIF-67 and MZnZIF-67, respectively) (Figure S1). For ZIF-67, 448 mg of $Co(NO_3)_2.6H_2O$ and 991 mg of 2-MIM were dissolved in 30 mL of methanol. $Co(NO_3)_2.6H_2O$ solution was passed through one inlet and 2-MIM solution through the other inlet with a flow rate of 80 µL/min. The product solution was collected from the outlet, washed with methanol, and dried overnight. For Zn-ZIF-67, $Co(NO_3)_2.6H_2O$ salt and zinc nitrate salt were taken in the molar ratio of

1:0.4 (582 mg:238 mg); the salts were dissolved in methanol. Similar to ZIF-67 microfluidic synthesis, salt solution and linker solution were pumped at 80 µL/min (flow rate was chosen based on synthesis time and yields obtained), and the product was collected at the outlet (as shown in Figure 1). The product was washed multiple times with ethanol and dried overnight [16,18,29]. The concentrations of materials were chosen based on the assumption that there should be a minimum 'build-up' of material in the channel to prevent its failure, while also considering a reasonable yield of material produced. The flow rate was chosen after conducting a time versus chip failure (blockages, leakages) probability after performing five trials of flow analysis for 30 min at 40, 60, 80, and 100 µL/min. It was observed that a flow rate of 80 µL/min produced the lowest number of chip failures for a decent amount of material produced.

2.3.2. Drug Loading

Diclofenac sodium (DS) was chosen as the model drug. It shows a characteristic peak at around 278 nm under UV–visible spectrophotometry. Various drug concentrations, namely 40, 50, 60, 70, 80, 90, and 100 ppm, were prepared by mixing the drug in de-ionized (DI) water.

2.3.3. MOF–Alginate Bead Synthesis

The bead formation method involves the addition of 5 mg of MOF to 2.5 mL of 90 ppm drug solution followed by agitation for 2 h for adsorption. The resultant solution was mixed with 2.5 mL of 10% (by w.t.) sodium alginate (SA) solution to prepare a 5% SA solution. This solution was pumped at 100 µL/min to produce small-sized drops which were dropped into a linker solution containing 5% $CaCl_2$ (by w.t.) (Figure S2). The beads were kept in the linker solution for 6 h for optimal linkage. The beads were then washed with water and dried in an oven overnight. The drying consequently reduced the size and weight of the beads to approximately 1 mm and 1 mg in diameter and weight, respectively, which theoretically should contain 25 µg of MOF per bead.

2.4. Characterization

All the materials (ZIF-67, MZIF-67, and MZnZIF-67) were analysed with various analytical techniques to understand their physical and chemical properties. The surface morphology of ZIF-67, MZIF-67, and MZnZIF-67 was determined using the field-emission scanning electron microscopic (FESEM) technique (HITACHI SU-70). Brunauer, Emmett, and Teller (BET) analysis was carried out to determine the multipoint surface area, and a microporous (MP) study was conducted to calculate the total pore volume. An adsorption–desorption study was performed using N2 at liquid nitrogen temperature ($-196\ ^\circ$C) on Belsorp-Max (M/s. Microtarc BEL, Osaka, Japan). During the sample analysis process, all the materials were subjected to a degassing process to expel the moisture content present at $100\ ^\circ$C for 2 h. To study the structural features, powder X-ray diffraction (XRD) patterns for the samples were recorded on an Ultima-IV X-ray diffractometer (M/s. Rigaku Corporation, Tokyo, Japan) with Ni-filtered Cu Kα radiation (1 = 1.5406 A°) with a 2θ scan speed of 2 degrees/min and a scan range of 5 to 80 degrees at 40 KV and 30 Ma. Zetasizer Nano ZS—ZEN3600 was used to perform zeta analysis of the samples at 7 pH. X-ray photoelectron spectroscopy (XPS) was used to analyse the chemical states of the elements in the sample using VG multi-lab 2000, which was operated at 3.125 meV using an Al-Kα as the energy source. The MTT assay was carried out using a microplate reader (Molecular Devices SpectraMax ABS Plus, TCL Asset Group Inc. Concord, Canada).

2.5. Adsorption Studies

Adsorption studies were performed by changing the two main reaction parameters, such as concentration and time of contact. The concentration studies were performed by adding the respective MOF powders to solutions with different drug concentrations, namely 40, 50, 60, 70, 80, 90, and 100 ppm, followed by shaking for 2 h. After 2 h, the solutions were centrifuged at 3000 rpm to separate the MOF powders and the supernatant was collected. The time studies were performed by adding MOF powders to 90 ppm solutions which were left for adsorption for up to 10, 20, 30, 40, 50, and 60 min, after which the solutions were centrifuged at 3000 rpm to separate the MOF powders and the supernatant was collected. The supernatant was analysed using a UV–visible spectrometer. The drug loading percentage (C_l was calculated using the following equation:

$$C_l = \frac{(C_i - C_e)}{C_i} \times 100$$

where C_i is the initial concentration and C_e is the drug concentration after adsorption.

2.6. Drug Release Studies

Drug release studies were performed by the addition of MOF powders and beads to phosphate-buffered saline (PBS, pH-7.4) for a time period of 3, 24, 48, 72, 96, 120, or 144 h, after which the supernatant was collected and analysed using a UV–visible spectrometer. The drug release percentage (C_r) was calculated using the following equation:

$$C_r = \frac{C_e}{C_i} \times 100$$

where C_i is the initial concentration and C_e is the drug concentration after release.

2.7. Cytotoxicity Studies

A 48 h MTT assay was conducted with MCF-7 breast cancer cells in a 96-well plate (200 µL). The materials were dispersed (partially dissolved) in a PBS–ethanol mixture with a volumetric ratio of 1:0.05 of PBS to ethanol to enhance the materials' solubility. A 100 µL solution of an equal population (10,000 cells) of MCF-7 cells in DMEM (with the addition of streptomycin to prevent bacterial growth) was filled in each well, and each plate (one for 24 h and the other for 48 h) was incubated for 24 h. After 24 h, each well was noted, and control populations along with 10 µL of solution with a known concentration of the materials were added to the wells (same volume of blank solution was added to the control wells). The endpoint masses of materials in the wells were 0.003 mg, 0.004 mg, 0.005 mg, 0.006 mg, and 0.007 mg for ZIF-67 and 0.0075 mg, 0.01 mg, 0.0125 mg, 0.015 mg, and 0.0175 mg for MZIF-67 and MZnZIF-67. The different concentrations were taken due to the low solubility/dispersibility of conventional ZIF-67 in PBS. After 24 h and 48 h, a 10 µL solution of MTT reagent (concentration of 5 mg/mL) was added to each well and left for 3 h for the cell–reagent interaction to take place [30]. After three hours, DMSO was added to the wells, the absorbance was measured using a microplate reader at 540 nm, and the OD values were noted. The assay could not be performed for the beads as even after the addition of streptomycin the solution in the wells became turbid and produced improper OD values.

3. Results

3.1. Characterization

The FESEM images of the single crystals of the MOF variants can be seen with their respective size bars in Figure 2. The comparison indicates that the crystal sizes of the batch produced and the microfluidically synthesized MOF variants have similar dimensions with a total length of 500 nm (with edge lengths of approximately 250 nm). The images show the crystals to have a similar structure as well as a cubic symmetry, indicating the formation of ZIF-67.

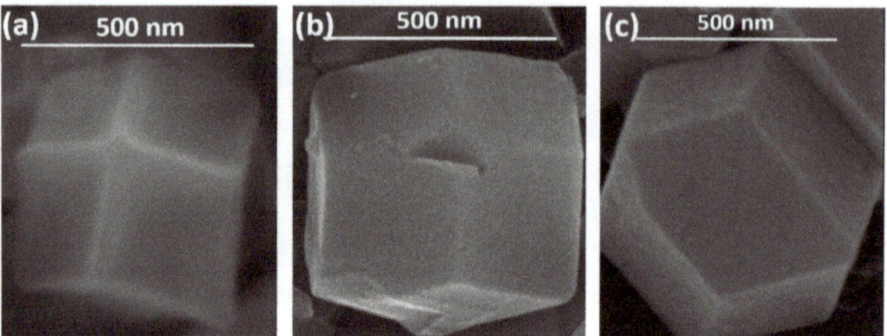

Figure 2. FESEM images of (**a**) ZIF-67, (**b**) MZIF-67, and (**c**) MZnZIF-67.

After the synthesis of MOF-incorporated alginate beads, FESEM images of the beads were obtained, as shown in Figure 3. The alginate cross-sectional surface can be seen in beads devoid of MOF, as shown in Figure 3a, whereas in other images (Figure 3b–d), the MOF crystals (as small clusters) are seen embedded in the alginate macro-structure. The images are meant to show the presence of crystals in the alginate structure, i.e., clusters of smaller crystals are specifically shown (the scale bar is for reference only).

Figure 3. FESEM images of (**a**) ALG, (**b**) ALG_ZIF-67, (**c**) ALG_MZIF-67, and (**d**) ALG_MZnZIF-67.

The XRD plot of the MOF variants is shown in Figure 4. It can be observed from the plot that the peaks of the synthesized materials match the simulated ZIF-67 XRD pattern. A comparison of the major peaks ((110), (211), (222)) of all three variants shows that the batch and microfluidically synthesized variants have peaks lying at the same 2θ values. All the major peaks are singular in nature, indicating the absence of bi-phasic or two different materials, especially in the case of MZnZIF-67. The zeta potential was determined to be −12.8 mV, −14.9 mV, and −14.1 mV for ZIF-67, MZIF-67, and MZnZIF-67, respectively.

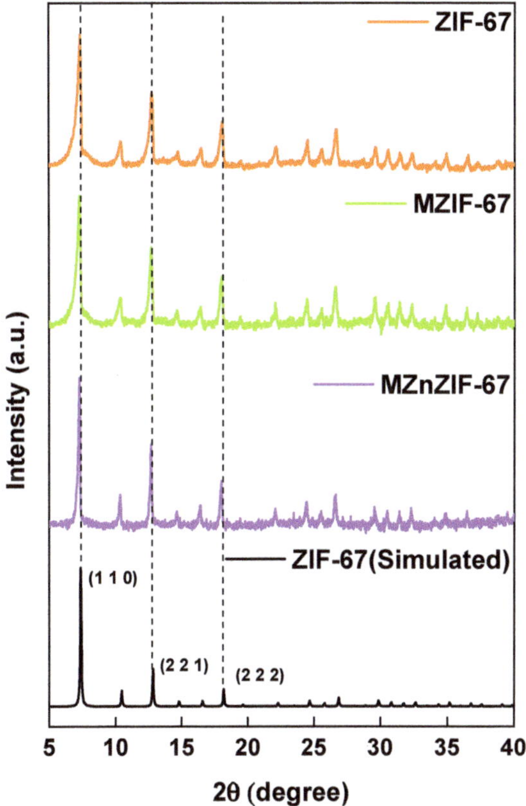

Figure 4. XRD patterns of different MOF materials.

BET isotherms of the materials are plotted and shown in Figure 5. All materials show a Type-I isotherm, indicating that the materials are microporous in nature, owing to their very high surface area. The pore diameters of the materials have similar dimensions, but the overall surface area of batch-synthesized ZIF-67 is higher than that of microfluidically synthesized variants (shown in Table S1). The FTIR spectra of all materials are shown in Figure 5d, with the dotted lines marking the peaks. The peak at 431 cm^{-1} represents the presence of the Co-N bond (and Zn-N in the case of MZnZIF-67), whereas peaks at 760, 1140, and 1438 cm^{-1} indicate the stretching and bending modes of the imidazole ring. The stretching mode of C–H from the aromatic ring and the aliphatic chain in 2-MIM are also described by peaks at 2948 cm^{-1} and 2880 cm^{-1}.

X-ray photoelectron spectroscopy (XPS) is a surface analysis technique that describes the elements present as well as their oxidation states and binding information. As can be observed in Figure 6a and Table S2, the elemental composition of the surface of Alg_MZnZIF-67 consists of carbon present in a C1s state with the de-convolution of its peak indicating peaks at 284.65 eV (C–C/C-H), 286.75 eV (C–O/C-O-C), and 289.27 eV (COO-), as well as oxygen in an O1s state with the de-convoluted peaks at 531.73 eV (O-C=O) and 533.05 eV (C-O) [31,32].

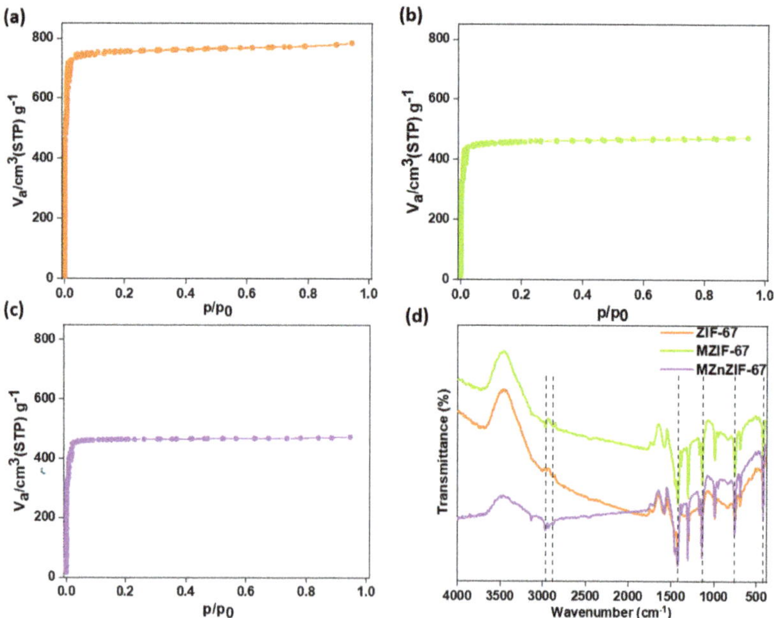

Figure 5. BET plots of (**a**) ZIF-67, (**b**) MZIF-67, and (**c**) MZnZIF-67 and (**d**) FTIR plots for the synthesized materials.

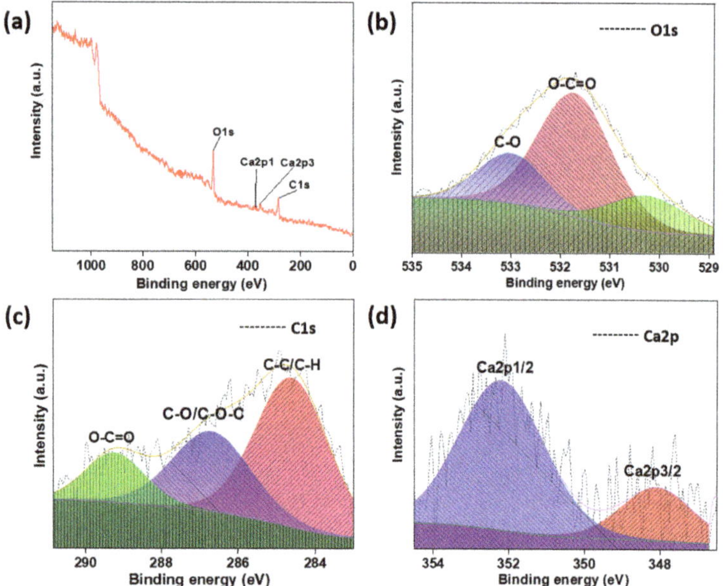

Figure 6. The XPS (**a**) survey spectra of Alg_MZnZIF-67, (**b**) high-resolution spectra of O1s, (**c**) high-resolution spectra of C1s, and (**d**) high-resolution spectra of Ca2p.

3.2. Adsorption Studies

Concentration and Time Studies

Concentration-based adsorption studies are generally performed to observe the maximum adsorption a material (in this case MOFs) can reach over a long period. It can be seen from Figure 7a that the MZIF-67 variant shows the highest drug adsorption across all concentrations of the drug, followed by MZnZIF-67 and ZIF-67. All three variants, MZIF-67, MZnZIF-67, and ZIF-67, reach their highest adsorption capacities of 78%, 63%, and 52%, respectively, at higher concentrations of 80 to 100 ppm (later, 100 ppm was chosen to perform time studies).

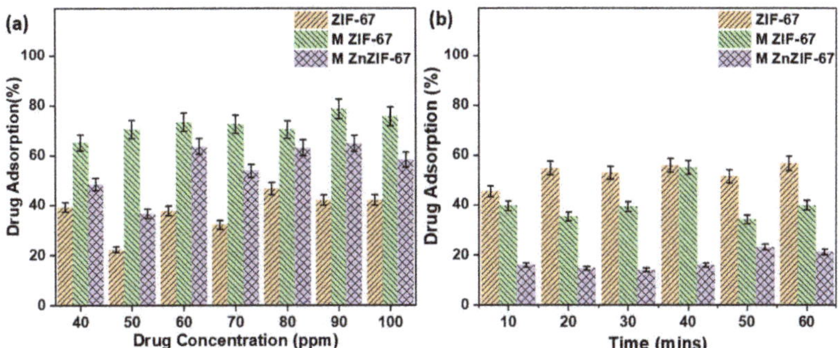

Figure 7. The plot of drug adsorption (%) with respect to (**a**) drug concentration (ppm) and (**b**) time (minutes) for the synthesized MOFs.

3.3. Drug Release Studies

The release percentage of the drug in PBS over a period of 3 to 144 h (0 to 6 days) for all MOF and MOF in alginate variants is shown in Figure 8. It can be seen that the powder variants (ZIF-67, MZIF-67, and MZnZIF-67) release most of the adsorbed drug within the first 24 h (1 day), whereas the alginate variants (Alg ZIF-67, Alg MZIF-67, and Alg MZnZIF-67) release their drug content slowly over a period of 144 h (6 days). The drug release of different materials starts at different values because they adsorbed different amounts of the drug, as can be seen from Figure 7a.

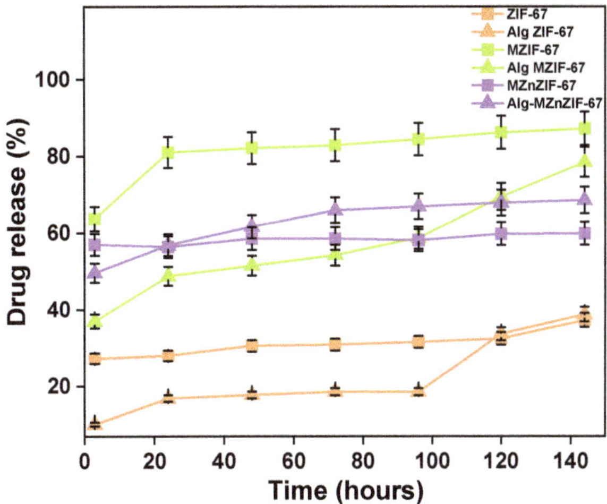

Figure 8. Plot for drug release over time of different MOF and MOF in alginate variants.

3.4. Cytotoxicity Studies

The results of the MTT assays performed on the materials are shown in Figure 9. As can be seen in Figure 9c, the ZIF-67 variant has an erratic OD value trend with comparative cytotoxicity between 24 and 48 h showing a sudden increase from 24 to 48 h. On the other hand, it can be observed from Figure 9a,b that the assays of materials MZIF-67 and MZnZIF-67 show an expected (mostly gradual, non-erratic) trend in OD values, with comparative cytotoxicity between 24 and 48 h showing small increase from 24 to 48 h.

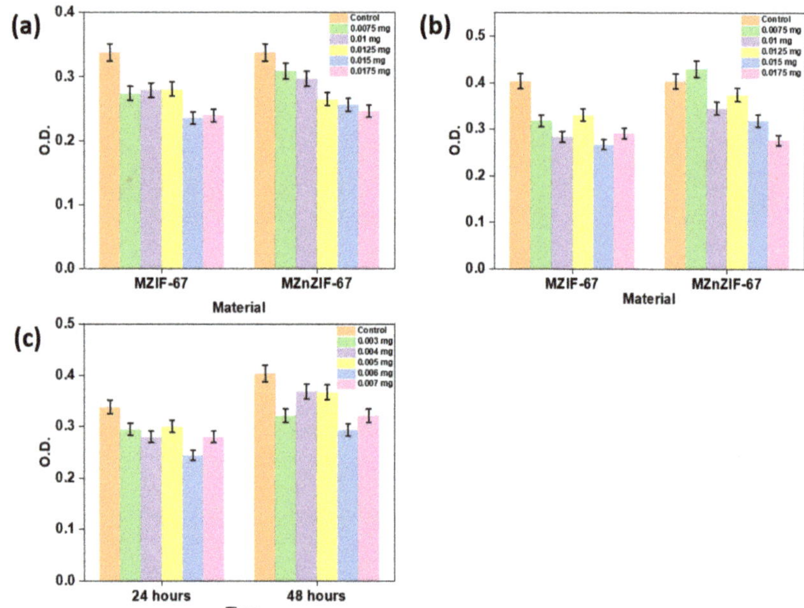

Figure 9. The plot of the OD values with respect to (**a**) materials (MZIF-67 and MZnZIF-67) at different concentrations for 24 h assay, (**b**) materials (MZIF-67 and MZnZIF-67) at different concentration for 48 h assay, and (**c**) ZIF-67 at different concentration for 24 and 48 h assay.

4. Discussion

4.1. Characterization

The FESEM images (Figure 2) prove that the synthesis of similarly sized crystals of MOFs can be achieved by microfluidic means in a shorter time than by the batch synthesis techniques. The observed size distribution of crystals achieved by both techniques was also comparable, while when embedded in the alginate beads (Figure 3), the distribution of these crystals seems to be in small clusters, owing to the inability of the MOF to be evenly dispersed in the highly viscous alginate (in water) solution. These small clusters are dispersed evenly in the alginate structure, not affecting the overall distribution/dispersion of the MOF crystals in the alginate microstructure. The presence of matching sharper peaks seen in the XRD plots (Figure 4) indicates the formation of highly/purely crystalline expected materials, which in turn dictates that microfluidic processes can reproduce such materials with ease. The decrease in surface areas (surface parameters calculated using the instrument software), as seen from the BET plots (Figure 5), might be due to changes in surface morphology/structure or the introduction of surface moieties due to microfluidic synthesis conditions and reduced synthesis time [33,34]. The high surface area and large pore volume may help in the loading of a large amount of drug molecules, but the same effect may be produced even at a lower surface area along with the presence of certain surface moieties or morphology. This might be supported by the fact that the zeta potential of the microfluidic variants is more negative than that of the bulk variant. The FTIR plots (Figure 5) also confirm the presence of the expected functional groups and thus further confirm the formation of the expected products. The XPS spectra indicate the presence of elements and bonds expected in alginate but the absence of any elements and bonds associated with the MOF embedded in the alginate matrix [35]. It can be gathered from the XPS plots that the metal salt (Ca^{2+} present in Ca2p state) used as a linker is present in the matrix, whereas the metals used for the MOF (Co^{2+}, Zn^{2+}) are not present, indicating that the MOF is properly embedded into a properly linked matrix. Similarly, Na^{2+} is also not present, which indicates two facts, namely, (i) the alginate matrix has linked properly such that no sodium alginate molecules are present and (ii) there is no leakage of sodium diclofenac from the MOF onto the surface of the beads.

4.2. Adsorption Studies and Drug Release Studies

As seen from Figures 7 and 8 and Table S1, three possible reasons can be drawn as to why microfluidic ZIF variants show higher adsorption than the batch-synthesized variant: (a) the stability of the batch-synthesized ZIF-67 in water was observed to be lower than that of the microfluidic variants, which may lead to a decrease in adsorption over time; (b) the faster microfluidic synthesis might lead to the introduction of defects and vacant sites, which might lead to the exposure of functional groups which may increase the adsorption capacity; and (c) a smaller pore size (for microfluidic ZIF-67) suggests that there is a possibility of the drug molecule occupying pores much better than the bigger pores of the batch-synthesized ZIF-67 (as drug molecules may move back and forth from the pore, constantly). The oscillatory nature of the drug adsorption values may be due to the constant adsorption and desorption of the drug molecule from the MOF's surface due to the processes of diffusion and dissolution [36]. This is also supported by the trend seen in Figure 8, where these MOFs release the majority of the drug molecules almost instantly.

The drug release profiles (Figure 8) of the various materials indicate that the MOF powders (specifically ZIF-67 and MZIF-67) have an erratic, non-uniform release or sudden pre-release where the release of drug molecules may be caused by the breakdown of the structure in the PBS within a short period of time. The MOF-in-alginate beads have a slow overall uniform release over 6 days (and also have a slower release in starting compared to their MOF-only counterparts) due to the preservation of the MOF structure in the alginate matrix and the slow breakdown of alginate in physiological conditions. Among all the MOF-in-alginate variants, the Alg-MZnZIF-67 variant shows the best uniform release, which may be due to the stability of MZnZIF-67 in PBS being higher than that of other variants due to the incorporation of zinc along with cobalt in the MOF structure [29].

4.3. Cytotoxicity Studies

The ZIF-67 has a highly toxic effect on the cells even at very low concentrations compared to the microfluidically synthesized materials, and the erratic OD values may support its unstable nature in aqueous solutions as it might be rapidly degrading in the solution, thus making it unsuitable as a drug carrier. The assays of materials MZIF-67 and MZnZIF-67 show an expected (mostly gradual, non-erratic) trend in OD values, which may support that the materials are stable in aqueous medium for at least up to 48 h. Among these two materials, MZnZIF-67 shows a much more stable trend of increasing cytotoxic effect with an increase in material amount, with the MZIF-67 having lower OD values than those of MZnZIF-67 for the same corresponding amount of material.

5. Conclusions

It can be gathered from the aforementioned data that MOFs can be used for drug delivery purposes after appropriate modifications. The results show that the microfluidic method produced homogenous MOF particles similar to batch/bulk processes in a short amount of time, as well as improved the particles' adsorption properties. All three synthesized variants follow Freundlich isotherm, which indicates multi-layered adsorption, and follow the second-order kinetics as well. The microfluidically synthesized variants show a higher adsorption capacity as the synthesis may have added or left some positive functional groups on the surface of the MOF particles (which might have also led to a decrease in surface area; this requires further investigation) which is supported by the fact that many MOF structures adsorb organic molecules via electronic interactions and hydrogen bonding as well as π-π stacking, i.e., diclofenac molecules may interact with positive surface moieties to be adsorbed. Even though such materials are unsuitable for drug delivery purposes due to the presence of 'toxic' ions such as cobalt and being unstable under physiological conditions, they can be further improved by combining other metals such as zinc to greatly improve their stability and further reduce toxicity (as can be seen by the MTT assay studies) (as shown in Figure S3). As the studies were performed on cancer cells and showed some cytotoxic effect towards these cells, this also supports the possibility of the use of such toxicity in a favourable way, as if the particle can be inserted or directed towards an affected area, the release of toxic ions can help kill the invading/infected moiety (bacteria, fungus, etc.) as well as destroy cancerous tissue while delivering the relevant medication. Subsequently, the incorporation of such materials in a biopolymer (sodium alginate) matrix helps further protect the delivery agent (MOFs) (slowing down the framework's (MOF's) breakdown) and the drug, thus making them biocompatible, safe, and easy to transport. These beads slowly release the drug over a long time of up to 6 days. MOFs possess many physiochemical properties that make them ideal drug delivery agents, but their qualities can be further improved by the utilization of different synthesis techniques and combinations of different metals in the structure as well as the incorporation of such materials in a biopolymer, as shown in this work.

Supplementary Materials: The following are available online at https://www.mdpi.com/article/10.3390/bios13070737/s1, Figure S1. Setup for microfluidic MOF synthesis; Figure S2. Schematic of setup for MOF-in-alginate bead synthesis; Figure S3. Stability of microfluidically synthesized MOF variants in PBS over a period of 5 days; Table S1. BET isotherm parameters of different MOFs; Table S2. The binding energies and elemental composition from the XPS of Alg_MZnZIF-67.

Author Contributions: Conceptualization, methodology, software, writing—original draft preparation, A.B.; formal analysis, data curation, writing—review and editing, visualization, V.H., K.V.A., K.K., D.S., and V.K.N.; investigation and validation, S.T.M., W.-S.H., and H.-Y.J.; supervision, project administration, resources, funding acquisition, M.D.K. All authors have read and agreed to the published version of the manuscript.

Funding: This research was funded by Department of Science and Technology (DST), India (DST/BDTD/EAG/2019), Department of Science and Technology (DST), India (DST/TDT/DDP-31/2021), and Vision Group for Science and Technology (VGST), India (VGST/K-FISTL1/GRD No. 1053/2021-22), along with Jain University, Minor Project (JU/MRP/Uni/3/2022) for the financial provision and support.

Institutional Review Board Statement: Not applicable.

Informed Consent Statement: Not applicable.

Data Availability Statement: The data presented in this study are available on request from the corresponding author.

Acknowledgments: We sincerely acknowledge Jain University, India, for providing infrastructure and various facilities for this work. Krishna Kant also acknowledges the European Union's Horizon 2020 research and innovation program under the Marie Sklodowska-Curie grant agreement no. 894227.

References

1. Uthappa, U.T.; Brahmkhatri, V.; Sriram, G.; Jung, H.-Y.; Yu, J.; Kurkuri, N.; Aminabhavi, T.M.; Altalhi, T.; Neelgund, G.M.; Kurkuri, M.D. Nature engineered diatom biosilica as drug delivery systems. *J. Control. Release* **2018**, *281*, 70–83. [CrossRef]
2. Pérez-Herrero, E.; Fernández-Medarde, A. Advanced targeted therapies in cancer: Drug nanocarriers, the future of chemotherapy. *Eur. J. Pharm. Biopharm.* **2015**, *93*, 52–79. [CrossRef]
3. Wilczewska, A.Z.; Niemirowicz, K.; Markiewicz, K.H.; Car, H. Nanoparticles as drug delivery systems. *Pharmacol. Rep.* **2012**, *64*, 1020–1037. [CrossRef]
4. Bhat, S.; Uthappa, U.T.; Sadhasivam, T.; Altalhi, T.; Soo Han, S.; Kurkuri, M.D. Abundant cilantro derived high surface area activated carbon (AC) for superior adsorption performances of cationic/anionic dyes and supercapacitor application. *Chem. Eng. J.* **2023**, *459*, 141577. [CrossRef]
5. Kuruppathparambil, R.R.; Jose, T.; Babu, R.; Hwang, G.-Y.; Kathalikkattil, A.C.; Kim, D.-W.; Park, D.-W. A room temperature synthesizable and environmental friendly heterogeneous ZIF-67 catalyst for the solvent less and co-catalyst free synthesis of cyclic carbonates. *Appl. Catal. B Environ.* **2016**, *182*, 562–569. [CrossRef]
6. Hegde, V.; Uthappa, U.T.; Suneetha, M.; Altalhi, T.; Soo Han, S.; Kurkuri, M.D. Functional porous Ce-UiO-66 MOF@Keratin composites for the efficient adsorption of trypan blue dye from wastewater: A step towards practical implementations. *Chem. Eng. J.* **2023**, *461*, 142103. [CrossRef]
7. Raj, A.; Rego, R.M.; Ajeya, K.V.; Jung, H.-Y.; Altalhi, T.; Neelgund, G.M.; Kigga, M.; Kurkuri, M.D. Underwater oleophobic-super hydrophilic strontium-MOF for efficient oil/water separation. *Chem. Eng. J.* **2023**, *453*, 139757. [CrossRef]
8. Lawson, H.D.; Walton, S.P.; Chan, C. Metal–Organic Frameworks for Drug Delivery: A Design Perspective. *ACS Appl. Mater. Interfaces* **2021**, *13*, 7004–7020. [CrossRef] [PubMed]
9. Maranescu, B.; Visa, A. Applications of Metal-Organic Frameworks as Drug Delivery Systems. *Int. J. Mol. Sci.* **2022**, *23*, 4458. [CrossRef]
10. Yan, C.; Jin, Y.; Zhao, C. Environment Responsive Metal–Organic Frameworks as Drug Delivery System for Tumor Therapy. *Nanoscale Res. Lett.* **2021**, *16*, 140. [CrossRef]
11. Yang, J.; Wang, H.; Liu, J.; Ding, M.; Xie, X.; Yang, X.; Peng, Y.; Zhou, S.; Ouyang, R.; Miao, Y.J.R.a. Recent advances in nanosized metal organic frameworks for drug delivery and tumor therapy. *RSC Adv.* **2021**, *11*, 3241–3263. [CrossRef]
12. Wu, M.-X.; Yang, Y.-W. Metal–Organic Framework (MOF)-Based Drug/Cargo Delivery and Cancer Therapy. *Adv. Mater.* **2017**, *29*, 1606134. [CrossRef] [PubMed]
13. Al Sharabati, M.; Sabouni, R.; Husseini, G.A. Biomedical Applications of Metal & minus;Organic Frameworks for Disease Diagnosis and Drug Delivery: A Review. *Nanomaterials* **2022**, *12*, 277. [PubMed]
14. Bieniek, A.; Terzyk, A.P.; Wiśniewski, M.; Roszek, K.; Kowalczyk, P.; Sarkisov, L.; Keskin, S.; Kaneko, K. MOF materials as therapeutic agents, drug carriers, imaging agents and biosensors in cancer biomedicine: Recent advances and perspectives. *Prog. Mater. Sci.* **2021**, *117*, 100743. [CrossRef]
15. Saeb, M.R.; Rabiee, N.; Mozafari, M.; Mostafavi, E. Metal-Organic Frameworks (MOFs)-Based Nanomaterials for Drug Delivery. *Materials* **2021**, *14*, 3652. [CrossRef]
16. Bibi, S.; Pervaiz, E.; Ali, M. Synthesis and applications of metal oxide derivatives of ZIF-67: A mini-review. *Chem. Pap.* **2021**, *75*, 2253–2275. [CrossRef]
17. Kolmykov, O.; Commenge, J.-M.; Alem, H.; Girot, E.; Mozet, K.; Medjahdi, G.; Schneider, R. Microfluidic reactors for the size-controlled synthesis of ZIF-8 crystals in aqueous phase. *Mater. Des.* **2017**, *122*, 31–41. [CrossRef]
18. Zhang, M.; Yu, Z.; Sun, Z.; Wang, A.; Zhang, J.; Liu, Y.-Y.; Wang, Y. Continuous synthesis of ZIF-67 by a microchannel mixer: A recyclable approach. *Microporous Mesoporous Mater.* **2021**, *327*, 111423. [CrossRef]
19. Bazban-Shotorbani, S.; Gavins, F.; Kant, K.; Dufva, M.; Kamaly, N. A Biomicrofluidic Screening Platform for Dysfunctional Endothelium-Targeted Nanoparticles and Therapeutics. *Adv. NanoBiomed Res.* **2022**, *2*, 2270011. [CrossRef]
20. Bendre, A.; Bhat, M.P.; Lee, K.-H.; Altalhi, T.; Alruqi, M.A.; Kurkuri, M. Recent developments in microfluidic technology for synthesis and toxicity-efficiency studies of biomedical nanomaterials. *Mater. Today Adv.* **2022**, *13*, 100205. [CrossRef]
21. Liu, Y.; Yang, G.; Hui, Y.; Ranaweera, S.; Zhao, C.-X. Microfluidic Nanoparticles for Drug Delivery. *Small* **2022**, *18*, 2106580. [CrossRef]
22. Ma, Z.; Li, B.; Peng, J.; Gao, D. Recent Development of Drug Delivery Systems through Microfluidics: From Synthesis to Evaluation. *Pharmaceutics* **2022**, *14*, 434. [CrossRef] [PubMed]
23. Rahme, K.; Dagher, N. Chemistry Routes for Copolymer Synthesis Containing PEG for Targeting, Imaging, and Drug Delivery Purposes. *Pharmaceutics* **2019**, *11*, 327. [CrossRef] [PubMed]
24. Sun, X.; Hu, D.; Yang, L.Y.; Wang, N.; Wang, Y.-G.; Ouyang, X.k. Efficient adsorption of Levofloxacin from aqueous solution using calcium alginate/metal organic frameworks composite beads. *J. Sol-Gel Sci. Technol.* **2019**, *91*, 353–363. [CrossRef]
25. Wang, L.; Xu, H.; Gao, J.; Yao, J.; Zhang, Q. Recent progress in metal-organic frameworks-based hydrogels and aerogels and their applications. *Coord. Chem. Rev.* **2019**, *398*, 213016. [CrossRef]
26. Zheng, X.; Zhang, Y.; Wang, Z.; Wang, Y.; Zou, L.; Zhou, X.; Hong, S.; Yao, L.; Li, C. Highly effective antibacterial zeolitic imidazolate framework-67/alginate fibers. *Nanotechnology* **2020**, *31*, 375707. [CrossRef]
27. Zhu, H.; Zhang, Q.; Zhu, S. Alginate Hydrogel: A Shapeable and Versatile Platform for In Situ Preparation of Metal–Organic Framework–Polymer Composites. *ACS Appl. Mater. Interfaces* **2016**, *8*, 17395–17401. [CrossRef] [PubMed]

28. Bhat, M.P.; Kurkuri, M.; Losic, D.; Kigga, M.; Altalhi, T. New optofluidic based lab-on-a-chip device for the real-time fluoride analysis. *Anal. Chim. Acta* **2021**, *1159*, 338439. [CrossRef] [PubMed]
29. Qian, X.; Ren, Q.; Wu, X.; Sun, J.; Wu, H.; Lei, J. Enhanced Water Stability in Zn-Doped Zeolitic Imidazolate Framework-67 (ZIF-67) for CO_2 Capture Applications. *Chemistryselect* **2018**, *3*, 657–661. [CrossRef]
30. Mosmann, T. Rapid colorimetric assay for cellular growth and survival: Application to proliferation and cytotoxicity assays. *J. Immunol. Methods* **1983**, *65*, 55–63. [CrossRef]
31. Huang, Q.; Liu, S.; Li, K.; Hussain, I.; Yao, F.; Fu, G. Sodium Alginate/Carboxyl-Functionalized Graphene Composite Hydrogel Via Neodymium Ions Coordination. *J. Mater. Sci. Technol.* **2017**, *33*, 821–826. [CrossRef]
32. Zhou, F.; Feng, X.; Yu, J.-G.; Jiang, X. High performance of 3D porous graphene/lignin/sodium alginate composite for adsorption of Cd(II) and Pb(II). *Environ. Sci. Pollut. Res.* **2018**, *25*, 15651–15661. [CrossRef] [PubMed]
33. Andrew Lin, K.-Y.; Yang, H.; Lee, W.-D. Enhanced removal of diclofenac from water using a zeolitic imidazole framework functionalized with cetyltrimethylammonium bromide (CTAB). *RSC Adv.* **2015**, *5*, 81330–81340. [CrossRef]
34. Sun, X.; Keywanlu, M.; Tayebee, R. Experimental and molecular dynamics simulation study on the delivery of some common drugs by ZIF-67, ZIF-90, and ZIF-8 zeolitic imidazolate frameworks. *Appl. Organomet. Chem.* **2021**, *35*, e6377. [CrossRef]
35. Jejurikar, A.; Seow, X.T.; Lawrie, G.; Martin, D.; Jayakrishnan, A.; Grøndahl, L.J.J.o.M.C. Degradable alginate hydrogels crosslinked by the macromolecular crosslinker alginate dialdehyde. *J. Mater. Chem.* **2012**, *22*, 9751–9758. [CrossRef]
36. Molavi, H.; Moghimi, H.; Taheri, R.A. Zr-Based MOFs with High Drug Loading for Adsorption Removal of Anti-Cancer Drugs: A Potential Drug Storage. *Appl. Organomet. Chem.* **2020**, *34*, e5549. [CrossRef]

 *biosensors*

Article

Quantitative, Temperature-Calibrated and Real-Time Glucose Biosensor Based on Symmetrical-Meandering-Type Resistor and Intertwined Capacitor Structure

Yangchuan Ma [†], **Tian Qiang** *, **Minjia Gao** [†], **Junge Liang and Yanfeng Jiang**

Department of Electronic Engineering, School of Internet of Things Engineering, Jiangnan University, Wuxi 214122, China; 6201924123@stu.jiangnan.edu.cn (Y.M.); 6201924078@stu.jiangnan.edu.cn (M.G.); jgliang@jiangnan.edu.cn (J.L.); yanfeng_jiang@yahoo.com (Y.J.)
* Correspondence: qtknight@jiangnan.edu.cn
† These authors contributed equally to this work.

Abstract: Here, we propose a glucose biosensor with the advantages of quantification, excellent linearity, temperature-calibration function, and real-time detection based on a resistor and capacitor, in which the resistor works as a temperature sensor and the capacitor works as a biosensor. The resistor has a symmetrical meandering type structure that increases the contact area, leading to variations in resistance and effective temperature monitoring of a glucose solution. The capacitor is designed with an intertwined structure that fully contacts the glucose solution, so that capacitance is sensitively varied, and high sensitivity monitoring can be realized. Moreover, a polydimethylsiloxane microfluidic channel is applied to achieve a fixed shape, a fixed point, and quantitative measurements, which can eliminate influences caused by fluidity, shape, and thickness of the glucose sample. The glucose solution in a temperature range of 25–100 °C is measured with variations of 0.2716 Ω/°C and a linearity response of 0.9993, ensuring that the capacitor sensor can have reference temperature information before detecting the glucose concentration, achieving the purpose of temperature calibration. The proposed capacitor-based biosensor demonstrates sensitivities of 0.413 nF/mg·dL^{-1}, 0.048 nF/mg·dL^{-1}, and 0.011 pF/mg·dL^{-1}; linearity responses of 0.96039, 0.91547, and 0.97835; and response times less than 1 second, respectively, at DC, 1 kHz, and 1 MHz for a glucose solution with a concentration range of 25–1000 mg/dL.

Keywords: biosensor; microfluidic channel; symmetrical meandering type resistor; intertwined capacitor; temperature calibration

Citation: Ma, Y.; Qiang, T.; Gao, M.; Liang, J.; Jiang, Y. Quantitative, Temperature-Calibrated and Real-Time Glucose Biosensor Based on Symmetrical-Meandering-Type Resistor and Intertwined Capacitor Structure. *Biosensors* **2021**, *11*, 484. https://doi.org/10.3390/bios11120484

Received: 10 November 2021
Accepted: 26 November 2021
Published: 28 November 2021

Publisher's Note: MDPI stays neutral with regard to jurisdictional claims in published maps and institutional affiliations.

1. Introduction

A biosensor is an instrument that is sensitive to biological substances and converts their concentration into electrical signals for detection. It uses immobilized biologically sensitive materials as identification elements, including enzymes, antibodies, antigens, microorganisms, cells, tissues, nucleic acids, etc. A biosensor device or system combines appropriate physical/chemical transducers with signal amplification [1–3]. In regard to a glucose biosensor, the detection sample is focused on the concentration of a glucose solution. Real-time and early detection of glucose concentration has become significant in clinical diagnoses and for assessing treatment progress. Diabetes is a global health issue affecting millions of people [4]. Diabetes is closely related to insulin concentration and glucose levels must be monitored often to avoid complications caused by blood sugar fluctuations [5]. Diabetic macroangiopathy and atherosclerosis due to diabetes can cause cerebrovascular, ischemic heart, peripheral arterial, and other vascular diseases, which are important causes of death in diabetic patients [6].

Currently, more and more researchers have become interested in glucose biosensors due to the need for early detection of glucose levels. On the basis of the physical/chemical transducers of a biosensor, different types of biosensors have been achieved, such as electrochemical, optical, field effect transistor (FET), piezoelectric, microwave, and capacitor biosensors. Among them, an electrochemical biosensor realizes the selective detection of biomarkers by adding specific enzymes to electrodes [7,8]. The electrodes can be used to obtain the voltage–current curves and resistance of a tested solution at different concentrations [9,10]. An electrochemical biosensor is simple to use and cost effective [11]; however, there are still some shortcomings, such as the introduction of external media that slows the sensor's response and decreases its reliability, and the need to replace it in a cycle of about half a year depending on the use environment. In addition, another factor restricting the application of electrochemical sensors is that the electrolyte needs to be replenished regularly and testing tips are not reusable, which may lead to an extra cost. In regard to optical biosensors, non-contact and non-destructive measurement methods can be realized based on optical principles [12,13], and they have excellent color recognition performance for test solutions added with fluorescent markers [14,15]. However, the measurement system constituted by optical devices is relatively complicated, and normal detection usually requires a long calibration and stabilization time, and it is susceptible to the influence of ambient light that causes detection errors. In regard to FET biosensors, the use of specific nanomaterials makes an FET biosensor highly sensitive, selective, and it has commercial potential [16,17]. It can be used to detect various biological or chemical molecules and to detect protein molecules as well as other ions in the physiological environment [18,19]. Nevertheless, FET biosensors are commonly fabricated using an oxide as the sensing layer. This causes the sensor to have monotonic drift and, at the same time, the response will also cause fluctuations. The use of active devices such as logic gates and transistors will increase energy consumption and will have high performance requirements for the FET chips. In regard to piezoelectric biosensors, although the sensitivity of piezoelectric biosensors is relatively high. It can detect small physiological changes similar to sweat, and the piezoelectric signal containing the change information in the concentration of a small molecule can also be obtained [20,21]; however, the selectivity and linearity of this kind of sensor still need to be improved. It is still difficult to distinguish small changes in a substance, and whether they cause discomfort in the human body remains to be discussed. In regard to microwave biosensors, they can detect and characterize the dielectric properties in materials with high Q-factors, narrow resonance, low insertion loss, and high sensitivity [22,23]. Microwave biosensors can translate a variation in dielectric properties of adjacent materials into quantifiable electrical signals such as resonant frequency and resonant amplitude in a remote non-contact manner [24]. However, a lack of accurate model characterization has restricted the simulation of microwave biosensors; the amount, position, and shape of a tested biomarker sample are important issues for accurate and quantitative detection of the biomarker solution, let alone the ambient temperature influence on detection. Recently, capacitor sensors have been especially applied for detecting humidity with high sensitivity, short response/recovery time, and good thermal stability [25–27]. As compared with other types of humidity sensors, it is very competitive. In addition, a capacitor-based biosensor could be a promising candidate in the biosensor research field due to its quick response time, real-time detection, excellent design flexibility, compact chip dimension, cost effectiveness, easy measurement process, convenient integration with matching circuits, etc.

In this study, we propose a resistor-based temperature sensor and a capacitor-based biosensor which can measure the temperature and concentration of a biomarker solution in real time. The concentration range of the glucose solution is 25–1000 mg/dL, which is injected into the microfluidic cavity and placed on the top of the sensors. Then, the signal is provided through an LCR meter, while capacitance and resistance are recorded. The proposed biosensor applies a polydimethylsiloxane (PDMS) quantitative cavity structure with fixed volume and fixed test points, which is convenient for quantitative detection of glucose solution samples, and can eliminate the interference and influence caused by the fluidity, shape, and thickness of glucose solution samples during the test, and therefore achieves accurate measurements of the biomarker solution. Moreover, the amount of solution required for the measurements is quite small and only 1.806 μL of glucose solution is required to complete a measurement. Commonly used and low-cost medical syringes can be used to transfer and inject the biomarker solution for measurement, without using expensive and strict quantitative pipettes. The sensor response of the proposed work is real time, i.e., after the glucose solution is introduced, the resistance and capacitance can be directly affected in less than 1 s, and the value can be read out immediately through a low-cost LCR meter instead of an expensive and complex vector network analyzer. The concentration range of the glucose solution tested in this study includes the normal range of diabetic patients currently being tested in clinical practice.

2. Fabrication Process

The proposed biosensor is fabricated on a glass substrate by micro-/nano processing technology, as shown in Figure 1. To begin, before the growth of the metal structure on the glass substrate, an atomic force microscope (AFM) image of bare glass substrate is analyzed, as shown in Figure 1(c−i). A root mean square (RMS) value of 6.74 nm is obtained, which is not good enough for the following seed metal to stick on the surface of glass, because metal that locates on a glass surface usually suffers from a peel-off issue due to the self-tension of the seed metal. Therefore, in the case of such an issue, first, the glass substrate is polished with an SPM solution (sulfuric/peroxide mi) in order to remove impurities and organic pollutants and to enhance the roughness of the surface, targeting to improve the adhesion of the following metal growth. Then, the seed metal layer of Ti/Au with a thickness of 20/80 nm is prepared by the sputtering method. An RMS value of 0.47 nm is obtained for the surface morphology of the Ti/Au seed metal, as shown in Figure 1(c−ii). In order to further ensure adhesion between the seed metal and the following photoresist as well as electroplated metal, the O_2 etching method is applied on the surface of the seed metal and an improvement of 0.12 nm RMS value is achieved, as shown in Figure 1(c−iii). Following this, we spin a 6 μm thick photoresist on the seed metal, form the designed pattern through a photomask, and then develop the place where we are going to make the target metal through a lift-off machine. Subsequently, a 5 μm thick Cu/Au metal layer is electroplated at the notch. Finally, all photoresists are washed out with acetone solution, and the surplus seed metal layer is etched through the reactive ion etching method, the metal part of the sensor is finished without any peel-off issues between the glass surface and seed metal, between the seed metal and electroplated metal, and between the seed metal and photoresist. A scanning electron microscope (SEM) image of the cross-section of metal is demonstrated in Figure 1(d−i) with a thickness of 4.7 μm. A decrease in thickness occurs mainly due to the control tolerance of electroplating time and reactive ion etching during the isolation process of the seed metal. The details of the fabrication process of the metal structure for the resistor and capacitor are shown in Figure 1a. The fabrication of the PDMS quantitative cavity is based on a reverse molding process using a SU-8 negative photoresist on silicon substrate. Firstly, the silicon substrate is cleaned by acetone. Then, a 200 nm thick silica (SiO_2) passivation layer is deposited over the silicon substrate using plasma-enhanced chemical vapor deposition. Next, we spin a SU-8 photoresist on the passivation layer and form the required SU-8 structure through the photomask, and then wash away the excess SU-8 photoresist through developer. The SEM image shows the SU-8's line pattern on the Si substrate, as shown in Figure 1(d−ii). Its height and width are measured through a surface profiler; a height of approximately 105 μm can be obtained along with a width of approximately 106 μm, as shown in Figure 1(d−iii). Finally, the PDMS microfluidic channel is obtained by a reverse molding process after the SU-8 and PDMS are separated, as shown in Figure 1b. The biosensor and PDMS quantitative cavity are bonded together after the plasma bonding process and the thermal treatment to complete the whole fabrication. When the glucose solution is put into the PDMS quantitative cavity, the solution will directly contact the resistor and capacitor to change the resistance and capacitance.

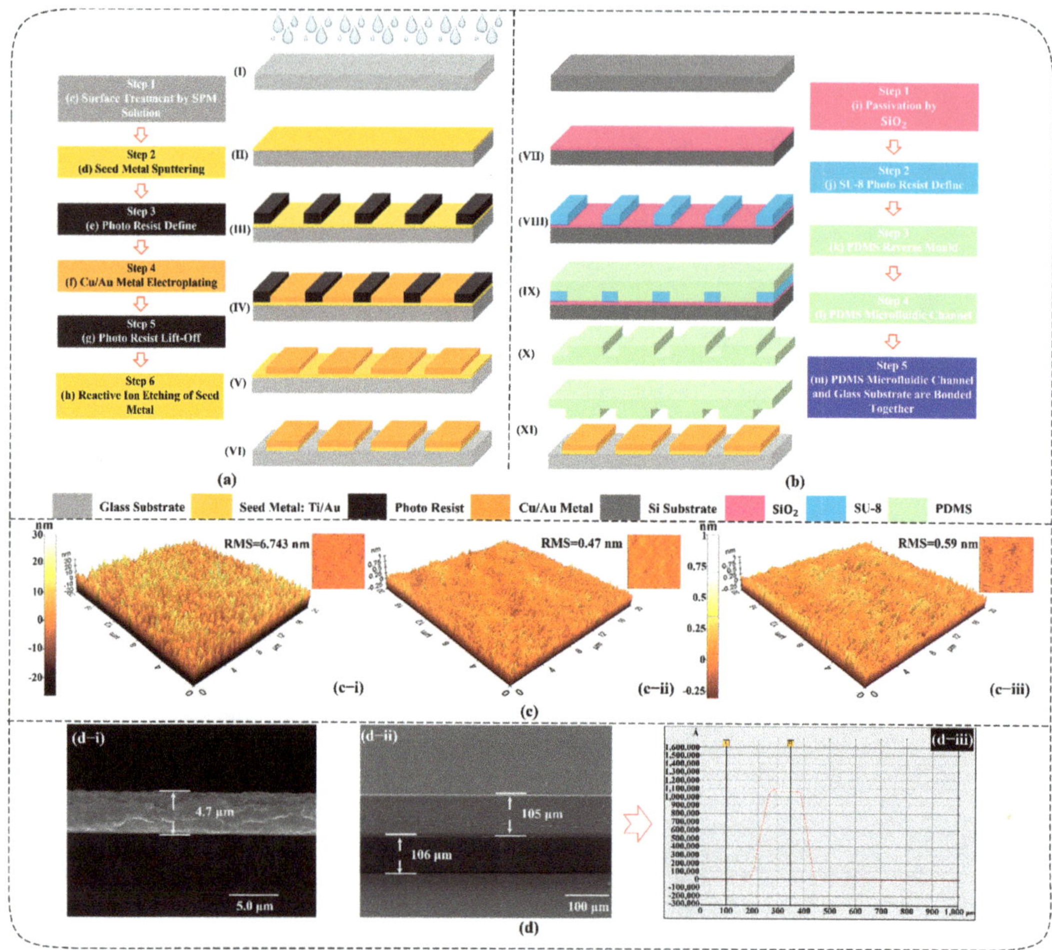

Figure 1. (**a**) Fabrication process of the metal structure for the resistor and capacitor; (**b**) fabrication process of the PDMS microfluidic channel; (**c**) morphological analysis; (**c–i**) surface profile of bare glass substrate; (**c–ii**) seed metal; (**c–iii**) seed metal after O_2 etching; (**d**) morphological analysis; (**d–i**) SEM image of metal on glass substrate; (**d–ii**) SU-8 line pattern on Si substrate; (**d–iii**) height and width measurements of the SU-8 line pattern.

3. Method and Analysis

A complete biosensor system should be comprised of four parts, which are the substance, as well as the detection, transduction, and signal conditioning components [28]. We are mainly focused on the transduction components based on our research background on RF/microwave engineering and micro-/nano fabrication technology for research in the field of potential glucose biosensing applications. The operation mechanism for the proposed work is on the basis of the RF/microwave theory and realistic glucose biosensing application. We describe a new concept for transducers with microfluidic devices. Our design is mainly composed of three parts, which are a temperature sensor, glucose biosensor, and a PDMS microfluidic channel.

Ambient temperature is an important factor that influences the properties of a glucose solution, especially the dielectric property. The dielectric property, which is the complex relative permittivity (ε_s), is the key factor for the capacitor or microwave biosensor, and a variation in the dielectric property could result in a capacitance or frequency change in the biosensor. The mathematical equation is given below:

$$\varepsilon_s = \varepsilon_s' - j\varepsilon_s'' \tag{1}$$

Note that, as the temperature of the glucose solution increased, the imaginary part of the complex relative permittivity of glucose ε_s'' decreases more rapidly than that of real part ε_s' [29]; the real and imaginary parts of the dielectric constant of glucose solution at 5 GHz can vary from 78.0 to 76.2 and from 15.7 to 12.4, respectively, when the ambient temperature is changed from 32 °C to 42 °C. Even for a microwave biosensor in [30], temperature is considered to be a critical parameter to scrutinize for glucose sensing application. Temperature variations from 10 °C to 50 °C have been carried out to study the temperature effect on resonating frequency for different glucose concentrations. Therefore, temperature information is needed so that the concentration of the glucose solution is tested with calibration of the ambient temperature influence.

In this study, based on a resistor and capacitor structure, a temperature sensor and a glucose biosensor are proposed. In order to increase the contact area with the glucose solution, a symmetrical meandering type resistor and an intertwined capacitor are designed, as shown in Figure 2. Figure 2a shows the layout model established using an advanced design system; both sensors are integrated on a glass substrate with a dimension of 19.4 mm × 8.0 mm × 4.6 mm. The integrated microfluidic channel is demonstrated in Figure 2(a−i) with detailed dimensions. Figure 2(b−i) and (c−i) show the detailed schematics of the capacitor and resistor inserted with the glucose sample, separately. They all adopt the structure of a winding line, which can achieve a high number of turns in a compact area so that more inter-turn gaps can be formed, laying the foundation for a highly sensitive sensing response with the biomarker solution. The metal structure for the resistor and capacitor includes an input and an output port, respectively, two 1 mm × 1 mm align key modules. The input and output ports are used for the probe measurement of the LCR meter. An align key module is used for precise alignment when bonding with the quantitative channel structure of the PDMS, since R can change with a change in ambient temperature, resulting in total resistance variation in the equivalent circuit of the resistor-based temperature sensor, as shown in Figure 2(b−ii). Similarly, C_c and C_g can change with different glucose concentrations, resulting in total capacitance variation in the equivalent circuit of the capacitor glucose biosensor, as shown in Figure 2(c−ii).

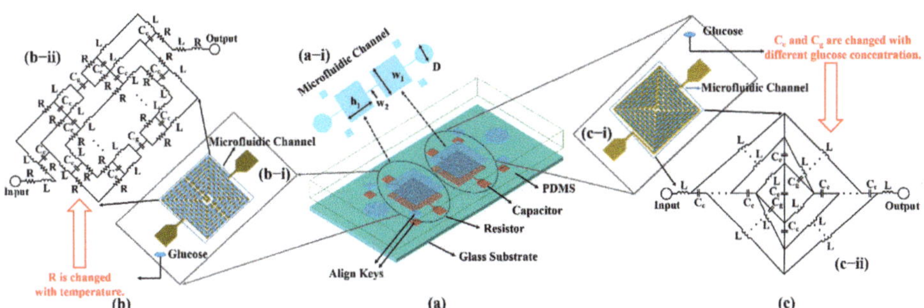

Figure 2. (a) Layout of the designed sensor: (**a–i**) shows the PDMS microfluidic channel structure with dimensions in mm, i.e., D = 2.6, w_1 = 4.3, w_2 = 0.1, and h_1 = 4.2. (**b**) The temperature sensor based on resistor structure: The inset image (**b–i**) shows the microfluidic channel inserted with glucose solution; the inset image (**b–ii**) demonstrates the equivalent circuit of the temperature sensor. (**c**) The glucose biosensor based on capacitor structure: The inset image (**c–i**) shows the microfluidic channel inserted with glucose solution; the inset image (**c–ii**) demonstrates the equivalent circuit of glucose biosensor.

The equivalent resistance of the resistor can be calculated using the equation below [31]:

$$R = R_{sh} \frac{[h{\cdot}N + d(N+1) + 2(b+a)]}{w} \tag{2}$$

where R_{sh} is the sheet resistance of the metal layer used for the resistor design; h represents the width of the winding line resistor; N represents the order of the winding line resistor; d represents the gap of the winding line resistor; a and b represent the size of the winding line at the input and output ports, respectively. The resistance is affected by the temperature of the glucose solution. When the temperature rises, due to an increase in the outermost electron energy of the atom, it can move irregularly and freely, so that the metal can conduct electricity. However, in addition to the free electrons, the atoms in the metal also vibrate. The higher the temperature, the stronger the vibration, which increases the probability of collision between the free electrons and atoms, hinders the directional movement of electrons, and leads to an increase in metal resistance, that is, the value of R_{sh} increases, and metal commonly performs a positive temperature coefficient at a certain temperature range. The relationship between the resistance and temperature is given as follows [32]:

$$R(T) = A + Ce^{BT} \tag{3}$$

where T is temperature (K); R(T) is the resistance (Ω); and A, B, and C are constants whose values are determined by conducting experiments at two temperatures and solving the equations simultaneously. As we calculated based on Equation (3), the resistance increases exponentially with increasing temperature for a metal with a positive temperature coefficient, therefore we use Cu/Au to fabricate the resistor structure and take advantage of a fraction of the exponential curve at a certain temperature range, in which the curve almost performs similar to a linear characteristic. The relationship between resistance and glucose solution temperature can be used to perform the temperature sensor response.

The capacitance of the biosensor is affected by the concentration of glucose solution, the equivalent capacitance of the capacitor can be calculated using the equation below [33]:

$$C = \left[\varepsilon_0\varepsilon_{sub} \left(\frac{1+\varepsilon_s}{2} \right) \frac{K\sqrt{1-k^2}}{K(k)} + \varepsilon_0\frac{t}{a} + \left\{ \frac{K(k)}{\varepsilon_0 K\sqrt{1-k^2}} \right\}^{-1} \right] L_C \tag{4}$$

where L_c is the coupled meandered-line length, t is the thickness of the metal, a is the distance between the winding line resistors, ε_0 (equals to 8.854×10^{-12}) denotes the free space permittivity, ε_s denotes the permittivity of the glucose solution, and ε_{sub} (equals to 4.1) denotes the permittivity of glass substrate; k (=a/b) and K(k) are the elliptic integrals of the first kind. Since the viscous effect increases as the concentration of the glucose solution increases, resulting in increased relaxation times and correspondingly decreased dielectric constants and increased loss factor [34,35], as a consequence, the capacitance of the proposed sensing capacitor in this study is expected to be maximized and minimized when the glucose level in the glucose sample is minimized and maximized at 25 mg/dL and 1000 mg/dL, respectively. Therefore, the relationship between capacitance and glucose concentration can be used to perform the biosensor response.

The equivalent circuits of the biosensor structures are depicted in Figure 2b,c, respectively. They represent the series/parallel combination of the inductor and capacitor or resistor which are equivalent to electrodes' conductive paths and the gap between them. In the presented equivalent structures, L represents inductance, R represents resistance, C_g represents gap capacitance, and C_c represents coupling capacitance. In the equivalent circuit of the resistor, because input and output ports are connected with each other, the inductor and resistor play the main role in the circuit, and the influence of coupling capacitance can be ignored. At the same time, because temperature has a significant influence on the resistance of metal, the resistor can be used to calibrate the temperature of the glucose solution in this design. In the equivalent circuit of the capacitor, because the input and output ports are disconnected, the inductor and capacitor play the main role in the circuit, and the influence of resistance can be ignored. When the glucose solution is introduced into the PDMS microfluidic channel, the solution fills all gaps, and the dielectric constant of the glucose solution with different concentrations are different, which will affect the coupling capacitance and gap capacitance. Therefore, in this design, the capacitor can be used to measure the concentration of the glucose solution. Moreover, the PDMS microfluidic cavity is designed to ensure that the glucose solution is in full contact with the metal structure at a fixed position and a fixed shape, so that the error caused by the different positions can be avoided and the error of solution quantity caused by the direct dropping can be eliminated. The inset image, as shown in Figure 2(a−i), shows the structure and dimension of the PDMS microfluidic channel. The height of the cavity is 0.1 mm, and the depth of the whole PDMS is 3.8 mm. The quantitative cavity structure includes a capacitor cavity and a resistor cavity, which are connected by a microchannel with a height of 0.1 mm and a width of 0.1 mm. A through hole with a diameter of 2.6 mm can be obtained at each end of the channel through a hole punch, separately, which is prepared for the joint with the plastic hose in the later experiment, so as to ensure a complete fit with the plastic hose without liquid leakage. Furthermore, there are four calibration modules around the PDMS quantitative cavity structure, which correspond to the metal structure alignment modules, respectively.

4. Experimental Setup

The proposed biosensor is fabricated on a glass substrate with a dielectric constant and loss tangent of 4.1 and 0.08, respectively. The thicknesses of the glass substrate and metal layer are 0.8 mm and 5 μm, respectively. The glucose solution interacts with the biosensor through the PDMS microfluidic cavity. An LCR meter (HIOKI IM3536) is used to measure the capacitance and resistance as the sensor response. Figure 3a shows the schematic diagram of the test setup. After the LCR meter is connected to the sensor, when the glucose solution is put into the PDMS microfluidic cavity, the changes in resistance and capacitance of the sensor can be observed. Such a measurement is completed only with a solution volume of 1.806 μL, which is cost effective for the tested biomarker solution.

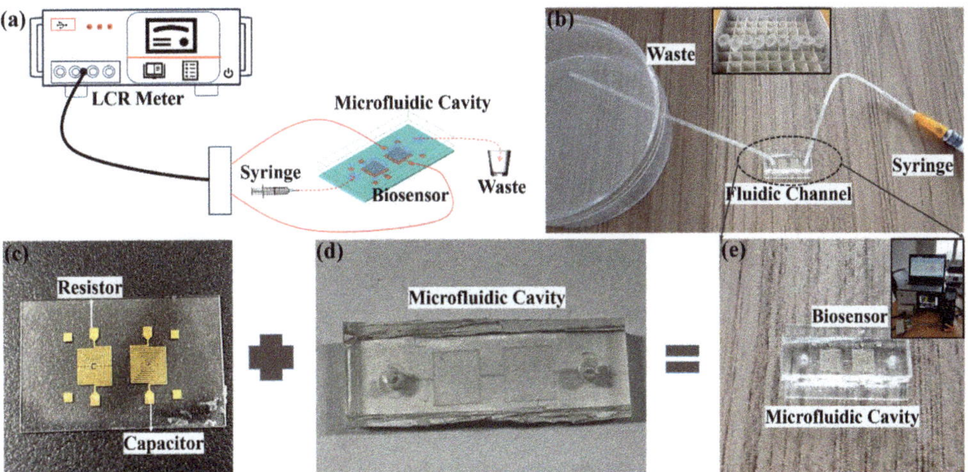

Figure 3. (**a**) Schematic diagram of the sensor and experimental setup; (**b**) measurement view of the solution injected into the microfluidic channel, the inset image is glucose sample; (**c**) fabricated sensor with metal structure; (**d**) fabricated PDMS microfluidic channel; (**e**) proposed sensors integrated with microfluidic channel, inset image is the LCR meter that records the experimental data.

In order to verify the experimental results, we used a glucose sample to measure the sensor response. The overall measurement process is shown in Figure 3b, after connecting the two interfaces of the LCR meter to both ports of the capacitor or resistor. A syringe is used to inject the biomarker solution into the tube and the PDMS microfluidic channel. Then, the glucose solution is in contact with the resistor and capacitor in the microfluidic cavity. At the same time, the LCR meter can display the resistance and capacitance in real time. Finally, the used solution is stored in a beaker after passing through the biosensor. In this experiment, the glucose solution is used with concentrations of 25, 50, 100, 200, 300, 400, 500, 600, 800, and 1000 mg/dL, as shown in Figure 3b. They are prepared using a quantized mixture of D-(+)-glucose powder (Aladdin Bio-Chem Technology Co., LTD) and deionized (DI) water. The fabricated sensor is shown in Figure 3e, which mainly includes three parts: the glass substrate, the metal structure of the sensor, and the PDMS quantitative cavity structure. The glass substrate has the advantages of smooth surface, easy PDMS bonding property, and transparency, which make it a strong candidate for a microfluidic-based application, as shown in Figure 3c. The metal structure for the resistor and capacitor is designed with a single layer structure which is easy to process and low in cost. The metal structure for the resistor and capacitor at high frequency is more easily affected by the concentration of the biomarker solution, the structure is shown in Figure 3c. The PDMS materials have uniform texture, few bubbles, strong bonding with glass, easy molding with the SU-8 photoresist mold, soft texture, and they are easy to combine with flexible electronic applications. The PDMS material after the plasma bonding process can be firmly and tightly combined with the glass substrate, and the area where the measurement port is located is reserved around the PDMS material, as shown in Figure 3d. An LCR meter is used to input signals of different frequencies and measures the corresponding capacitance and resistance, as shown in Figure 3e. During the experiment, the same concentration of glucose solution is measured three times for temperature sensor measurements and biosensor measurements, separately. After each measurement, another syringe is used to inject DI water to clean the glucose solution, and air is injected through different syringes to dry the DI water. Before each measurement, it should be ensured that there is no residual solution in the biosensor module, and the channel is dry and free from blockage.

5. Results and Discussion

We injected glucose solutions with different concentrations into the microfluidic cavity and recorded the response of the temperature sensor with an LCR meter. A remarkably quick response time of less than 1 s was realized when the tested glucose was injected into the shape-fixed, position-fixed, and volume-fixed PDMS microfluidic structures, which is shown in the supplementary video in Supplementary Materials. Next, we measured the resistance using a temperature sensor. Figure 4a shows the resistance measurement histogram of the glucose solution in the temperature range of 25–100 °C. Figure 4b shows the linear relationship between the temperature of the glucose solution and the change in resistance realized by the resistor-based temperature sensor. This relationship can be expressed as:

y_1 [Ω] = 99.14418 + 0.27157 × x [°C], a good correlation with the linear fit of R^2 = 0.99931, relative standard deviation (RSD) $\leqq$ 0.49558, sensitivity is 0.272 Ω/°C, where y_1 is the resistance of the resistor-based temperature sensor, and x is the temperature of the glucose solution. The regression analysis data show that the temperature of the glucose solution has a good linear correlation with the resistance of the resistor-based temperature sensor, that is, it can measure the temperature of the glucose solution before the capacitor-based biosensor measures the concentration, and therefore assists in calibrating the influence of solution temperature.

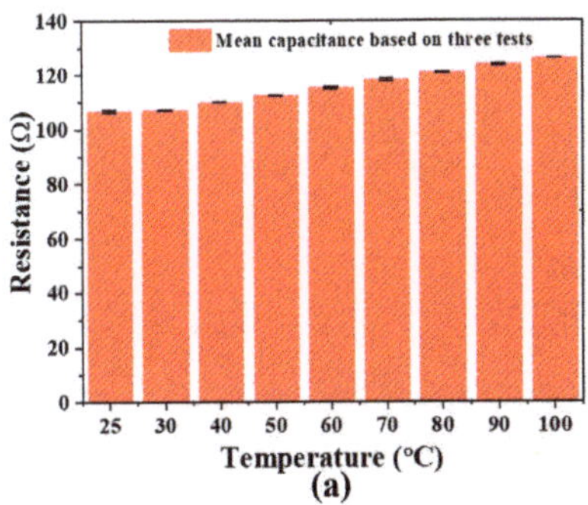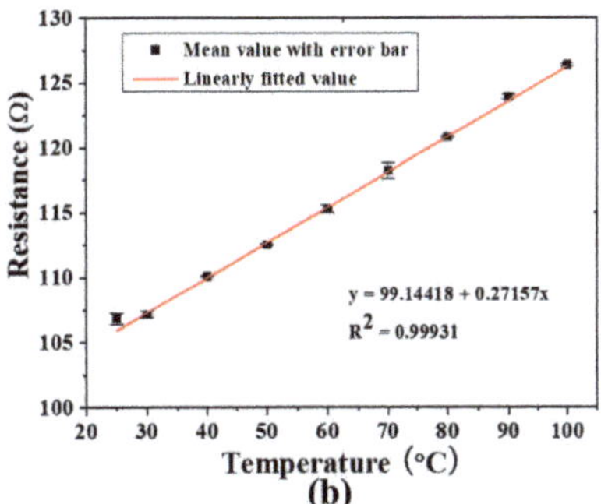

Figure 4. (**a**) The average value of the resistor biosensor response to the glucose solution at different temperatures; (**b**) regression analysis of resistance (n = 9) at different temperatures.

Under the condition of ambient relative humidity ranging from 47.4% to 48.6%, the capacitances of the capacitor-based biosensor for bare chip and DI water are measured three times under different input frequencies, as shown in Table 1. It can be seen that with an increase in input frequency, the measured capacitances show a decreasing trend in the above two cases. Such a phenomenon is mainly caused by a decrease in the effective dielectric constant of the glucose solution when the input frequency increases [35]. The measured capacitances of the bare-chip capacitor at DC, 1 kHz, and 1 MHz are far different from the capacitance measured after passing glucose solution. Therefore, in the actual measurements, the influence of bare chip and DI water on the concentration of glucose solution can be ignored.

Table 1. Performances of the different measured capacitance for the bare chip and under DI water for the PDMS cavity-based DC/1 kHz/1 MHz capacitor-based biosensor.

[1] S.F	Item	Capacitance (pF/nF)					RSD (%)
		1st Test	2nd Test	3rd Test	[2] Mean	Mean ± RSD ([3] C_{av})	
DC	Bare Chip	28.023 nf	28.99 nf	28.877 nf	28.630 nf	28.63 nf ± 1.847	1.847
	DI water	938.945 nf	957.672 nf	980.598 nf	959.072 nf	959.072 nf ± 2.175	2.175
1 kHz	Bare Chip	3.224 pf	3.262 pf	3.244 pf	3.243 pf	3.243 pf ± 0.586	0.586
	DI water	64.375 nf	65.174 nf	66.37 nf	65.306 nf	65.306 nf ± 1.537	1.537
1 MHz	Bare Chip	2.424 pf	2.393 pf	2.405 pf	2.407 pf	2.407 pf ± 0.649	0.649
	DI water	23.483 pf	23.543 pf	23.528 pf	23.518 pf	23.518 pf ± 0.133	0.133

[1] S.F, signal frequency; [2] mean = average of the three experiments; [3] C_{av}, final average capacitance.

The measured sensing responses of the capacitor-based biosensor to different concentrations of glucose solution are shown in Figure 5. Figure 5a–c, respectively, shows the measurement histograms of the glucose solution in the concentration range of 25–1000 mg/dL based on the capacitor-based biosensor at DC, 1 kHz, and 1 MHz signal frequencies. Figure 5d–f shows the linear relationship between the glucose solution concentrations and capacitance change realized by the capacitor-based biosensor with input signals of DC, 1 kHz, and 1 MHz. The corresponding linear regression equations are as follows:

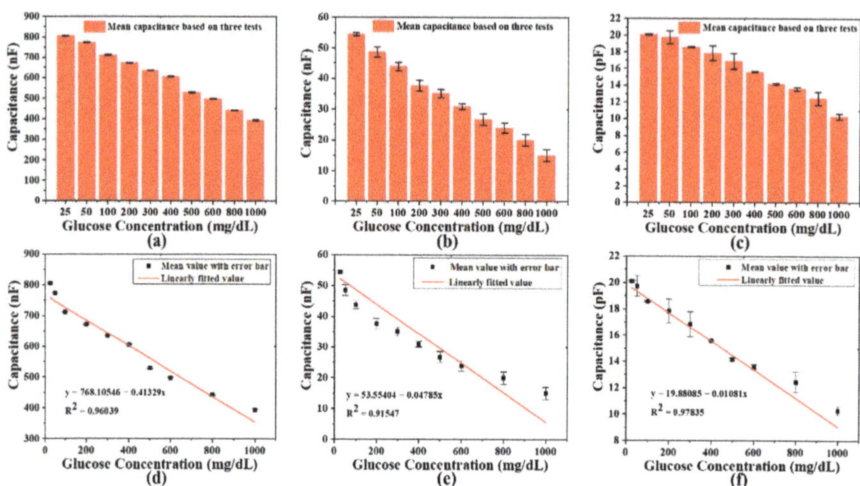

Figure 5. The average of three sensing responses of the capacitor-based biosensor to different concentrations of glucose solution at: (**a**) DC signal frequency; (**b**) 1 kHz signal frequency; (**c**) 1 MHz signal frequency. Regression analysis is performed for different glucose concentrations (n = 10) at: (**d**) DC signal; (**e**) 1 kHz signal; (**f**) 1 MHz signal.

y_2 [nF] = 768.10546 − 0.41329 × x [mg/dL], a correlation with the linear fit of R^2 = 0.96039, RSD $\leq$ 2.63909%, sensitivity is 0.413 nF/mg·dL^{-1};

y_3 [nF] = 53.55404 − 0.04785 × x [mg/dL], a correlation with the linear fit of R^2 = 0.91547, RSD $\leq$ 1.97151%, sensitivity is 0.048 nF/mg·dL^{-1};

y_4 [pF] = 19.88085 − 0.01081 × x [mg/dL], a correlation with the linear fit of R^2 = 0.97835, RSD $\leq$ 0.96428% sensitivity is 0.011 pF/mg·dL^{-1}; where y_2, y_3 and y_4 are the capacitances of the capacitor-based biosensor at DC, 1 kHz, and 1 MHz; x is the concentration of glucose solution in the range of 25–1000 mg/dL. The regression analysis reveals a good linear correlation between glucose concentration and capacitance. In addition, it shows that the capacitance can be used as a parameter for glucose concentration detection, and the limit of detection (LOD) values of the capacitor-based biosensor are 67.236 mg/dL, 17.724 mg/dL, and 2.944 mg/dL at the input signal frequencies of DC, 1 kHz, and 1 MHz, respectively. The RSD of the mathematical parameters can be obtained based on the statistical data of multiple measurements of the capacitance, which are less than 3%, 2%, and 1% at the signal frequencies of DC, 1 kHz, and 1 MHz, respectively, indicating that the capacitance of the capacitor-based biosensor is quite stable in the process of multiple measurements.

Since environmental temperature is a key factor that influences the measurement accuracy of a glucose biosensor, we performed a temperature calibration test based on the 1 MHz capacitive biosensor and measured the sensing response at a temperature range between 30 °C and 80 °C. As shown in Figure 6a, there was almost no change in the bare chip capacitance, which proved the stability of our fabricated capacitor. In addition, with an increase in temperature, the capacitor-based biosensor injected with a similar concentration of glucose sample shows a decreasing capacitance, such a phenomenon is mainly due to the higher temperature that gives more energy to the glucose molecules, leading to more active molecule movements, and thus a higher viscosity within a fixed volume of glucose sample. Therefore, a lower dielectric constant can be achieved and, finally, results in a decrease in capacitance as long as the temperature is increasing. Furthermore, the slope value (I tan θ I) and maximum variation (ΔC) of the capacitance fitting curve at different glucose concentrations are calculated, as shown in Figure 6b. The I tan θ I and ΔC are both decreasing when the glucose concentration is increasing, while the temperature is also increasing. It can be concluded that a higher concentration of the glucose sample suffers less from temperature influence as comparing with a lower glucose concentration.

Such a phenomenon is mainly due to the fact that the higher concentration glucose sample possesses more glucose molecules within a fixed sample volume, and therefore the glucose molecules have limited space to be active, resulting in a relatively lower viscous effect and a smaller variation in $|\tan\theta|$ and ΔC. Therefore, consistent with our original design intention, it is necessary to design a temperature sensor before measuring capacitance to result in more accurate measurements of glucose concentration.

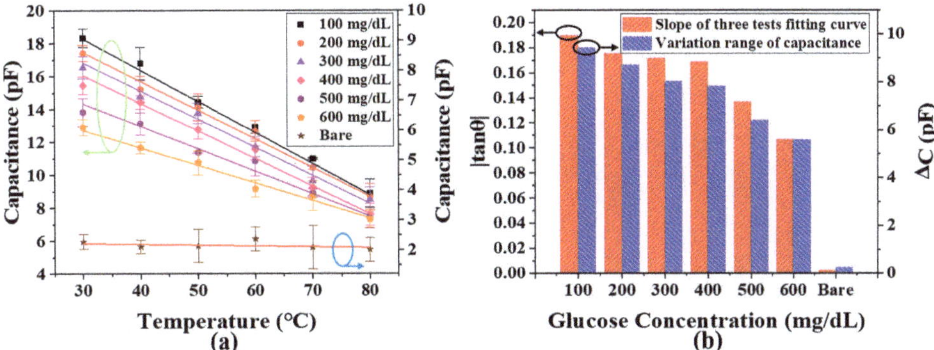

Figure 6. (a) The response of the bare chip and capacitor to different concentrations of glucose solution at the temperature range of 30–80 °C; (b) the change of slope for the fitting curve between different concentrations of glucose solution and capacitance, and the change range of capacitance with temperature.

Figure 7 shows the current density of the proposed capacitor-based biosensor. As shown in Figure 7a, a dark blue color can be observed for the proposed biosensor at both DC and 1 kHz, even though the maximum current density at this scale is only 8.952×10^{-9} A/m, indicating that a few of the input signals couple between the conductor gaps. However, the simulation results of the current density for the proposed biosensor at 1 MHz shows a dark red color with a maximum value of 4.000×10^{-3} A/m at the scale bar, as shown in Figure 7b, indicating that the input signal not only concentrates on the conductor itself, but also has strong coupling between the conductor gaps. Under such a condition, the tested glucose solution that flows in the conductor gap could be in full contact with the electric field at stable ambient conditions in the microfluidic cavity, providing stable mathematical statistics of R^2, RSD, and LOD. Moreover, signal with higher frequency possesses smaller coverage area as compared with lower frequencies such as DC and 1 kHz, thus, smaller coverage area benefits from fewer ambient noise interferences, which also verifies the excellent properties of R^2, RSD, and LOD of a 1 MHz biosensor. Among these three frequencies, the measurement results of 1 MHz show relatively excellent linearity, sensitivity, LOD, and RSD. Therefore, we can consider this frequency in practical glucose biosensor applications. A performance comparison between this study and previously reported studies is summarized in Table 2 in terms of sensor structure, sensitivity (in molar values), detection range (in molar values), detection limit (in molar values), temperature calibration, response time, and quantitative measurement. It can be inferred that the proposed work is excellent, especially for the functions of temperature calibration, quick response time, and quantitative measurement. Although the microwave biosensor mentioned in [36] also achieved the abovementioned functions, it suffered from relatively low sensitivity and higher sample consumption. Moreover, the proposed biosensor in this study possessed a wide detection range which covered glucose measurements in the application of diabetes testing, as well as applications in some fruits and sweet foods. In regard to the detection limit, a moderate performance was obtained with a value of 0.1636 mmol/L at 1 MHz; such a value was not excellent enough as compared with electrochemical, SERS, optoelectronic, and LFA biosensors. Furthermore, since the proposed biosensor was in different sensitivity units as comparing with the reported electrochemical, SERS, optoelectronic, and LFA biosensors, proper comparisons could not be made. Few of the reported studies focused on a capacitor-based biosensor; among the studies, humidity sensors were quite popular. However, this study focused on a biosensor test based on the structure of the capacitor, providing a new method for glucose testing.

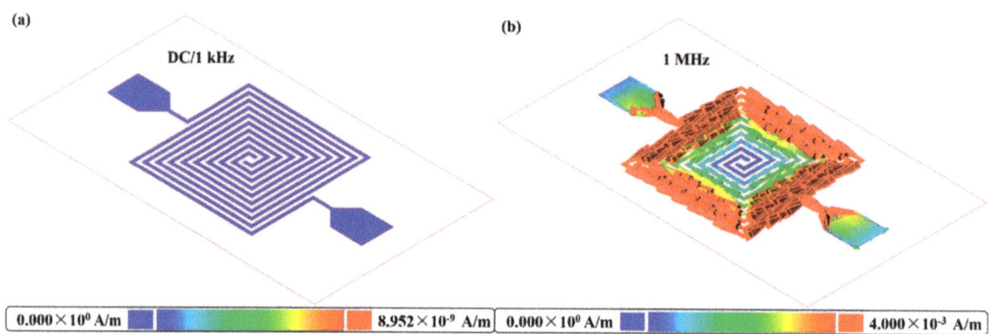

Figure 7. Current density diagram of the capacitor-based biosensor: (**a**) At DC and 1 kHz signal frequency; (**b**) at 1 MHz signal frequency.

Table 2. Performance comparison with other previously reported methods.

Refs.	Sensor Structure	Sensitivity		Detection Range	Detection Limit	Temperature Calibration	Real Time	Quantitative Test
[37]	Electrochemical biosensor	65.6 μA/mmol·L^{-1}·cm^2		0.0003–2.1 mmol/L	0.3 μmol/L	No	No	No
[38]	Electrochemical biosensor	1.41 μA/mmol·L^{-1}		0.1–25 mmol/L	25 μmol/L	No	No	Yes
[39]	Electrochemical based nonenzymatic biosensor	1467.32 μA/mmol·L^{-1}·cm^2		0.005–5.89 mmol/L	0.012 μmol/L	No	No	No
[36]	Microwave biosensor	0.00144 dB/mmol·L^{-1}		0–22.22 mmol/L	-	Yes	Yes	Yes 3.9 μL
[40]	SERS biosensor	2350 a.u./mol·L^{-1}		0–1 mol/L	0.01 mol/L	No	No	No
[41]	Optoelectronic biosensor	13.6 uA mM^{-1} cm^{-2}		0–11 mmol/L	0.015 mmol/L	No	No	No
[42]	LFA biosensor	-		0–5.56 mmol/L	0.0128 mmol/L	No	No	No
This work	Capacitor-based biosensor	DC	2.574 nF/mmol·L^{-1}	1.39–55.56 mmol/L	3.735 mmol/L	Yes	Yes	Yes 1.806 μL
		1kHz	0.864 nF/mmol·L^{-1}		0.9847 mmol/L			
		1MHz	0.198 pF/mmol·L^{-1}		0.1636 mmol/L			

6. Conclusions

Our study investigated a resistor-based temperature sensor and capacitor-based biosensor combining with a PDMS microfluidic channel, functioning for real-time, quantitative, and temperature-calibrated glucose detection. The resistor and capacitor were designed with symmetrical meandering and intertwined structure, respectively, so that compact chip size of integrated sensors could be achieved. Moreover, we managed to apply a PDMS microfluidic channel with fixed shape, fixed volume, and fixed test position, and integrated it with the metal structure of the glass substrate. The results from this study suggest that the resistor-based and capacitor-based biosensor responds rapidly and linearly to a range of temperatures (25–100 °C) and glucose solution concentrations (25–1000 mg/dL) with good sensitivity, excellent stability, and low cost. Owing to the outstanding performance of our developed temperature-calibrated glucose biosensor, a new method is proposed for clinical detection of diabetes in early stage, which is of great medical significance for early detection of diabetes.

Supplementary Materials: The following are available online at https://www.mdpi.com/article/10.3390/bios11120484/s1, Video S1: Measurement Step: LCR Meter + Proposed Biosensor + Syringe + Tube + Connection Line + Beaker.

Author Contributions: Y.M. and M.G. contributed equally to this work. Conceptualization, Y.M. and T.Q.; methodology, Y.M.; software, Y.M.; validation, Y.M. and M.G.; formal analysis, Y.M. and M.G.; investigation, Y.M. and M.G.; resources, Y.M. and T.Q.; data curation, Y.M. and M.G.; writing—original draft preparation, Y.M.; writing—review and editing, Y.M., T.Q., J.L. and Y.J.; visualization, Y.M., M.G. and T.Q.; supervision, T.Q.; project administration, T.Q.; funding acquisition, T.Q. All authors have read and agreed to the published version of the manuscript.

Funding: This research is supported by National Natural Science Foundation of China (grant no. 61801146), Project funded by China Postdoctoral Science Foundation (grant no. 2021M691284) and the Fundamental Research Funds for the Central Universities (grant no. JUSRP12026).

Institutional Review Board Statement: Not applicable.

Informed Consent Statement: Not applicable.

Data Availability Statement: Not applicable.

Conflicts of Interest: The authors declare no conflict of interest.

References

1. Li, H.Y.; Lin, H.Y.; Lv, W.X.; Gai, P.P.; Li, F. Equipment-free and visual detection of multiple biomarkers via an aggregation induced emission luminogen-based paper biosensor. *Biosens. Bioelectron.* **2020**, *165*, 112336. [CrossRef]
2. Bai, Y.; Xu, T.; Zhang, X. Graphene-Based Biosensors for Detection of Biomarkers. *Micromachines* **2020**, *11*, 60. [CrossRef]
3. Gou, Y.; Liu, J.; Sun, C.; Wang, P.; You, Z.; Ren, D. Inertial-Assisted Immunomagnetic Bioplatform towards Efficient Enrichment of Circulating Tumor Cells. *Biosensors* **2021**, *11*, 183. [CrossRef]
4. Kim, J.; Campbell, A.S.; Wang, J. Wearable non-invasive epidermal glucose sensors: A review. *Talanta* **2018**, *177*, 163–170. [CrossRef]
5. Zatterale, F.; Longo, M.; Naderi, J.; Raciti, G.A.; Desiderio, A.; Miele, C.; Beguinot, F. Chronic Adipose Tissue Inflammation Linking Obesity to Insulin Resistance and Type 2 Diabetes. *Front. Physiol.* **2020**, *10*. [CrossRef]
6. Katakami, N. Mechanism of Development of Atherosclerosis and Cardiovascular Disease in Diabetes Mellitus. *J. Atheroscler. Thromb.* **2018**, *25*, 27–39. [CrossRef]
7. Khan, N.I.; Maddaus, A.G.; Song, E. A Low-Cost Inkjet-Printed Aptamer-Based Electrochemical Biosensor for the Selective Detection of Lysozyme. *Biosensors* **2018**, *8*, 7. [CrossRef] [PubMed]
8. Govindasamy, M.; Wang, S.F.; Subramanian, B.; Ramalingam, R.J.; Al-Lohedan, H.; Sathiyan, A. A novel electrochemical sensor for determination of DNA damage biomarker (8-hydroxy-2′-deoxyguanosine) in urine using sonochemically derived graphene oxide sheets covered zinc oxide flower modified electrode. *Ultrason. Sonochem.* **2019**, *58*, 104622. [CrossRef] [PubMed]
9. Wang, Y.H.; Wang, F.T.; Han, Z.W.; Huang, K.J.; Wang, X.M.; Liu, Z.H.; Wang, S.Y.; Lu, Y.F. Construction of sandwiched self-powered biosensor based on smart nanostructure and capacitor: Toward multiple signal amplification for thrombin detection. *Sens. Actuators B Chem.* **2020**, *304*, 127418. [CrossRef]
10. Meng, W.; Wen, Y.; Dai, L.; He, Z.; Wang, L. A novel electrochemical sensor for glucose detection based on Ag@ZIF-67 nanocomposite. *Sens. Actuators B Chem.* **2018**, *260*, 852–860. [CrossRef]
11. Manavalan, S.; Ganesamurthi, J.; Chen, S.M.; Veerakumar, P.; Murugan, K. A robust Mn@FeNi-S/graphene oxide nanocomposite as a high-efficiency catalyst for the non-enzymatic electrochemical detection of hydrogen peroxide. *Nanoscale* **2020**, *12*, 5961–5972. [CrossRef]
12. Jahn, I.J.; Grjasnow, A.; John, H.; Weber, K.; Popp, J.; Hauswald, W. Noise Sources and Requirements for Confocal Raman Spectrometers in Biosensor Applications. *Sensors* **2021**, *21*, 5067. [CrossRef] [PubMed]
13. Tao, W.; Song, Y.; Singhal, N.; McGoverin, C.; Vanholsbeeck, F.; Swift, S. A novel optical biosensor for in situ and small-scale monitoring of bacterial transport in saturated columns. *J. Environ. Manag.* **2021**, *289*, 112452. [CrossRef]
14. Williams, R.M.; Lee, C.; Heller, D.A. A Fluorescent Carbon Nanotube Sensor Detects the Metastatic Prostate Cancer Biomarker uPA. *ACS Sens.* **2018**, *3*, 1838–1845. [CrossRef] [PubMed]
15. Severi, C.; Melnychuk, N.; Klymchenko, A.S. Smartphone-assisted detection of nucleic acids by light-harvesting FRET-based nanoprobe. *Biosens. Bioelectron.* **2020**, *168*, 112515. [CrossRef]

16. Selvarajan, R.S.; Rahim, R.A.; Majlis, B.Y.; Gopinath, S.C.B.; Hamzah, A.A. Ultrasensitive and Highly Selective Graphene-Based Field-Effect Transistor Biosensor for Anti-Diuretic Hormone Detection. *Sensors* **2020**, *20*, 2642. [CrossRef]

17. Tsang, D.K.H.; Lieberthal, T.J.; Watts, C.; Dunlop, I.E.; Ramadan, S.; Hernandez, A.E.d.R.; Klein, N. Chemically Functionalised Graphene FET Biosensor for the Label-free Sensing of Exosomes. *Sci. Rep.* **2019**, *9*, 13946. [CrossRef] [PubMed]

18. Mei, J.; Li, Y.-T.; Zhang, H.; Xiao, M.-M.; Ning, Y.; Zhang, Z.-Y.; Zhang, G.-J. Molybdenum disulfide field-effect transistor biosensor for ultrasensitive detection of DNA by employing morpholino as probe. *Biosens. Bioelectron.* **2018**, *110*, 71–77. [CrossRef]

19. Baldacchini, C.; Montanarella, A.F.; Francioso, L.; Signore, M.A.; Cannistraro, S.; Bizzarri, A.R. A Reliable BioFET Immunosensor for Detection of p53 Tumour Suppressor in Physiological-Like Environment. *Sensors* **2020**, *20*, 6364. [CrossRef]

20. Pohanka, M. Piezoelectric biosensor for the determination of Tumor Necrosis Factor Alpha. *Talanta* **2018**, *178*, 970–973. [CrossRef] [PubMed]

21. Zhang, W.L.H.; Zhang, L.L.; Gao, H.L.; Yang, W.Y.; Wang, S.; Xing, L.L.; Xue, X.Y. Self-Powered Implantable Skin-Like Glucometer for Real-Time Detection of Blood Glucose Level In Vivo. *Nano-Micro Lett.* **2018**, *10*, 32. [CrossRef] [PubMed]

22. Ebrahimi, A.; Scott, J.; Ghorbani, K. Microwave reflective biosensor for glucose level detection in aqueous solutions. *Sens. Actuators A-Phys.* **2020**, *301*, 111662. [CrossRef]

23. Bahar, A.A.M.; Zakaria, Z.; Arshad, M.K.M.; Isa, A.A.M.; Dasril, Y.; Alahnomi, R.A. Real Time Microwave Biochemical Sensor Based on Circular SIW Approach for Aqueous Dielectric Detection. *Sci. Rep.* **2019**, *9*, 5467. [CrossRef] [PubMed]

24. Alibakhshikenari, M.; Virdee, B.S.; Shukla, P.; Parchin, N.O.; Azpilicueta, L.; See, C.H.; Abd-Alhameed, R.A.; Falcone, F.; Huynen, I.; Denidni, T.A.; et al. Metamaterial-Inspired Antenna Array for Application in Microwave Breast Imaging Systems for Tumor Detection. *IEEE Access* **2020**, *8*, 174667–174678. [CrossRef]

25. Yu, X.; Chen, X.D.; Ding, X.; Zhao, X. A High-Stability Quartz Crystal Resonator Humidity Sensor Based on Tuning Capacitor. *IEEE Trans. Instrum. Meas.* **2018**, *67*, 715–721. [CrossRef]

26. Liang, J.G.; Wang, C.; Yao, Z.; Liu, M.Q.; Kim, H.K.; Oh, J.M.; Kim, N.Y. Preparation of Ultrasensitive Humidity-Sensing Films by Aerosol Deposition. *ACS Appl. Mater. Interfaces* **2018**, *10*, 851–863. [CrossRef] [PubMed]

27. Chappanda, K.N.; Chaix, A.; Surya, S.G.; Moosa, B.A.; Khashab, N.M.; Salama, K.N. Trianglamine hydrochloride crystals for a highly sensitive and selective humidity sensor. *Sens. Actuators B Chem.* **2019**, *294*, 40–47. [CrossRef]

28. Lee, H.-J.; Yook, J.-G. Recent research trends of radio-frequency biosensors for biomolecular detection. *Biosens. Bioelectron.* **2014**, *61*, 448–459. [CrossRef]

29. Bababjanyan, A.; Melikyan, H.; Kim, S.; Kim, J.; Lee, K.; Friedman, B. Real-Time Noninvasive Measurement of Glucose Concentration Using a Microwave Biosensor. *J. Sens.* **2010**, *2010*. [CrossRef]

30. Kumar, A.; Wang, C.; Meng, F.Y.; Zhou, Z.L.; Zhao, M.; Yan, G.F.; Kim, E.S.; Kim, N.Y. High-Sensitivity, Quantified, Linear and Mediator-Free Resonator-Based Microwave Biosensor for Glucose Detection. *Sensors* **2020**, *20*, 4024. [CrossRef] [PubMed]

31. Murji, R.; Deen, M.J. A scalable meander-line resistor model for silicon RFICs. *IEEE Trans. Electron Devices* **2002**, *49*, 187–190. [CrossRef]

32. Lall, M.P. P. Resistors. In *Electrical Engineering Handbook*; Dorf, R.C., Ed.; CRC Press: Boca Raton, FL, USA, 1997; pp. 5–15.

33. Endres, H.E.; Drost, S. Optimization of the Geometry of Gas-Sensitive Interdigital Capacitors. *Sens. Actuators B Chem.* **1991**, *4*, 95–98. [CrossRef]

34. Yoon, G. Dielectric properties of glucose in bulk aqueous solutions: Influence of electrode polarization and modeling. *Biosens. Bioelectron.* **2011**, *26*, 2347–2353. [CrossRef] [PubMed]

35. Topsakal, E.; Karacolak, T.; Moreland, E.C. Glucose-dependent dielectric properties of blood plasma. In Proceedings of the 2011 3th URSI General Assembly and Scientific Symposium, Istanbul, Turkey, 13–20 August 2011; pp. 1–4.

36. Jang, C.; Park, J.K.; Lee, H.J.; Yun, G.H.; Yook, J.G. Non-Invasive Fluidic Glucose Detection Based on Dual Microwave Complementary Split Ring Resonators with a Switching Circuit for Environmental Effect Elimination. *IEEE Sens. J.* **2020**, *20*, 8520–8527. [CrossRef]

37. Yoon, H.; Nah, J.; Kim, H.; Ko, S.; Sharifuzzaman, M.; Barman, S.C.; Xuan, X.; Kim, J.; Park, J.Y. A chemically modified laser-induced porous graphene based flexible and ultrasensitive electrochemical biosensor for sweat glucose detection. *Sens. Actuators B Chem.* **2020**, *311*, 127866. [CrossRef]

38. Cao, L.; Han, G.-C.; Xiao, H.; Chen, Z.; Fang, C. A novel 3D paper-based microfluidic electrochemical glucose biosensor based on rGO-TEPA/PB sensitive film. *Anal. Chim. Acta* **2020**, *1096*, 34–43. [CrossRef]

39. Ahmad, R.; Khan, M.; Mishra, P.; Jahan, N.; Ahsan, M.A.; Ahmad, I.; Khan, M.R.; Watanabe, Y.; Syed, M.A.; Furukawa, H.; et al. Engineered Hierarchical CuO Nanoleaves Based Electrochemical Nonenzymatic Biosensor for Glucose Detection. *J. Electrochem. Soc.* **2021**, *168*, 017501. [CrossRef]

40. Sung, C.-J.; Chao, S.-H.; Hsu, S.-C. Rapid Detection of Glucose on Nanostructured Gold Film Biosensor by Surface-Enhanced Raman Spectroscopy. *Biosensors* **2021**, *11*, 54. [CrossRef] [PubMed]

41. Yang, W.; Xu, W.; Zhang, N.; Lai, X.; Peng, J.; Cao, Y.; Tu, J. TiO_2 nanotubes modified with polydopamine and graphene quantum dots as a photochemical biosensor for the ultrasensitive detection of glucose. *J. Mater. Sci.* **2020**, *55*, 6105–6117. [CrossRef]

42. Ki, H.; Jang, H.; Oh, J.; Han, G.-R.; Lee, H.; Kim, S.; Kim, M.-G. Simultaneous Detection of Serum Glucose and Glycated Albumin on a Paper-Based Sensor for Acute Hyperglycemia and Diabetes Mellitus. *Anal. Chem.* **2020**, *92*, 11530–11534. [CrossRef]

 *biosensors*

Article

RFID-Based Microwave Biosensor for Non-Contact Detection of Glucose Solution

Minjia Gao [†], Tian Qiang [*], Yangchuan Ma [†], Junge Liang and Yanfeng Jiang

Department of Electronic Engineering, School of Internet of Things Engineering, Jiangnan University, Wuxi 214122, China; 6201924078@stu.jiangnan.edu.cn (M.G.); 6201924123@stu.jiangnan.edu.cn (Y.M.); jgliang@jiangnan.edu.cn (J.L.); yanfeng_jiang@yahoo.com (Y.J.)
* Correspondence: qtknight@jiangnan.edu.cn
† These authors contributed equally to this work.

Abstract: Due to the increasing number of diabetic patients, early monitoring of glucose levels is particularly important; therefore, glucose biosensors have attracted enormous attention from researchers. In this paper, we propose a glucose microwave biosensor based on RFID and achieve a non-contact measurement of the concentration of glucose solutions. The Reader is a complementary split-ring resonator (CSRR), and the Tag is comprised of a squared spiral capacitor (SSC). A poly-dimethylsiloxane microfluidic quantitative cavity with a volume of 1.56 µL is integrated on the Tag to ensure that the glucose solution can be accurately set to the sensitive area and fully contacted with the electromagnetic flux. Because the SSC exhibits different capacitances when it contacts glucose solutions of different concentrations, changing the resonant frequency of the CSRR, we can use the relationship to characterize the biosensing response. Measurement results show that bare CSRR and RFID-based biosensors have achieved sensitivities of 0.31 MHz/mg·dL^{-1} and 10.27 kHz/mg·dL^{-1}, and detection limits of 13.79 mg/dL and 1.19 mg/dL, respectively, and both realize a response time of less than 1 s. Linear regression analysis of the abovementioned biosensors showed an excellent linear relationship. The proposed design provides a feasible solution for microwave biosensors aiming for the non-contact measurement of glucose concentration.

Keywords: RFID; non-contact measurement; microwave biosensor; complementary split-ring resonator; glucose concentration

Citation: Gao, M.; Qiang, T.; Ma, Y.; Liang, J.; Jiang, Y. RFID-Based Microwave Biosensor for Non-Contact Detection of Glucose Solution. *Biosensors* **2021**, *11*, 480. https://doi.org/10.3390/bios11120480

Received: 18 October 2021
Accepted: 24 November 2021
Published: 26 November 2021

Publisher's Note: MDPI stays neutral with regard to jurisdictional claims in published maps and institutional affiliations.

1. Introduction

Diabetes is a chronic metabolic disease characterized by elevated blood sugar levels, which can cause severe damage to body organs such as the heart, blood vessels, eyes, kidneys, and nerves over time [1,2]. Therefore, the early detection and screening of diabetic patients are very important. It is crucial to develop a method to realize early detection of glucose for the treatment and management of diabetes [3].

Nowadays, biosensing detection conversion plays a very important role in the sensing system because of several key factors such as portability, accuracy, and low cost [4]. Compared with traditional glucose detection methods, such as fluorescent and colorimetric [5,6], the latest biosensors convert glucose solutions of different concentrations into corresponding output electrical signals with higher accuracy, better linearity, as well as quicker response times [7,8]. Since the test solution usually exhibits different conductivity, refractive index, and dielectric constant under different concentration conditions, biosensors are usually categorized into electrochemical biosensors, optical biosensors, and microwave biosensors [9–11]. Electrochemical biosensors are widely accepted for the detection of glucose solutions due to their excellent selectivity, high sensitivity, simple operation, and effective cost. The operating mechanism is to detect biomarkers by adding specific enzymes to the electrodes [7,12,13]. However, the introduction of foreign media, such as some small molecules that serve as channels between enzymes and motor electrodes, slows the response, reduces the performance, and ultimately leads to deteriorating reliability. Moreover, another key factor limiting the application of electrochemical sensors is that the electrolytes need to be replenished regularly, which increases subsequent costs. With regard to optical biosensors, non-contact and non-destructive measurement methods can be realized based on optical principles and have excellent color recognition performance for the test solution added with fluorescent markers [14–16]. However, due to the complex measurement system of optical devices, long calibration and stabilization time, and the detection deviation caused by the influence of ambient light, it cannot be popularized on a large scale. Recently, microwave biosensors, which are considered to be a promising and competitive candidate for the realization of the third-generation glucose biosensor, have attracted great attention [17–19]. Microwave biosensors rely on unique advantages such as having lower performance degradation during service time when compared with other types of biosensors, are not sensitive to environmental conditions (such as external light, noise interference), and can maintain stability in long-term complex environments [20]. Microwave biosensors can perform rapid detection of biomarkers in real time without the need for pre-stabilization time. More importantly, microwave biosensors can realize a non-contact detection of solution concentration [21].

In this work, we propose a biosensor based on the concept of RFID, which consists of three parts, namely the Reader, Tag, and PDMS-based microchannel. Benefiting from the novel biosensor design with complementary split-ring resonator (CSRR) and squared spiral capacitor (SSC) structure, it was found that the glucose concentration is proportional to the resonant frequency of the biosensor, which implies that the resonant frequency can be applied to map the different glucose levels, indicating the feasibility of the proposed biosensor in glucose detection. Moreover, the sensing Tag was made on a glass substrate, which is a biologically compatible substrate and is quite promising in biological fields. The microfluidic channel based on PDMS can simulate biological structures and even biological tissues. Furthermore, the volume of test solution required was very low (1.56 μL). The Reader could receive microwave signals from Tag; therefore, the purpose of non-contact detection and monitoring of blood glucose concentration can be achieved. Finally, the proposed biosensor based on RFID with a fast response time, robust and accurate test results, and excellent non-contact performance were realized, which opens up new routes for many applications such as the non-contact detection of glucose concentration.

2. Materials and Methods

2.1. Biosensor Design

An RFID-based microwave biosensor with microchannels was designed and simulated using a high-frequency structure simulator. The simulated Reader structure, along with its lumped circuit model, is presented in Figure 1a. The Reader is a CSRR structure with a high Q-factor, which is used for accumulating and transferring the energy to the Tag. Electromagnetic resonators demonstrated for the sensing of liquid samples are usually designed in the microwave regime. Geometric sizes associated with this frequency regime are easy to handle and therefore suitable for portable applications. Going further down in size and operating at mm wavelength and THz frequencies has the drawback of reduced reliability since, at high frequencies, the effective ε_r of the liquid samples will be changed due to increased absorption.

The Reader was fabricated on a printed circuit board substrate with a dielectric constant of 9.6, loss tangent of 0.003, and a thickness of 1.524 mm. The CSRR structure was made of 35 μm copper. In order to prevent the copper from oxidation and to make it convenient for subsequent soldering, a surface tin sinking process was used. The structure and size of the proposed structure is shown in Figure 1b, in which $l_1 = 11.3$ mm, $w_1 = 1.5$ mm, $l_2 = w_2 = 9.2$ mm, $l_3 = w_3 = 6.4$ mm, $g_2 = g_3 = g_4 = 0.2$ mm, and $l_4 = 1.2$ mm. Figure 1c shows the equivalent circuit of CSRR, which is a conductive circuit path with three gaps on it. The electrical behavior of the CSRR can be modeled with an equivalent resistor R, capacitor C, and inductor L. The gap capacitance C_g, substrate capacitance C_{sub}, and coupling capacitance C_c are shown in the below equation,

$$C = C_g + C_{sub} + C_c \tag{1}$$

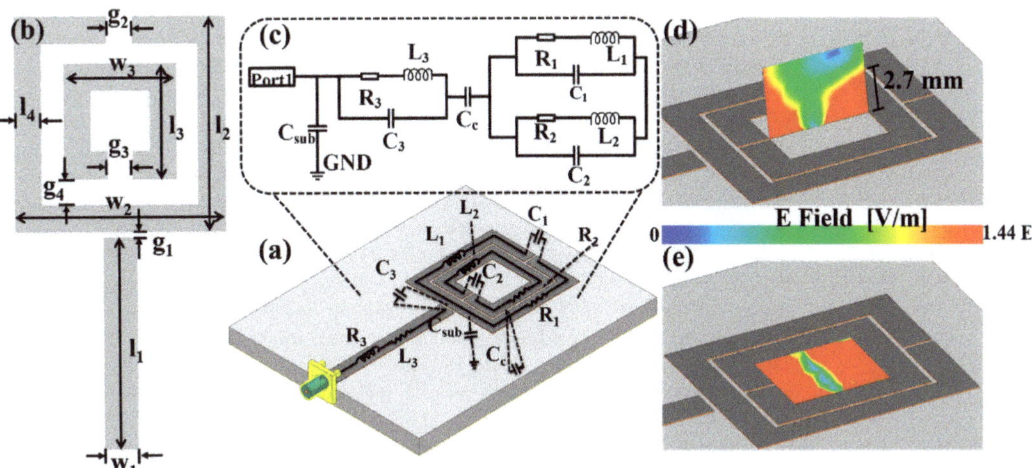

Figure 1. (**a**) Structure and size of the proposed Reader, (**b**) the equivalent circuit of the proposed split-ring structure, (**c**) the proposed Reader structure model, (**d**) the electric field intensity distribution on the vertical, and (**e**) horizontal planes in the middle area of the Reader.

The *RLC* circuit model parameters are inherent parameters of CSRR geometry, which is defined by gap width (g_1, g_2, g_3), conductive path width (w_1, w_2, w_3), the distance between two conductive paths (g_4), and the length of conductor (l_1, l_2, l_3). Analytical models are developed in this work linking the *RLC* quantities to the geometry parameters of the CSRR. Following these models, the resonant frequency f_0 and Q factor of the CSRR can be derived based on the below equations [22]:

$$f_0 = \frac{1}{2\pi\sqrt{L(C_g + C_{sub} + C)_c}} \tag{2}$$

$$Q = \frac{1}{R}\sqrt{\frac{L}{C_g + C_{sub} + C_c}} \tag{3}$$

Compared with C_g, C_{sub} is too small and is ignored; C_c is a constant. Therefore, although a higher C_g can potentially concentrate the electric field in a single region of the CSRR and increase the sensitivity, based on Equation (3), a high total C can cause a deterioration in Q, which presents a design trade-off. The nominal Q of the Reader simulated, as such, can achieve a value of 46. Moreover, Q is observed to decrease with further dielectric loading, especially in the context of sensing with liquid samples. Therefore, in order to maintain the high Q of the biosensor, a CSRR structure with optimized key dimensions and good performance was adopted, which is illustrated in Figure 1d,e. It is obvious to see from the schematic diagram that there is an electric field distribution in the middle area of the Reader whether it is vertical or horizontal, which also proves that this area is sensitive and can radiate the signal upwards about 2.7 mm, and such a radiation distance is enough for energy transmission.

The 3D structure of the RFID-based biosensor is illustrated in Figure 2a. The microchannel, Tag, and Reader are located from top to bottom. The design of the Reader has been explained above. The specific structure of the designed Tag is shown in Figure 2(a-i). The specific dimensions are described as follows, $l_5 = 3.9$ mm, $w_5 = 4$ mm, $g_5 = w_4 = 0.1$ mm, and $l_6 = w_6 = 0.9$ mm. The Tag was designed with two metal wires meandered into a squared spiral structure. Four square metal blocks were prepared around the SSC. This is to align with the designed microchannel, which is depicted in Figure 2(a-ii), and the dimensions are $l_7 = 4.3$ mm, $w_7 = 4.2$ mm, $l_8 = w_8 = 1$ mm, $d_1 = 2.6$ mm, respectively. Its middle structure can just cover the area of the Tag. Four calibration modules were slightly larger than the metal block of the Tag in order to facilitate subsequent alignment and bonding. The Tag was fabricated through a Ti/Au (20/80 nm) sputtering process followed by a Cu (5 μm) plating process and made on a 0.8 mm glass substrate (SCHOTT B 270). The usage of a glass substrate can not only lower the cost but also reduce the dielectric loss of the substrate [22]. The microchannel was made of PDMS because of its good biological properties and metal compatibility. The PDMS microchannel was attached to the glass substrate by plasma treatment and hotplate heating, and the 3D simulation diagram is shown in Figure 2c when the RFID chip is placed in the middle of the Reader. Figure 2b displays the biosensor equivalent circuit, the specific structure of the Tag is shown in Figure 2(b-i), and the corresponding equivalent circuit model is represented in Figure 2(b-ii). As depicted in Figure 2d, the electromagnetic signal passes through the Reader, and the energy can pass through the glass substrate and radiate to the Tag. The electromagnetic signal propagates between the metal lines of the Tag, and its direction will change with the change of the phase. At the resonant frequency of 3.77 GHz, the direction of the electric field lines at a certain phase is shown in Figure 2(d-i). If there is a biological medium between the metal wires, the electromagnetic signals will interact with them, as shown in Figure 2(d-ii).

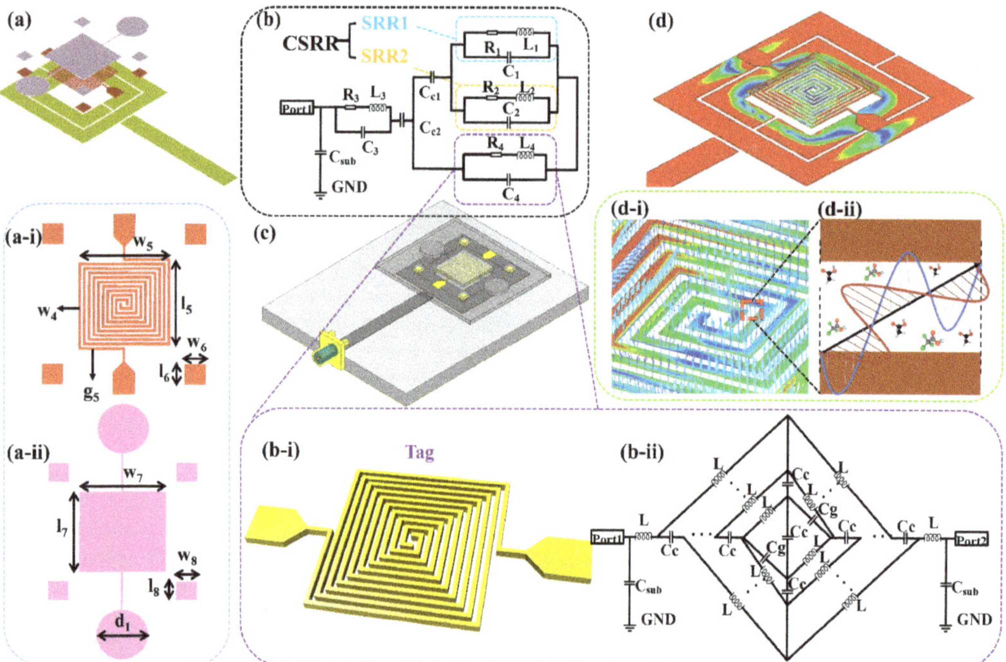

Figure 2. (**a**) Proposed 3D structure of the biosensor and the structure and size of (**a-i**) Tag and (**a-ii**) microchannel, (**b**) the equivalent circuit model of the biosensor, and (**b-i**) Tag structure and its (**b-ii**) equivalent circuit model. (**c**) The model of the proposed biosensor, (**d**) the electric field distribution of Reader and Tag at resonant frequency, and (**d-i**) Tag's electric field coupling model including (**d-ii**) the schematic diagram of the electromagnetic wave acting inside the metal wire.

The electromagnetic wave radiated by the Reader strongly couples with the SSC Tag, and the simulated current density of the biosensor at the resonant frequency is illustrated in Figure 2d. Therefore, the SSC Tag is able to transform a small change in the permittivity caused by the changing of glucose concentration into effective capacitance variations. Accordingly, the CSRR transforms the shift in resonant frequency. The capacitance of SSC is based on the finger width (*WF*), finger length (*LF*), finger gap (*GF*), the gap between electrode and feed line, and feed line width. The 3D schematic, dimension marker, and equivalent circuit of the proposed SSC structure are shown in Figure 2(b-i, a-i, b-ii). The equivalent circuit of the SSC is represented with series/parallel combinations of inductor and capacitor according to design conductive path, coupling, and gap capacitance, respectively. In the presented equivalent structure, L represents inductance, C_g represents gap capacitance, and C_c represents coupling capacitance. As for conventional SSC models, C_g is considered as an essential factor in deciding overall device capacitance. However, after the advancement of micro-fabrication techniques, C_c becomes a significant parameter for capacitive variation. The overall circuit capacitance of SSC can also be represented based on an analytical semiempirical equation as [23,24],

$$C_{SSC} = \left[\varepsilon_0 \varepsilon_{sub} \left(\frac{1+\varepsilon_s}{2} \right) \times \frac{K\left(\sqrt{1-K^2}\right)}{K(k)} + \varepsilon_0 \varepsilon_s \frac{t}{a} + \left\{ \frac{K(k)}{\varepsilon_0 \varepsilon_s K\left(\sqrt{1-k^2}\right)} \right\}^{-1} \right] \times LF \quad (4)$$

$$K = \left(1 + \frac{2 \times WF}{2 \times GF + WF} \right) \times \left(\sqrt{\frac{1}{1 + \frac{2 \times WF}{GF}}} \right) \quad (5)$$

where ε_0 represents the effective permittivity of free space ($\varepsilon_0 = 8.854 \times 10^{-12}$ F/m), ε_{sub} represents substrate permittivity, and ε_s is the permittivity of the glucose sample. K represents the first elliptical integral, k represents the ratio of the finger width and finger gap (WF/GF), and LF represents the length of the coupled meandered line. Equation (4) represents the device capacitance and is directly proportional to the complex permittivity of testing materials. In a testing experiment, the complex permittivity of the tested solution will change with the change of the solution concentration that correspondingly alters the overall device capacitance.

Since glucose solutions with different shapes, volumes, and test positions could lead to a detection accuracy issue, PDMS microchannel was applied to achieve the shape setting (fixed surface area and thickness), volume setting, and test position setting of the tested sample for a quantitative measurement with only 1.56 µL (4.0 mm × 3.9 mm × 0.1 mm), which can eliminate the influence caused by shape, volume, and test position of the glucose sample during the test, ensuring the accuracy of measurement results.

2.2. Biosensor Operating Mechanism

According to the Debye dispersion model, a more general complex quantity expression for the glucose sample can be expressed using the dielectric constant (ε_s') and loss factor (ε_s'') as follows [24],

$$\varepsilon_s = \varepsilon_s' + j\varepsilon_s'' = \left[\frac{(\varepsilon_{sub}-\varepsilon_\infty)}{1+\omega^2\tau^2} + \varepsilon_\infty \right] + j\left[\frac{(\varepsilon_{sub}-\varepsilon_\infty)\omega\tau}{1+\omega^2\tau^2} \right] \quad (6)$$

The above equation is an analytical semiempirical equation, which was used to study the influence of the sample glucose solution on permittivity. Due to the compositional characteristics of the molecule, when glucose anhydrase is dissolved in water, the monosaccharide molecules ($C_6H_{12}O_6$) contain a higher number of -OH groups compared with the -H bands, which leads to less available water to interact with the alternating current (AC) field. This explains why the dielectric constant of the water-glucose solution was lower than water. Furthermore, the molecular weight of glucose is as high as 180.2 kDa, and it is ten times the molecular weight of water molecules. Thus, the glucose molecule is much heavier than the water molecule, which helps to explain the dielectric mechanism of the water-glucose solution. Firstly, the large size caused a more pronounced viscous effect, leading to a difficult rotation with the AC field. Secondly, a large dipole moment is unable to provide the molecule with enough compactness to facilitate reorientation with the AC field. According to Equation (6), the viscous effect is directly proportional to the concentration level of glucose. As a consequence, different glucose solutions lead to differences in their dielectric constants. The differences in dielectric constants correspondingly change the capacitance of the SSC Tag, and the change in the capacitance of SSC influences the capacitance of the overall RFID biosensor, which causes the reflection parameters of the CSRR resonator and ultimately reflects the deviation of the resonant frequency.

2.3. Microwave Detection Methods

The proposed biosensor was constructed from a CSRR Reader and an SSC Tag integrated with a PDMS microchannel. The change of Tag capacitance results in a resonant frequency shift of the Reader. To interrogate the SSC Tag without contact, the Reader was coupled with the Tag. Through monitoring the reflection coefficient S_{11} of the Reader, the measured biosensor response could be obtained. Before conducting the experiment, an SMA needed to be soldered to the Reader; the fabricated Reader is shown in Figure 3a. Before aligning and placing the Tag in the middle area of the Reader, the microchannel was fixed on the Tag. PDMS (Slygard 184, Dow Corning, USA) was mixed with 10% by weight of curing agent. The mixture was poured onto an SU-8 mold and was degassed in a vacuum chamber. After that, it was cured and solidified at 90 °C for 30 min on a hotplate. The formed PDMS microchannels were then cut and punctured to create the inlet/outlet holes; the structure and size can be observed from Figure 3d. After perforating the microchannel, it was aligned and bonded with the Tag on the glass substrate. The fabricated Tag with dimension marker on the glass substrate is shown in Figure 3b. Inserting the catheter into the previously punched hole and assembling the various parts together, the proposed biosensor was obtained, as shown in Figure 3c.

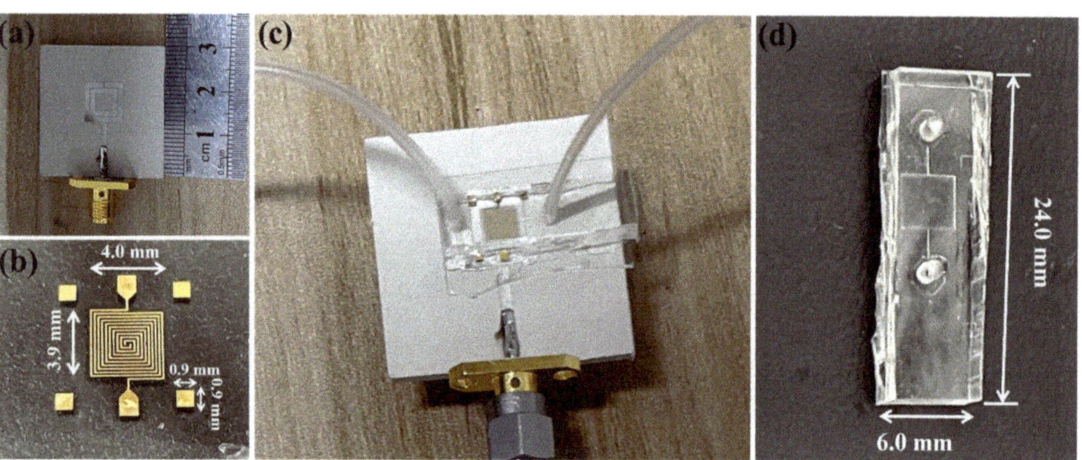

Figure 3. The fabricated RFID chip, including (**a**) Reader, (**b**) Tag, (**c**) RFID chip, and (**d**) microchannel.

In order to measure the microwave response of the proposed biosensor, a measurement platform was successfully constructed, consisting of a vector network analyzer (VNA), an under test biosensor, a quantitative micropipette, and a temperature and humidity sensor (Figure 4a). Ten different glucose samples were prepared, ranging from 25 to 600 mg/dL (25, 50, 100, 150, 200, 250, 300, 400, 500 and 600 mg/dL). These samples were composed of a mixture of glucose anhydrose and deionized (DI) water. The VNA was initially calibrated using a mechanical calibration kit with open, short, and load testing capabilities. After calibration, the VNA was set to 1 GHz bandwidth to measure the reflection coefficient (S_{11}) over 1001 equally spaced data points between 3.5 GHz and 4.5 GHz. Before injecting the glucose solutions of different concentrations into the catheter through a syringe, we first tested and collected the relevant microwave parameters of the bare Reader and the bare chip for subsequent comparative analyses. All the experimental sample solutions were under the experimental conditions of temperature and humidity ranging from 29.2 °C to 30.6 °C and 66.9% RH to 68.4% RH, respectively. This was to reduce the influence of environmental factors on the final experimental results. The 3D schematic with dimension marker of the biosensor is shown in Figure 4b. Inside the red line is the designed biosensor, and the enlarged image of the biosensor structure is shown in Figure 4c.

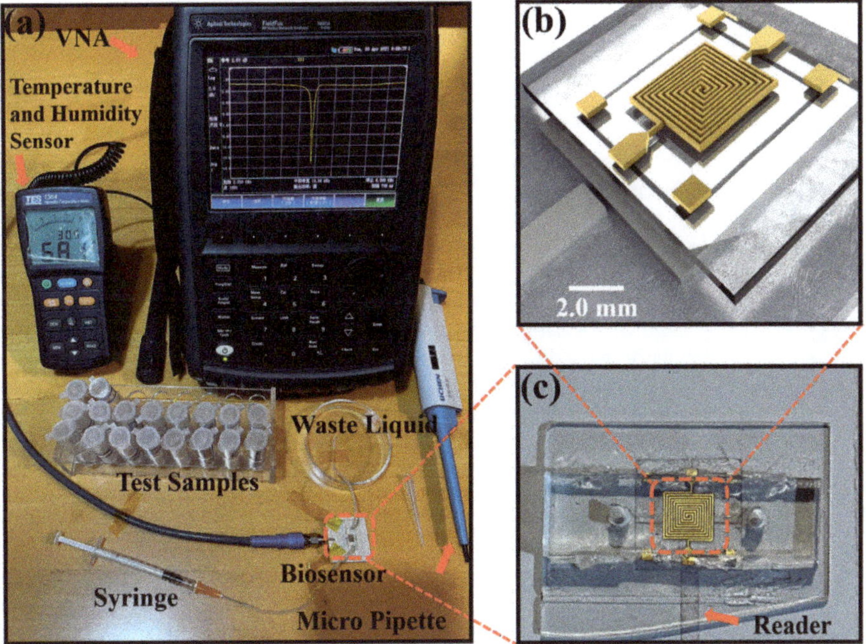

Figure 4. (**a**) Proposed experimental setup for chip measurement of detecting variable glucose level. (**b**) The 3D schematic diagram of RFID chip. (**c**) The fabricated biosensor.

3. Results and Discussion

3.1. Liquid Dielectric Loading

Measured spectra for unloaded and DI water-loaded microchannels are shown in Figure 5. When the Tag is in the sensing area of the Reader, results demonstrate that the resonant frequency of the biosensor will shift approximately 150 MHz. This is because whether RFID is made of a glass substrate or metal, they all have a dielectric constant, which greatly enhances the capacitance of the Reader capacitor and causes a significant drop in the resonant frequency. The offset for the RFID biosensor occurs because the DI water is completely filled in the 10 μm metal line gap, the original air medium is replaced by the DI water medium, so the effective dielectric constant is changed, leading to a change in capacitance, which causes the resonant frequency shifting of the entire system. If the intertwined metal wire of the Tag is unrolled, the sensing capacitance of the Tag will increase compared to the sensing capacitance of the Reader, and when a larger sensing capacitance exists, the change in the dielectric constant will be more obvious. This is why the RFID-based biosensor in Figure 5 changes more when DI water is added to it than just to a bare Reader.

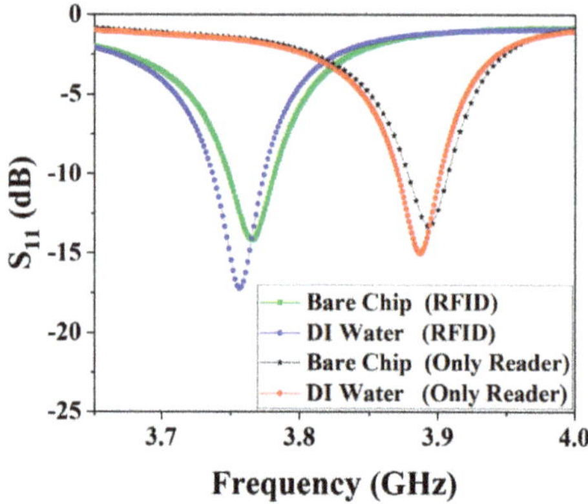

Figure 5. Measured reflection spectra (S_{11}) for Reader and RFID-based biosensor. Microchannels are kept empty and filled with DI water, respectively, while simultaneous measurements are conducted.

3.2. Glucose Solution Loading

Glucose solutions with various concentrations (25–600 mg/dL) were introduced by syringe through the microchannels to measure the Reader and RFID separately. In the experiment, the temperature and humidity changes of 1.4 °C and 1.5% RH had a negligible influence on the dielectric constant (real part and imaginary part < 1) of the glucose solution. Therefore, the effect of temperature and humidity on the capacitance was almost negligible, so the change in the resonant frequency was almost negligible. The gap of the outside split-ring was selected to test the sensitivity of the Reader. This is because, at the sensitive area observed at the resonant frequency, only the outside gap responded to glucose solutions of different concentrations. The experimental results of the Reader are shown in Figure 6a. The maximum offset can reach 200 MHz, which reflects the relatively high sensitivity of the Reader. The designed Tag structure was fixed in the middle area of the Reader, and the measurement results are shown in Figure 6b. The sensitivity of the test in direct contact with the Reader appeared low. This may be caused by two reasons. One reason is that after tin is deposited on the surface of the Reader, it hinders the radiation of electromagnetic flux to a certain extent, resulting in transmission loss, causing the frequency change to be insignificant. The second reason is due to the disadvantage of non-contact transmission and the natural non-contact transmission efficiency and sensitivity being lower. Five measurements at each concentration level were taken. Error bars were added to show relative standard deviations (RSD). As for the large scale of the error bars in Figure 6c,d, there may be individual data for the experimental operation tolerance, or DI-water may not be completely discharged from the microfluidic cavity; therefore, a relatively large difference was observed in the measurement results. Fit lines are added to the averages of the measured S_{11} of the bare Reader and RFID system in Figure 6c,d. During the experiment, due to the differences in position, shape, and thickness of the glucose sample, the experimental frequency results could have different deviations; therefore, due to these interference factors, 3.840 GHz was measured for the bare Reader, and such a value falls into the error bars of 100 and 200 mg/dL. In order to eliminate the influence of the abovementioned factors, we use a microfluidic cavity to fix the position, shape, and thickness of the glucose sample. However, due to a few shortcomings of our microfluidic cavity design, there are still some issues remaining to be solved that cause similar problems as the bare Reader did, which are explained as follows: the microfluidic cavity was designed with a square structure, and when the glucose solution was injected, the square structure is difficult to completely fill before the solution is discharged, which will cause a certain tolerance in the experimental results. Besides, in the measurement process, there will be solution residuals remaining in the microfluidic cavity after being rinsed with deionized water several times. It is difficult to completely remove the residual liquid when the gas is passed in such a square cavity. As a result, 3.840 GHz falls into the error bars of 100 and 200 mg/dL in Figure 6d as well. Compared with the results of a single CSRR, although the sensitivity of the biosensor after the composing RFID chip decreased, which may be caused by non-contact reasons, its linearity was greatly improved from $R^2 = 0.88471$ to $R^2 = 0.97185$ at the same time. The data reveals an absolute sensitivity to glucose concentration levels of 0.31 MHz/mg·dL^{-1} and 10.27 kHz/mg·dL^{-1}, the minimum detection limits of 13.79 mg/dL and 1.19 mg/dL, and a response time of less than 1 s, separately. Less than 1 s is required to finish the microwave signal sweep between 3.5 GHz to 4.5 GHz. A comparative analysis based on microwave biosensors is summarized in Table 1, where the detection method and some features of this work are superior to those achieved in previous studies. For example, the proposed design uses a multi-layer structure to achieve a non-contact detection of glucose solution. Although its sensitivity is relatively low compared with the method of direct contact with the detected solution, the sensitivity is relatively good in the non-contact detection method. Besides, the design uses a microfluidic cavity, which can achieve quantitative detection of the solution, reflecting its higher detection accuracy compared to other non-quantitative detection methods.

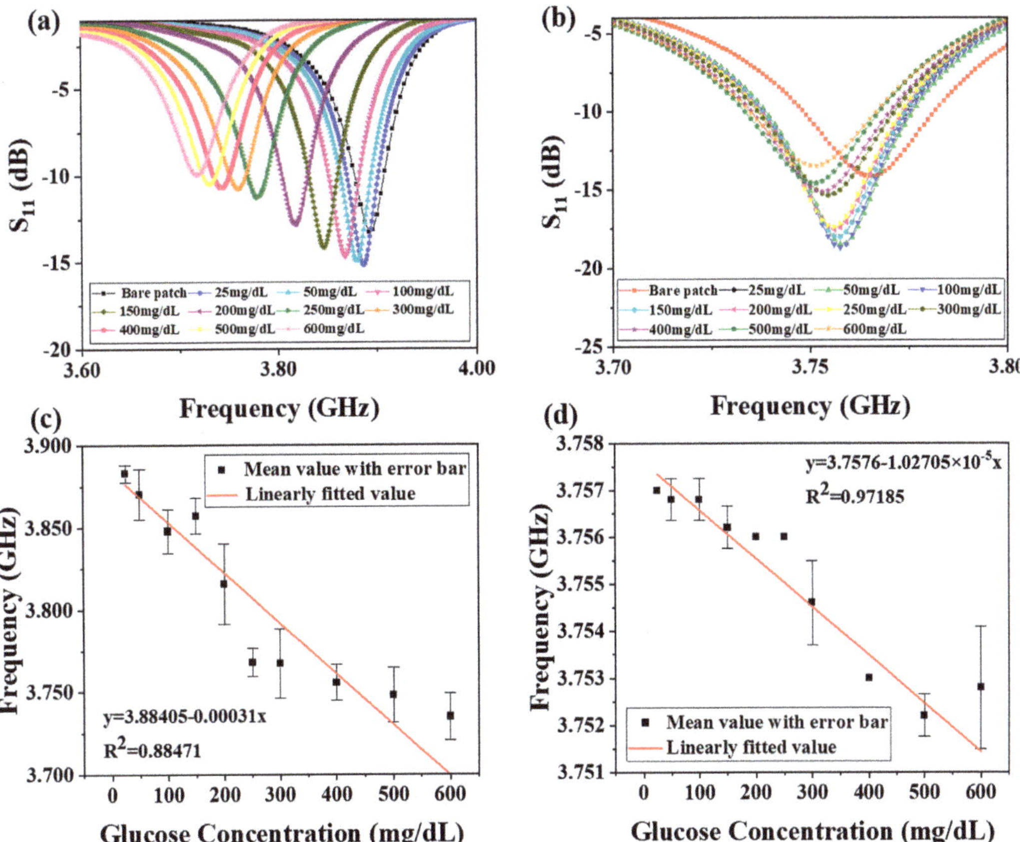

Figure 6. Measured S$_{11}$ for different concentrations of glucose sample solutions (25–600 mg/dL): (**a**) bare Reader and (**b**) RFID system. Linearly fitted resonant frequency, including the mean value of resonant frequencies with error bars. (**c**) Reader structure ($R^2 = 0.88471$, RSD < 1%) and (**d**) RFID system ($R^2 = 0.97185$, RSD < 1%).

Table 1. Performance comparison with other previously reported methods.

Ref.	Sensor structure	Sensitivity	Operating Frequency	Non-Contact	Quantitative Test
[18]	Triple-pole CSRR	0.062 dB/mg·dL^{-1}	2.3 GHz	No	Pipette required (600 μL)
[24]	IDC and spiral inductor	1.99 MHz/mg·dL^{-1}	2.45 GHz	No	Pipette required (5 μL)
[25]	Air bridge structure	1.08 MHz/mg·dL^{-1}	9.20 GHz	No	Pipette required (1 μL)
[26]	CSRR	5 kHz/mg·dL^{-1}	2.48 GHz	No	Channel required (0.637 mL)
[27]	Cylindrical dielectric resonator antenna sensor	2.81 kHz/mg·dL^{-1}	5.25 GHz	Yes	No
This work	**CSRR and SSC**	**10.27 kHz/mg·dL^{-1}**	**3.77 GHz**	**Yes**	**Yes (1.56 μL)**

4. Conclusions

This paper presents a non-contact inspired microwave resonator based on the RFID concept integrated with a microfluidic system developed for the sensing of glucose solutions. The overall system is comprised of a Reader, Tag, and microchannel, functioning for healthcare applications such as the detection of the blood level in the human body, etc. Our work provides an efficient solution for the non-contact or possibly short distance signal transmission and biomarker detection method and a feasible fabrication plan for the microwave sensor combined with metal-on-glass, PCB, and PDMS microfluidic channel has been made. The simulation and measurement results show its ability to sense changes in glucose concentrations around the sensitive area. The CSRR Reader can converge electromagnetic flux into the middle gap, and the condensed energy can penetrate the glass substrate to reach the special-customized SSC Tag, as well as the glucose solution. Glucose samples enter the microfluidic cavity through the microchannel and directly contact the Tag, causing the change of the SSC and, thus, changing its resonant frequency. The biosensing response was obtained based on the relationship between glucose concentration and resonant frequency. In general, the proposed microwave design based on the one-to-one correspondence with RFID function will greatly improve the detection speed and efficiency in the face of the complex and huge amount of data that may exist in the future, most importantly, providing an idea for short-distance non-intrusive testing, and some preparations and pavements for future in vitro experiments have been made.

Author Contributions: M.G. and Y.M. contributed equally to this work. Conceptualization, M.G. and T.Q.; methodology, M.G.; software, M.G.; validation, M.G. and Y.M.; formal analysis, M.G. and Y.M.; investigation, M.G. and Y.M.; resources, M.G. and T.Q.; data curation, M.G. and Y.M.; writing—original draft preparation, M.G.; writing—review and editing, M.G., T.Q., J.L., and Y.J.; visualization, M.G., Y.M. and T.Q.; supervision, T.Q.; project administration, T.Q.; funding acquisition, T.Q. All authors have read and agreed to the published version of the manuscript.

Funding: This research is funded by National Natural Science Foundation of China (Grant No. 61801146), Project funded by China Postdoctoral Science Foundation (Grant No. 2021M691284, No. 2021M691360), and the Fundamental Research Funds for the Central Universities (Grant No. JUSRP12026), and Suzhou Institute of Nano-Tech and Nano-Bionics Open Project (No. 20SZ04), Zhejiang Postdoctoral Research Foundation (No. ZJ2020101).

Institutional Review Board Statement: Not applicable.

Informed Consent Statement: Not applicable.

Data Availability Statement: Not applicable.

Conflicts of Interest: The authors declare no conflict of interest.

References

1. Cho, N.H.; Shaw, J.E.; Karuranga, S.; Huang, Y.; Fernandes, J.D.d.R.; Ohlrogge, A.W.; Malanda, B. IDF Diabetes Atlas: Global estimates of diabetes prevalence for 2017 and projections for 2045. *Diabetes Res. Clin. Pract.* **2018**, *138*, 271–281. [CrossRef] [PubMed]
2. Danaei, G.; Finucane, M.M.; Lu, Y.; Singh, G.M.; Cowan, M.J.; Paciorek, C.J.; Lin, J.K.; Farzadfar, F.; Khang, Y.-H.; Stevens, G.A.; et al. National, regional, and global trends in fasting plasma glucose and diabetes prevalence since 1980: Systematic analysis of health examination surveys and epidemiological studies with 370 country-years and 2.7 million participants. *Lancet* **2011**, *378*, 31–40. [CrossRef]
3. Fagherazzi, G.; Ravaud, P. Digital diabetes: Perspectives for diabetes prevention, management and research. *Diab. Metab.* **2019**, *45*, 322–329. [CrossRef] [PubMed]
4. Salek-Maghsoudi, A.; Vakhshiteh, F.; Torabi, R.; Hassani, S.; Ganjali, M.R.; Norouzi, P.; Hosseini, M.; Abdollahi, M. Recent advances in biosensor technology in assessment of early diabetes biomarkers. *Biosens. Bioelectron.* **2018**, *99*, 122–135. [CrossRef]
5. Liu, Y.; Deng, C.; Tang, L.; Qin, A.; Hu, R.; Sun, J.Z.; Tang, B.Z. Specific Detection of D-Glucose by a Tetraphenylethene-Based Fluorescent Sensor. *J. Am. Chem. Soc.* **2011**, *133*, 660–663. [CrossRef] [PubMed]
6. Dong, Y.-L.; Zhang, H.-G.; Rahman, Z.U.; Su, L.; Chen, X.-J.; Hu, J.; Chen, X.-G. Graphene oxide-Fe3O4 magnetic nanocomposites with peroxidase-like activity for colorimetric detection of glucose. *Nanoscale* **2012**, *4*, 3969–3976. [CrossRef]
7. Rahsepar, M.; Foroughi, F.; Kim, H. A new enzyme-free biosensor based on nitrogen-doped graphene with high sensing performance for electrochemical detection of glucose at biological pH value. *Sens. Actuators B Chem.* **2019**, *282*, 322–330. [CrossRef]
8. Lu, Z.; Wu, L.; Dai, X.; Wang, Y.; Sun, M.; Zhou, C.; Du, H.; Rao, H. Novel flexible bifunctional amperometric biosensor based on laser engraved porous graphene array electrodes: Highly sensitive electrochemical determination of hydrogen peroxide and glucose. *J. Hazard. Mater.* **2021**, *402*, 123774. [CrossRef]
9. Min, J.; Sempionatto, J.R.; Teymourian, H.; Wang, J.; Gao, W. Wearable electrochemical biosensors in North America. *Biosens. Bioelectron.* **2021**, *172*, 112750. [CrossRef]
10. Tseng, M.L.; Jahani, A.; Leitis, A.; Altug, H. Dielectric metasurfaces enabling advanced optical biosensors. *ACS Photonics* **2021**, *8*, 47–60. [CrossRef]
11. Mehrotra, P.; Chatterjee, B.; Sen, S. EM-Wave Biosensors: A Review of RF, Microwave, mm-Wave and Optical Sensing. *Sensors* **2019**, *19*, 1013. [CrossRef]
12. Xu, Z.; Han, R.; Liu, N.; Gao, F.; Luo, X. Electrochemical biosensors for the detection of carcinoembryonic antigen with low fouling and high sensitivity based on copolymerized polydopamine and zwitterionic polymer. *Sens. Actuators B Chem.* **2020**, *319*, 128253. [CrossRef]
13. Parlak, O.; Incel, A.; Uzun, L.; Turner, A.P.F.; Tiwari, A. Structuring Au nanoparticles on two-dimensional MoS2 nanosheets for electrochemical glucose biosensors. *Biosens. Bioelectron.* **2017**, *89*, 545–550. [CrossRef]
14. Zhang, Q.; Wu, Y.; Xu, Q.; Ma, F.; Zhang, C.-Y. Recent advances in biosensors for in vitro detection and in vivo imaging of DNA methylation. *Biosens. Bioelectron.* **2021**, *171*, 112712. [CrossRef]
15. Chen, C.; Wang, J. Optical biosensors: An exhaustive and comprehensive review. *Analyst* **2020**, *145*, 1605–1628. [CrossRef] [PubMed]
16. Yoo, S.M.; Lee, S.Y. Optical biosensors for the detection of pathogenic microorganisms. *Trends Biotechnol.* **2016**, *34*, 7–25. [CrossRef] [PubMed]
17. Jain, M.C.; Nadaraja, A.V.; Mohammadi, S.; Vizcaino, B.M.; Zarifi, M.H. Passive microwave biosensor for real-time monitoring of subsurface bacterial growth. *IEEE Trans. Biomed. Circuits Syst.* **2021**, *15*, 122–132. [CrossRef] [PubMed]
18. Omer, A.E.; Shaker, G.; Safavi-Naeini, S.; Alquie, G.; Deshours, F.; Kokabi, H.; Shubair, R.M. Non-invasive real-time monitoring of glucose level using novel microwave biosensor based on triple-pole CSRR. *IEEE Trans. Biomed. Circuits Syst.* **2020**, *14*, 1407–1420. [CrossRef]
19. Mohammadi, S.; Nadaraja, A.V.; Luckasavitch, K.; Jain, M.C.; June Roberts, D.; Zarifi, M.H. A Label-free, non-intrusive, and rapid monitoring of bacterial growth on solid medium using microwave biosensor. *IEEE Trans. Biomed. Circuits Syst.* **2020**, *14*, 2–11. [CrossRef]
20. Lee, H.-J.; Lee, J.-H.; Moon, H.-S.; Jang, I.-S.; Choi, J.-S.; Yook, J.-G.; Jung, H.-I. A planar split-ring resonator-based microwave biosensor for label-free detection of biomolecules. *Sens. Actuators B Chem.* **2012**, *169*, 26–31. [CrossRef]
21. Hosseini, N.; Baghelani, M. Selective real-time non-contact multi-variable water-alcohol-sugar concentration analysis during fermentation process using microwave split-ring resonator based sensor. *Sens. Actuators A Phys.* **2021**, *325*, 112695. [CrossRef]
22. Camli, B.; Altinagac, E.; Kizil, H.; Torun, H.; Dundar, G.; Yalcinkaya, A.D. Gold-on-glass microwave split-ring resonators with PDMS microchannels for differential measurement in microfluidic sensing. *Biomicrofluidics* **2020**, *14*, 054102. [CrossRef]
23. Kumar, A.; Wang, C.; Meng, F.-Y.; Liang, J.-G.; Xie, B.-F.; Zhou, Z.-L.; Zhao, M.; Kim, N.-Y. Aerosol deposited BaTiO3 film based interdigital capacitor and squared spiral capacitor for humidity sensing application. *Ceram. Int.* **2021**, *47*, 510–520. [CrossRef]
24. Kim, N.-Y.; Adhikari, K.K.; Dhakal, R.; Chuluunbaatar, Z.; Wang, C.; Kim, E.-S. Rapid, sensitive, and reusable detection of glucose by a robust radiofrequency integrated passive device biosensor chip. *Sci. Rep.* **2015**, *5*, 7807. [CrossRef]

25. Kim, N.Y.; Dhakal, R.; Adhikari, K.K.; Kim, E.S.; Wang, C. A reusable robust radio frequency biosensor using microwave resonator by integrated passive device technology for quantitative detection of glucose level. *Biosens. Bioelectron.* **2015**, *67*, 687–693. [CrossRef]
26. Ebrahimi, A.; Scott, J.; Ghorbani, K. Microwave reflective biosensor for glucose level detection in aqueous solutions. *Sens. Actuators A Phys.* **2020**, *301*, 111662. [CrossRef]
27. Iqbal, A.; Smida, A.; Saraereh, O.A.; Alsafasfeh, Q.H.; Mallat, N.K.; Lee, B.M. Cylindrical dielectric resonator antenna-based sensors for liquid chemical detection. *Sensors* **2019**, *19*, 1200. [CrossRef]

 *biosensors*

Article

Sequence-Independent DNA Adsorption on Few-Layered Oxygen-Functionalized Graphene Electrodes: An Electrochemical Study for Biosensing Application

Narges Asefifeyzabadi [1], Torrey E. Holland [2], Poopalasingam Sivakumar [2], Saikat Talapatra [2], Ishani M. Senanayake [1], Boyd M. Goodson [1] and Mohtashim H. Shamsi [1,*]

[1] School of Chemical and Biomolecular Sciences, Southern Illinois University, 1245 Lincoln Drive, Carbondale, IL 62918, USA; narges.asefifeyzabadi@siu.edu (N.A.); ishani.senanayake@siu.edu (I.M.S.); bgoodson@chem.siu.edu (B.M.G.)
[2] School of Physics and Applied Physics, Southern Illinois University, Carbondale, IL 62918, USA; torrey.holland@siu.edu (T.E.H.); psivakumar@siu.edu (P.S.); saikat@siu.edu (S.T.)
* Correspondence: mshamsi@siu.edu

Abstract: DNA is strongly adsorbed on oxidized graphene surfaces in the presence of divalent cations. Here, we studied the effect of DNA adsorption on electrochemical charge transfer at few-layered, oxygen-functionalized graphene (GO_x) electrodes. DNA adsorption on the inkjet-printed GO_x electrodes caused amplified current response from ferro/ferricyanide redox probe at concentration range 1 aM–10 nM in differential pulse voltammetry. We studied a number of variables that may affect the current response of the interface: sequence type, conformation, concentration, length, and ionic strength. Later, we showed a proof-of-concept DNA biosensing application, which is free from chemical immobilization of the probe and sensitive at attomolar concentration regime. We propose that GO_x electrodes promise a low-cost solution to fabricate a highly sensitive platform for label-free and chemisorption-free DNA biosensing.

Keywords: DNA biosensors; graphene electrodes; inkjet-printing; trinucleotide repeats; label-free; electrochemical biosensors

Citation: Asefifeyzabadi, N.; Holland, T.E.; Sivakumar, P.; Talapatra, S.; Senanayake, I.M.; Goodson, B.M.; Shamsi, M.H. Sequence-Independent DNA Adsorption on Few-Layered Oxygen-Functionalized Graphene Electrodes: An Electrochemical Study for Biosensing Application. *Biosensors* **2021**, *11*, 273. https://doi.org/10.3390/bios11080273

Received: 26 July 2021
Accepted: 12 August 2021
Published: 14 August 2021

Publisher's Note: MDPI stays neutral with regard to jurisdictional claims in published maps and institutional affiliations.

1. Introduction

Graphene comprises a single layer of graphite in which carbon atoms arrange themselves in a 2D hexagonal lattice with metal-like charge carrier properties [1,2]. The interfacial properties of graphene interface, e.g., charge transport, can be easily and sensitively modulated by interactions with biomolecules [3–5], Therefore, single and multilayered graphene and its derivatives (i.e., graphene oxide and reduced graphene) have been widely exploited as sensing platforms to detect DNA [6–8], proteins [9–12], and small molecules [13] using a variety of detection methods including optical [6,13], scanning probe [10], electrical [7,8], and electrochemical techniques [9,11,12].

To harness the properties of such interfaces for sensitive biosensing platforms, interactions of nucleic acids with graphene-based surfaces have been extensively studied in two regimes, i.e., physisorption and chemisorption [14]. The interfacing of DNA and graphene surfaces is often achieved by the simple mixing of DNA oligonucleotide solution with graphene solution (a liquid/liquid interface) to form DNA-graphene hybrids [15,16], and DNA self-assembly on graphene electrodes (a liquid/solid interface) [17]. Theoretical studies have confirmed π stacking interactions between hydrophobic DNA basepairs and graphitic carbon rings of graphene as a driving force of adsorption [18–20]. Nucleobases show differential interactions where base rings are parallel to the graphene surface, which maximizes π–π stacking [21]. Theoretical studies have ranked the adsorption energies of bases on graphene as G > A > T > C [18,19], whereas experimental studies involving isothermal titration calorimetry indicates that the trend follows as G > A > C > T [22].

Thus, non-electrostatic interactions dominate the binding, where purine bases bind more strongly than pyrimidines [18,22,23]. Moreover, not all bases in a duplex conformation are adsorbed, and a diverse range of DNA conformations are likely to exist, which may result in competitive binding between bases and graphene oxide (GO)—leading to partial denaturation of double-stranded DNA (dsDNA) and exposition of single-stranded DNA (ssDNA) regions [24]. Both DNA and GO carry negative charge; therefore, a high ionic strength (up to a few mM Mg^{2+} or $\geq$100 mM NaCl) is used to facilitate the adsorption that overcomes the kinetic barrier [15]. DNA conformation can also influence the adsorption process due to intramolecular basepairing. A solution-based study involving the adsorption of ssDNA on GO has revealed that shorter DNA adsorbs faster and tighter to the surface, whereas lower pH and higher ionic strength conditions favor the adsorption. Despite the expanded theoretical and experimental research in this area, the question of how DNA interaction with graphene-based materials such as oxidized graphene surfaces can modulate electrical and electrochemical charge transport across the interface—a liquid/solid interface—is underexplored.

For this study, we hypothesized that DNA adsorption on oxidized graphene interface may lead to unique electrochemical signatures, which eventually may lead to development of a simple electrochemical DNA biosensor. First, we prepared and thoroughly characterized a few-layered, oxygen-functionalized graphene ink (GO_x), followed by fabricating the GO_x electrodes on indium tin oxide (ITO) surfaces by inkjet printing, as illustrated in Scheme 1a. Then, we physisorbed the DNA sequences on the GO_x electrodes in high divalent cation concentration (20 mM Mg^{2+}), and monitored the electrochemical responses of the interface using differential pulse voltammetry, as depicted in Scheme 1b. The proposed sensing mechanism in the scheme explains the lower current obtained on the unmodified electrode due to repulsion between the negatively charged redox probe and negative O-groups on the GO_x surface. Then, the adsorption of the DNA sequence on a GO_x surface in presence of Mg^{2+} ions reduces the electrostatic repulsion, which leads to higher diffusion of the redox probe producing higher electrochemical current measured by DPV. In particular, we studied their electrochemical signature with respect to sequence type, conformation, concentration, length, and ionic strength in the presence of a soluble redox probe, $Fe(CN)_6^{3-/4-}$. Finally, we performed a proof-of-concept biosensing test that does not require chemical labelling and immobilization of the probe.

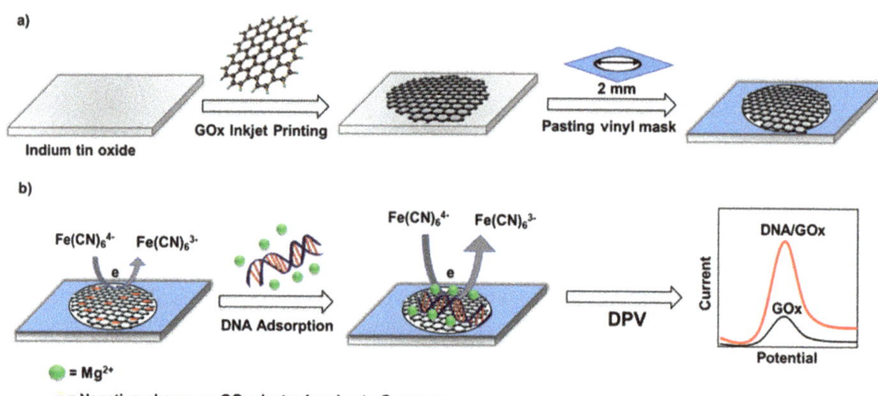

Scheme 1. (**a**) Illustrating fabrication of electrodes by inkjet printing of exfoliated few-layered oxygen functionalized graphene (GO_x) on indium tin oxide substrate, followed by adhesion of a vinyl mask selectively exposing a GO_x surface with 2 mm diameter. (**b**) Depiction of sensing mechanism: the lower current on the unmodified electrode is due to the repulsion between the negatively charged redox probe and the negative O-groups on the GO_x surface. Adsorption of DNA sequence on a GO_x surface in presence of Mg^{2+} ions reduces the electrostatic repulsion, which leads to higher diffusion of the redox probe producing higher electrochemical current measured by DPV.

2. Experimental

2.1. Materials

For the sequence type, trinucleotide repeats sequences (TNRs) were used with CGG, CAG, and GAA repeats. All the synthetic TNR oligos were purchased from Integrated DNA Technologies (Coralville, IA, USA). The different sequences and lengths of the TNRs are shown in Table S1. Graphite flakes, 99% carbon basis (-325 mesh particle size, $\geq$99%), ethyl cellulose with viscosity 4 cP (5%) in toluene/ethanol 80:20 (lit), and 48% ethoxyl, α-Terpineol (90%), tris(hydroxymethyl)aminomethane (Tris-ClO$_4$) were purchased from Sigma-Aldrich (St Louis, MO, USA). Ethanol (95%); cyclohexanone, K$_4$[Fe(CN)$_6$], K$_3$[Fe(CN)$_6$] (99+%), and phosphate-buffered saline (PBS) pH 7.4 (5$\times$ solution) were purchased from Fisher Scientific. Holey carbon-coated grids for electron microscopic studies were purchased from SPI Supplies (West Chester, PA, USA). Indium tin oxide coated substrate (ITO) with resistivity 4–8 Ω/sq was obtained from Delta Technologies (Loveland, CO, USA). An Electrochemical Workstation CHI 660E from CHI Instruments (Austin, TX, USA) was used for electrochemical characterization of the interface. A Metrohm Autolab potentiostat was used for differential pulse voltammetry measurements. An Ag/AgCl (with KCl solution) reference electrode and platinum wire auxiliary electrode was procured from Bioanalytical Systems Inc. (West Lafayette, IN, USA). Fujifilm Dimatix Materials Printer (DMP-2800) was used for printing on ITO substrate and purchased from Integrity Industrial Inkjet Integration (W. Lebanon, NH, USA). A NanoDrop One Spectrometer from Thermo Scientific was used for UV-Visible characterization of the few-layered graphene ink. A Graphtec Cutting Plotter (CE6000-40) was used to cut vinyl sheets to be used for covering the ITO substrates, while exposing a 2 mm diameter circle for studying the interface. Open-source software, Inkscape (Version 0.91), was used to design patterns for printing. A portable four-point probe test meter (HM21) from Jandel Engineering Limited (UK) was used for electrical characterization of the printed graphene ink. Raman characterization of graphene ink was performed using a Horiba iHR550 imaging spectrometer with near-infrared (NIR) excitation light source at a wavelength of ~785 nm (iBeam-Smalt-785-S-WS, TOPTICA Photonics). A Hitachi H-7650 transmission electron microscope operating at 60 kV and a Quanta 450 FEG (FEI) at the SIU Imaging Center were used for electron microscopic characterization of graphene ink and fabricated devices. Dynamic Light Scattering (DLS) measurements were performed using a DynaPro NanoStar purchased from Wyatt Technology Corporation.

2.2. Preparation of Ink

First, graphene sheets were exfoliated by ultrasonic exfoliation of graphite flakes in ethyl cellulose/ethanol mixture using an ultrasonication method as reported earlier [25,26], which gives a solid product of graphene/ethyl cellulose (Gr/EC). To prepare the graphene-based ink for inkjet printing, the Gr/EC powder was dispersed in an 83:17 cyclohexanone/terpineol mixture, sonicated for 2.5 h [25,26], and stored at room temperature. For inkjet printing, 3.5% ink concentration was prepared having 10 cP viscosity, which is suitable for inkjet printing using Dimatix printer.

2.3. Characterization of Ink

For ink characterization, a UV-vis spectrum scan of the ink was obtained between 200–800 nm by dropping a 5 µL aliquot of the ink on the NanoDrop Spectrometer. Transmission electron microscopic (TEM) images of the graphene nanosheets in the ink were obtained by dropping a small amount of the ink on a holey carbon coated grid followed by air drying before imaging. To measure the particle size of the ink, 200 µL of the sample was diluted up to 50 mL using absolute ethanol as a solvent. The mixture was sonicated for one minute. The DLS analysis was performed using 300 µL of the solution. This dilution factor allowed a mostly translucent solution to be analyzed. To confirm the type of graphene material, the ink was analyzed by Raman spectroscopy after annealing at 350 °C. The ink was directly analyzed on glass slides (Eisco microscope slides) using a modular Raman microscope system. A grating of 600 g/mm was used in the spectrometer, and a CCD camera on the Raman microscope was used to locate the ink deposition. After performing the characterization tools mentioned above, we identified our ink as few-layered oxygen functionalized graphene (GO_x).

2.4. Fabrication and Characterization of GO_x Electrodes

The GO_x ink was printed on ITO substrates using a Dimatix inkjet printer, employing parameters as previously reported [26]. For resistance measurements, line patterns were printed on ITO using printing cycles (3, 5, and 7) and sintered in a furnace for 30 min at 350 °C. Later, thicknesses of the patterned lines from different cycles were measured by scanning electron microscopy (SEM) before and after sintering.

Circular graphene patterns 2 mm in diameter were printed on the clean and dry ITO substrates for DNA adsorption studies. Following printing, thermal annealing of the electrodes was performed at various temperatures (250 °C to 350 °C) in air for 30 min. The working electrode area was separated from the rest of the ITO surface by pasting a vinyl sheet on the surface with a 2 mm-diameter hole to expose the inkjet-printed electrode while covering the rest of the electrically conducting area. The electrochemical impedance spectroscopy (EIS) was performed in the presence of 1 mM $K_4[Fe(CN)_6]/K_3[Fe(CN)_6]$ (1:1) prepared in PBS buffer (pH 7.2) at 0.3 V against Ag/AgCl reference (3 M KCl) using 5 mV amplitude and 100 kHz to 0.1 Hz frequency. Then, cyclic voltammetry with a potential window of 0−0.6 V was performed at the scan rates of 50–500 mV/s in 1 mM $K_4[Fe(CN)_6]/K_3[Fe(CN)_6]$ (1:1) prepared in PBS buffer (pH 7.2). The electrochemical measurements were recorded against an Ag/AgCl reference electrode and a platinum wire counter electrode.

2.5. Electrochemical Study of DNA/GO_x Interface

Solution hybridized double-stranded DNA (dsDNA) were prepared in 100 µM Tris buffer (pH 8.5) containing 200 mM NaCl and 20 mM $MgCl_2$. The mixture was heated to the melting temperature for 30 min and annealed at room temperature for 1 h. For electrochemical characterization of the DNA/GO_x interface, 5 µL aliquots of DNA solution were dropped on the GO_x electrode and incubated for 12–16 h at 4 °C. For concentration-dependence study, a range from 1 aM to 10 nM was used to adsorb dsDNA (dsCGG-8, dsCAG-8, and dsGAA-8; see Table S1) on the GO_x surface. For the length–dependence study, 5, 8, and 10 repeat lengths of 1 aM dsCGG were adsorbed on the GO_x surface. For surface hybridization detection, concentration of the physisorbed probe sequence (ssCGG-8) was first optimized, and later exposed to complementary GGC-8 target and noncomplementary sequences (NC-1 = TTC-8 and NC-2 = CAG-8) at 1 aM concentration. To confirm the ionic strength effect on the adsorption, EDTA was added to the DNA solution to chelate Mg^{2+} before the adsorption on the GO_x surface. Following the adsorption of the sequences for various variables, differential pulse voltammetry was performed with a potential range of 0.1–0.5 V and an amplitude of 0.05 V in 1 mM $Fe(CN)_6^{3-/4-}$ prepared in PBS buffer (pH 7.2).

3. Results and Discussion

Figure 1a illustrates the graphene-based ink preparation in an environmentally benign solvent as reported previously [25–27], which was found stable even after 15 months of storage at room temperature. The graphene-based ink was characterized by TEM, as shown in Figure 1b, indicating that the process yielded few-layered graphene nanosheets. The UV-vis characterization shown in Figure 1c confirms the formation of graphene sheets with a strong absorption peak at 275 nm due to excitation of π-plasmon of graphitic structure, which corroborates the previous reports [28–30]. The bands shown in the Raman spectrum in Figure 1d are similar to annealed exfoliated graphene comprising four bands, i.e., G band at ∼1590 cm^{-1}, 2D band at ∼2700 cm^{-1}, and disorder-related D and D′ peaks at ∼1350 and ∼1620 cm^{-1}, respectively [27]. The sharpness of the D and G bands are similar to the graphene-ethylene cellulose nanocomposite [27]. As previously indicated, a lower value of the I(D)/I(G) peak ratio suggests an increase in graphitization and lower defects, which can lead to better electrical conductivity [31]. Moreover, defects and oxidation might be introduced in the material during the sonication and annealing process, and may also arise from the smaller sizes of the sheets [27,32]. The I(D)/I(G) value for this ink is ~0.58, which indicates a moderate level of defects and oxidation during exfoliation and annealing [27], while the sharp peaks confirm the improved graphitization compared to typical graphene oxide (GO) material [33]. The composition of the ink was verified by EDS elemental analysis (Figure S1 in Supplementary Information) showing 56% C-atoms and 34% O-atoms after sintering the surface yielding a C/O ratio = 1.5, which confirms the partial oxidation of the ink during the formulation and annealing process. Figure 1e shows the particle size distribution of 82% of the particles with a diameter in the range of 250 to 700 nm, with an average diameter of 432 nm. The extensive characterization presented here suggests the ink material as few-layered oxygen-functionalized graphene ink (GO_x), where the presence of a moderate amount of oxygen functional groups on the surface may facilitate intermediate loadings and binding interactions for DNA [34].

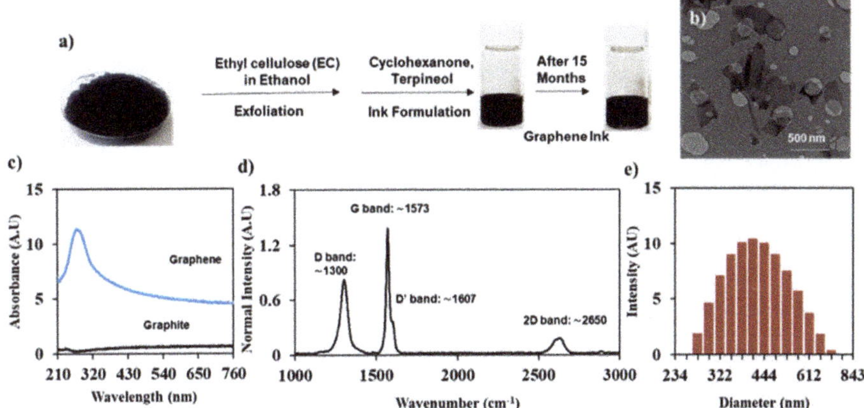

Figure 1. Synthesis and characterization of the few-layered graphene ink. (**a**) Exfoliation and formulation, (**b**) TEM image of exfoliated graphene sheets, (**c**) UV−vis spectrum of graphite and graphene ink, (**d**) Raman spectra of the ink, and (**e**) dynamic light scattering of the ink.

The GO_x ink was printed and characterized by various methods. We investigated the relationship between the printing cycles and pattern thickness by SEM shown in Figure 2. The cross-section images of 3, 5, and 7 printing cycles of GO_x patterns following sintering at 300 °C (Figure 2a) confirm the increase in thickness of the printed pattern with the number of printing cycles, i.e., 3 cycles (0.75 μm), 5 cycles (1.5 μm), and 7 cycles (2.4 μm). The morphology of the GO_x surface reveals a smooth surface before sintering, and the appearance of wrinkles after sintering. Such wrinkled roughness may endow a material with higher electrochemical current due to an increase in the surface area, as previously reported [35,36]. Then, we investigated the electrical and electrochemical behavior of the GO_x electrode. Figure 3a shows that the resistivity significantly decreased with high precision when the sintering temperature reached at 350 °C. Figure 3b indicates that the trend decreases with the printing cycles, where five cycles have significantly higher precision. Figure 3c shows the Nyquist form of electrochemical impedance spectroscopy (EIS) plots for the bare ITO substrate (see inset), GO_x ink before sintering, and GO_x ink after sintering (see inset). The charge transfer resistance of the GO_x ink before sintering (R_{ct} = 246 kΩ) is almost two orders of magnitude higher than that of bare ITO (R_{ct} = 9.5 kΩ). While the charge transfer resistance of the GO_x ink following sintering decreased (R_{ct} = 4.2 kΩ), which is less than 50% of the bare ITO resistance. This is an indication of substantially improved electrochemical property of the GO_x/ITO electrode after sintering. The modified Randle's equivalent circuit model used for fitting the Nyquist plot and the fitting values are shown in Figure S2 and Table S2, respectively. Following this characterization, we used the conditions of five printing cycles and sintering at 350 °C for 30 min to study the DNA/GO_x interface.

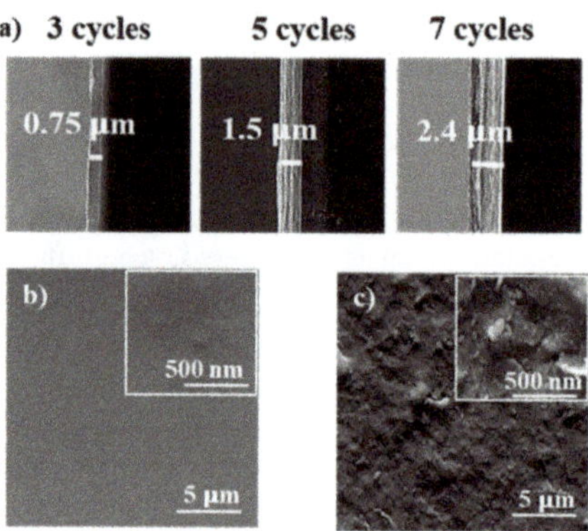

Figure 2. (**a**) SEM images of 3, 5, and 7 printed cycles of GO_x ink after sintering at 300 °C. (**b**) SEM images of top surface of the printed GO_x electrode before sintering and (**c**) after sintering.

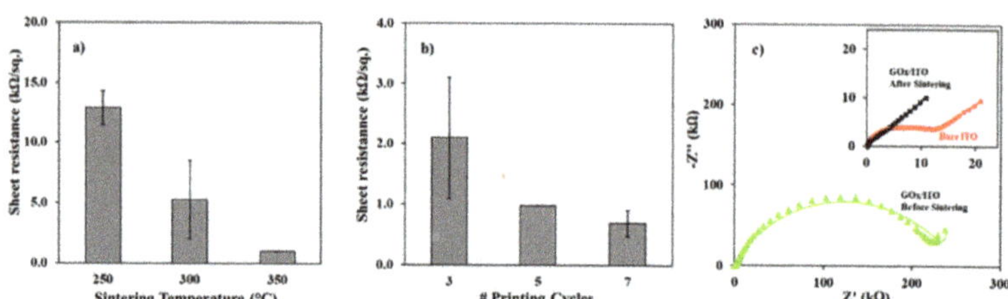

Figure 3. (**a**) Resistivity versus sintering temperature for 30 min following five printing cycles. (**b**) Resistivity versus number of printing cycles following sintering under fixed conditions (350 °C for 30 min). (**c**) Nyquist form of the EIS plot for bare ITO and GO_x/ITO (before and after sintering).

The electrochemical current of the surface-adsorbed dsCGG-8 (10 nM) was investigated on the GO_x electrode by voltametric techniques. Figure 4a shows cyclic voltametric curves before and after DNA modification of the GO_x electrode. The integrated peak current of the dsCGG-8/GO_x (12.5 µA) is almost twice the GO_x current (6.5 µA). The enhanced charge transport behavior of the DNA modified GO_x electrode was monitored over a range of 50–500 mV/s scan rates (Figure 4b), which follows the Randles–Sevcik equation model wherein peak current is proportional to square root of the scan rate [37]. However, Figure 4c shows that cathodic-anodic peaks separation (ΔE_p) increases with the scan rate (see cyclic voltamogram in Figure S3), which is an indicative of a quasi-reversible electron kinetics at the electrode surface and is typical to the DNA modified electrodes [38]. Figure 4d shows the cyclic voltammetry performed up to 50 scans for the DNA modified GO_x electrode, which presents no substantial change in the peak current and overpotential with the number of scans, thus it suggests a stable DNA/GO_x interface. We attribute this strong adsorption affinity to mild oxidation of the graphene sheets in the GO_x ink (C/O = 1.5), similar to recently reported strong adsorption of ssDNA and dsDNA on graphdiyne surface with a low oxidation degree [39]. Moreover, high cationic strength screens negative charge between the two materials and facilitates the adsorption. There is an intriguing question of whether this current enhancement is due to the DNA mediated electron transport or enhanced diffusion of the redox probe owing to high cationic strength environment. To test the effect, we added EDTA in the DNA solutions to chelate the Mg^{2+}, followed by incubation of the mixture with the GO_x electrode. Figure 5 shows the DPV current of ssCGG-8 and dsCGG-8 in the presence and absence of Mg^{2+} and compared with the unmodified GO_x electrode current. Evidently, the current in the absence of Mg^{2+} is very similar to background current or unmodified GO_x current (~22 µA). This result confirms the dual role of Mg^{2+}, i.e., promoting the DNA adsorption as suggested previously [15], and reducing the electrostatic repulsion between the redox probe and the electrode surface. It is important to note that DNA modified electrodes have shown higher resistance to charge transfer than that of unmodified electrode in low cationic strength due to lack of diffusion of the redox probe [40], which was improved after the addition of divalent cations [41–43]. Based on these results, we conclude that the high concentration of divalent cations in the environment screen the negative charge on both materials, which ultimately increases the diffusion of the redox probe, $Fe(CN)_6^{3-/4-}$, to the electrode surface leading to higher current response.

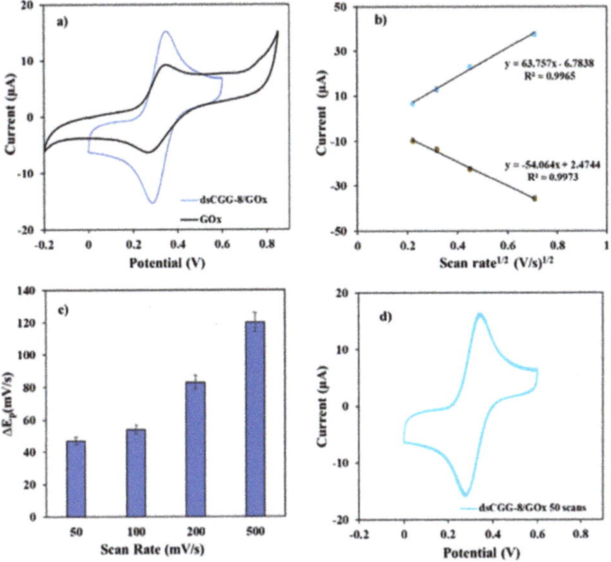

Figure 4. (a) Cyclic voltametric curves before and after adsorption of dsCGG-8 on GO_x. (b) Plot of cathodic and anodic voltametric current of dsCGG-8 as a function of square root of scan rate. (c) Cathodic–Anodic peaks separation (ΔE_p) versus scan rate for the dsCGG-8 modified electrodes. (d) 50 scans of CV curves of dsCGG-8 on GO_x electrode. The electrochemical measurements were performed in presence of 1 mM $Fe(CN)_6^{3-/4-}$ using Pt wire counter electrode and Ag/AgCl (3 M KCl) reference electrode. Scan rate for CV measurement is 100 mV/s. Concentration of dsCGG-8 was 10 nM.

Next, we tested the current responses of conformation, concentration, length, and sequence types of DNA using DPV. Figure 6a shows the DPV curves of GO_x, ssCGG-8, dsCGG-8, and a noncomplementary mixture of CGG-8/CAG-8 at 10 nM concentration. The current response of the double-stranded conformation (dsCGG-8) was higher than the single-stranded conformation (ssCGG-8), while the current response of the noncomplementary mixture showed similar current response as the single-stranded conformation, which corroborates the fact that noncomplementary strands remained unhybridized and in single-stranded conformation. Figure 6b shows the current response of dsCGG-8 over a range of 1 aM to 10 nM. The least-squares regression for dsCGG-8 formed a linear curve fit with a correlation coefficient $R^2 = 0.9785$ (Figure 6b). The detection limit signal, LOD_y, was obtained using LOD_y = (blank signal + 3*SD_{blank}) − intercept; where blank signal = 22 µA, 3*SD_{blank} = 3*2.5 µA, and intercept = 46.566 µA. Then, the 'theoretical' limit of detection concentration, LOD_x, was determined as 0.8 aM by calculating antilog ($LOD_y ÷$ sensitivity), where sensitivity or the slope of the curve 'm' is 0.9338. Figure 6c shows the current responses of dsCGG comprising 5, 8, and 10 dsCGG repeat units (i.e., 15, 24, and 30 nucleotide lengths respectively) at 1 aM concentration regime. The upward trend of the current response with respect to length indicates that the increase in DNA adsorption aided by Mg^{2+} facilitates the charge transport at the interface. This length-dependent current response may lead to sensitive and label-free discrimination of normal and abnormal lengths of DNA repeat sequences associated with neurodegenerative diseases, as reported recently on MoS_2 nanosheets surfaces [44]. Then, we tested the adsorption of various single-stranded, double-stranded, and noncomplementary sequences of same length (i.e., 8 trinucleotide repeats or 24 nucleotides) at 10 nM concentration. Figure 6d shows the current responses of trinucleotide repeat types CGG, CAG, and GAA (single-, double-, and noncomplementary) revealing the following trends. First, the current responses of double-stranded conformations, in all cases, are higher than their single strands. Second, the noncomplementary responses are similar to single-stranded conformation. Third, the current responses for all sequences are similar with respect to their conformation. Overall, we learn that the increase in DNA interaction with the surface in the high cationic environment increases the current response, which can be indicative of its conformation, concentration, and length. Despite the differential adsorption affinity of graphene-based materials for different nucleotide types, the DNA adsorption was sequence-independent. This result can be rationalized as the effect of high concentration of Mg^{2+} perhaps masks the sequence-dependent affinity between DNA and GO_x.

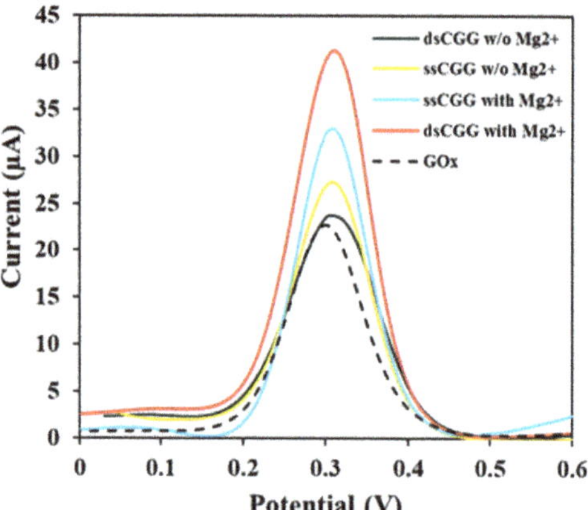

Figure 5. DPV curves of unmodified GO_x and after exposure to 10 nM ssCGG-8 and dsCGG-8 in presence (with) and absence (w/o) of Mg^{2+}.

Finally, we tested the performance of the interface for biosensing application. We optimized the adsorption of the probe sequence, ssCGG-8, which was later exposed to a complementary target and two different noncomplementary sequences. Figure 7a shows the response of ssCGG-8 probe sequence between 0.1 pM-10 µM concentrations. The maximum current of 32.3 ± 1.2 µA was observed at 1 nM concentration, which was assumed to be saturation of the surface with the probe strands. The current responses in Figure 7b show significant enhancement in the current following the surface hybridization reaction with the complementary target (CCG-8) at 1 aM concentration. The change in current ΔI, which is (target current−probe current), shown in Figure 7c, evidently distinguishes between the complementary target and the noncomplementary sequences (NC-1 = TTC-8 and NC-2 = CAG-8). One may notice that the current enhancement of surface hybridized complex is much higher than the solution hybridized complex at this concentration regime, which intrigued us to study the effect of target concentration, as shown in Figure 7d. Although current was enhanced following hybridization over a wide concentration range of the target, ΔI shows a downward trend with the target concentration. As mentioned above, competitive binding between nucleobases and graphene oxide (GO) exist, leading to partial denaturation of double-stranded DNA [24]. Based on this knowledge, we rationalize that not all target strands will hybridize on the surface, and unhybridized target strands may adsorb non-specifically on the pre-saturated interface, leading to a decrease in current by impeding diffusion of the redox probe. Nevertheless, further investigation is necessary to rationalize the high sensitivity at very low target concentration and the downward concentration trend. We propose that, in future applications, the strong adsorption of the probe in the presence of Mg^{2+} and intense surface hybridization signal would eliminate the steps involved in covalent immobilization of a probe (chemisorption) [14] or expensive modification of the probe with 1-pyrenebutanoic acid succinimidyl ester (PASE) linker for stronger π–π interactions with graphene surface [45]. In contrast to other electrochemical detection of TNRs [46,47], this label-free GO_x platform does not rely on guanine oxidation for sensing—therefore, a noninvasive method—and shows higher sensitivity down to the attomolar level compared to a recently reported solution-gated graphene transistor having detection limit only up to femtomolar range.

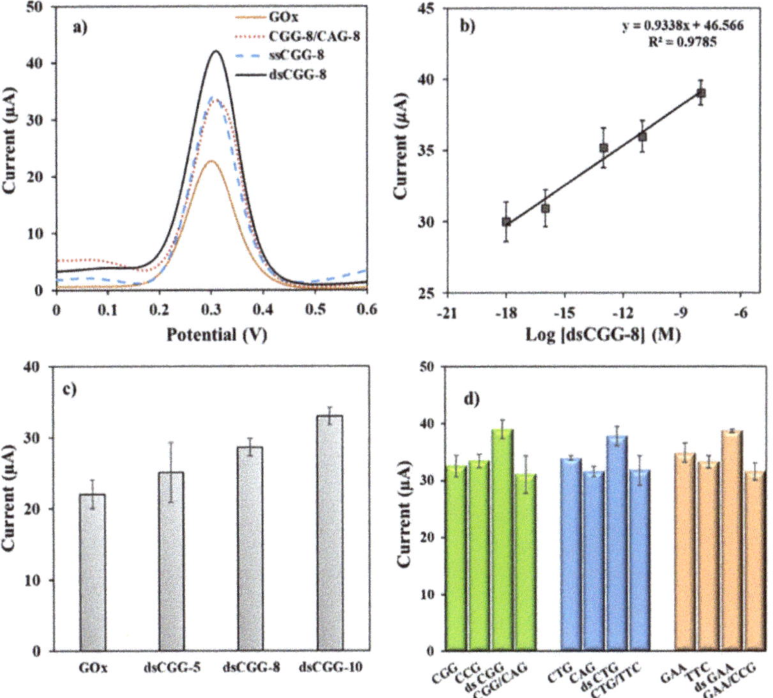

Figure 6. (**a**) Differential pulse voltamogram of 10 nM ssCGG-8, dsCGG-8, and noncomplementary mixture CGG-8/CAG-8. (**b**) Calibration curve dsCGG-8. (**c**) DPV current response of various lengths of dsCGG at 1 aM concentration. (**d**) DPV current response of various single-stranded, double-stranded, and noncomplementary sequences at 10 nM concentration. Error bars represent standard deviation for N $\geq$ 3.

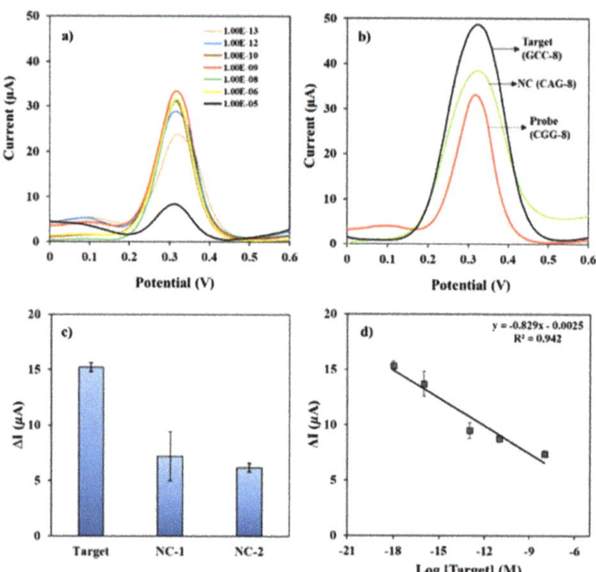

Figure 7. DPV current of (**a**) different concentration of probe sequence (ssCGG-8), (**b**) 1 nM probe, surface hybridized target (GGC-8) and noncomplementary (CAG-8) at 1 aM concentration. (**c**) Difference in current ΔI (target current − probe current) for complementary (CCG-8) and noncomplementary targets (NC-1 = TTC-8 and NC-2 = CAG-8), and (**d**) ΔI versus concentration of the surface hybridized CCG-8 [Target]. Hybridization time 1 h.

4. Conclusions

To best of our knowledge, this is the first comprehensive study on the electrochemical behavior of the DNA adsorbed, oxidized graphene electrodes (GO_x). The GO_x was synthesized and thoroughly characterized followed by fabrication of the GO_x electrodes on ITO surfaces. The DNA adsorption on the GO_x electrodes was performed in the high ionic strength (20 mM Mg^{2+} + 200 mM Na^+). The DNA-adsorbed GO_x surfaces produced significantly higher current due to enhanced diffusion of the soluble redox probe, which was rationalized as the effect of high concentration of divalent cation (Mg^{2+}) in the environment. The study reveals that increase in DNA interaction with the surface increase the current response which can be indicative of its conformation, concentration, and length. The detection limit of the dsCGG adsorption on the GO_x was 0.8 aM. The sequence-independent behavior was reasoned as the effect of high concentration of Mg^{2+} that neutralizes the negative charges on both materials and facilitates the adsorption of different types of sequences. Surface hybridization between the complementary sequences was distinguishable from the noncomplementary sequence, even at 1 aM concentration regime. Nevertheless, further study is suggested to investigate the difference in concentration trends of solution versus surface hybridization. We propose that the strong DNA adsorption in presence of Mg^{2+} and intense surface hybridization signal can develop into simple and label-free DNA biosensors on inkjet-printed GO_x devices.

Supplementary Materials: The following are available online at https://www.mdpi.com/article/10.3390/bios11080273/s1.

Author Contributions: Conceptualization: M.H.S.; Experimentation: N.A., T.E.H., I.M.S.; Formal analysis: N.A.; Supervision and Funding acquisition: M.H.S.; Resources (Raman and DLS): P.S. and B.M.G.; Writing and editing: S.T., B.M.G. and M.H.S. All authors have read and agreed to the published version of the manuscript.

Funding: This research was funded by NSF (CBET-1940716 and CHE-1905341) and NIH support (5R21CA220137). The APC charges were supported by NSF (CBET-1940716).

Institutional Review Board Statement: Not applicable.

Informed Consent Statement: Not applicable.

Data Availability Statement: Not applicable.

Acknowledgments: M.H.S. acknowledges N.S.F. (CBET-1940716) to support this research. B.M.G. acknowledges N.S.F. and N.I.H. support (CHE-1905341 and 5R21CA220137) to support Ishani M. Senanayake.

Conflicts of Interest: Authors do not have any financial/commercial conflict of interest related to this manuscript.

References

1. Wei, Z.; Barlow, D.E.; Sheehan, P.E. The Assembly of Single-Layer Graphene Oxide and Graphene Using Molecular Templates. *Nano Lett.* **2008**, *8*, 3141–3145. [CrossRef]
2. Katsnelson, M.I. Graphene: Carbon in two dimensions. *Mater. Today* **2007**, *10*, 20–27. [CrossRef]
3. Shan, C.; Yang, H.; Song, J.; Han, D.; Ivaska, A.; Niu, L. Direct Electrochemistry of Glucose Oxidase and Biosensing for Glucose Based on Graphene. *Anal. Chem.* **2009**, *81*, 2378–2382. [CrossRef]
4. Bo, Y.; Yang, H.; Hu, Y.; Yao, T.; Huang, S. A novel electrochemical DNA biosensor based on graphene and polyaniline nanowires. *Electrochim. Acta* **2011**, *56*, 2676–2681. [CrossRef]
5. Li, D.; Zhang, W.; Yu, X.; Wang, Z.; Su, Z.; Wei, G. When biomolecules meet graphene: From molecular level interactions to material design and applications. *Nanoscale* **2016**, *8*, 19491–19509. [CrossRef] [PubMed]
6. Lu, C.H.; Yang, H.H.; Zhu, C.L.; Chen, X.; Chen, G.N. A graphene platform for sensing biomolecules. *Angew. Chem. Int. Ed. Engl.* **2009**, *48*, 4785–4787. [CrossRef] [PubMed]
7. Dong, X.; Shi, Y.; Huang, W.; Chen, P.; Li, L.-J. Electrical Detection of DNA Hybridization with Single-Base Specificity Using Transistors Based on CVD-Grown Graphene Sheets. *Adv. Mater.* **2010**, *22*, 1649–1653. [CrossRef]
8. Lin, C.-T.; Loan, P.T.K.; Chen, T.-Y.; Liu, K.-K.; Chen, C.-H.; Wei, K.-H.; Li, L.-J. Label-Free Electrical Detection of DNA Hybridization on Graphene using Hall Effect Measurements: Revisiting the Sensing Mechanism. *Adv. Funct. Mater.* **2013**, *23*, 2301–2307. [CrossRef]
9. Xu, H.; Dai, H.; Chen, G. Direct electrochemistry and electrocatalysis of hemoglobin protein entrapped in graphene and chitosan composite film. *Talanta* **2010**, *81*, 334–338. [CrossRef] [PubMed]
10. Han, Y.; Li, H.; Jafri, S.H.M.; Ossipov, D.; Hilborn, J.; Leifer, K. Optimization and analysis of pyrene-maltose functionalized graphene surfaces for Con A detection. *Appl. Surf. Sci.* **2020**, *510*, 145409. [CrossRef]
11. Jampasa, S.; Siangproh, W.; Laocharoensuk, R.; Vilaivan, T.; Chailapakul, O. Electrochemical detection of c-reactive protein based on anthraquinone-labeled antibody using a screen-printed graphene electrode. *Talanta* **2018**, *183*, 311–319. [CrossRef] [PubMed]
12. Zhao, J.; Lv, Y.; Kang, M.; Wang, K.; Xiang, Y. Electrochemical detection of protein by using magnetic graphene-based target enrichment and copper nanoparticles-assisted signal amplification. *Analyst* **2015**, *140*, 7818–7822. [CrossRef] [PubMed]
13. Zhang, H.; Li, Y.; Su, X. A small-molecule-linked DNA–graphene oxide-based fluorescence-sensing system for detection of biotin. *Anal. Biochem.* **2013**, *442*, 172–177. [CrossRef] [PubMed]
14. Lopez, A.; Liu, J. Covalent and Noncovalent Functionalization of Graphene Oxide with DNA for Smart Sensing. *Adv. Intell. Syst.* **2020**, *2*, 2000123. [CrossRef]
15. Wu, M.; Kempaiah, R.; Huang, P.-J.J.; Maheshwari, V.; Liu, J. Adsorption and Desorption of DNA on Graphene Oxide Studied by Fluorescently Labeled Oligonucleotides. *Langmuir* **2011**, *27*, 2731–2738. [CrossRef]
16. Tikum, A.F.; Ko, J.W.; Kim, S.; Kim, J. Reduced Graphene Oxide-Oligonucleotide Interfaces: Understanding Based on Electrochemical Oxidation of Guanines. *ACS Omega* **2018**, *3*, 15464–15470. [CrossRef]
17. Huang, P.-J.J.; Liu, J. Molecular Beacon Lighting up on Graphene Oxide. *Anal. Chem.* **2012**, *84*, 4192–4198. [CrossRef]
18. Antony, J.; Grimme, S. Structures and interaction energies of stacked graphene–nucleobase complexes. *Phys. Chem. Chem. Phys.* **2008**, *10*, 2722–2729. [CrossRef]
19. Zeng, S.; Chen, L.; Wang, Y.; Chen, J. Exploration on the mechanism of DNA adsorption on graphene and graphene oxide via molecular simulations. *J. Phys. D Appl. Phys.* **2015**, *48*, 275402. [CrossRef]
20. Zhao, X. Self-Assembly of DNA Segments on Graphene and Carbon Nanotube Arrays in Aqueous Solution: A Molecular Simulation Study. *J. Phys. Chem. C* **2011**, *115*, 6181–6189. [CrossRef]
21. Vovusha, H.; Sanyal, B. Adsorption of nucleobases on 2D transition-metal dichalcogenides and graphene sheet: A first principles density functional theory study. *RSC Adv.* **2015**, *5*, 67427–67434. [CrossRef]
22. Varghese, N.; Mogera, U.; Govindaraj, A.; Das, A.; Maiti, P.K.; Sood, A.K.; Rao, C.N.R. Binding of DNA Nucleobases and Nucleosides with Graphene. *ChemPhysChem* **2009**, *10*, 206–210. [CrossRef] [PubMed]
23. Ortmann, F.; Schmidt, W.G.; Bechstedt, F. Attracted by Long-Range Electron Correlation: Adenine on Graphite. *Phys. Rev. Lett.* **2005**, *95*, 186101. [CrossRef] [PubMed]

24. Manohar, S.; Mantz, A.R.; Bancroft, K.E.; Hui, C.-Y.; Jagota, A.; Vezenov, D.V. Peeling Single-Stranded DNA from Graphite Surface to Determine Oligonucleotide Binding Energy by Force Spectroscopy. *Nano Lett.* **2008**, *8*, 4365–4372. [CrossRef] [PubMed]
25. Secor, E.B.; Prabhumirashi, P.L.; Puntambekar, K.; Geier, M.L.; Hersam, M.C. Inkjet Printing of High Conductivity, Flexible Graphene Patterns. *J. Phys. Chem. Lett.* **2013**, *4*, 1347–1351. [CrossRef] [PubMed]
26. Seo, J.-W.T.; Zhu, J.; Sangwan, V.K.; Secor, E.B.; Wallace, S.G.; Hersam, M.C. Fully Inkjet-Printed, Mechanically Flexible MoS2 Nanosheet Photodetectors. *ACS Appl. Mater. Interfaces* **2019**, *11*, 5675–5681. [CrossRef] [PubMed]
27. Liang, Y.T.; Hersam, M.C. Highly Concentrated Graphene Solutions via Polymer Enhanced Solvent Exfoliation and Iterative Solvent Exchange. *J. Am. Chem. Soc.* **2010**, *132*, 17661–17663. [CrossRef]
28. Wang, S.; Wang, C.; Ji, X. Towards understanding the salt-intercalation exfoliation of graphite into graphene. *RSC Adv.* **2017**, *7*, 52252–52260. [CrossRef]
29. Ji, Z.; Wu, J.; Shen, X.; Zhou, H.; Xi, H. Preparation and characterization of graphene/NiO nanocomposites. *J. Mater. Sci.* **2011**, *46*, 1190–1195. [CrossRef]
30. Bindumadhavan, K.; Srivastava, S.; Srivastava, I. Green Synthesis of Graphene. *J. Nanosci. Nanotechnol.* **2013**, *13*, 4320–4324.
31. Genc, R.; Alas, M.O.; Harputlu, E.; Repp, S.; Kremer, N.; Castellano, M.; Colak, S.G.; Ocakoglu, K.; Erdem, E. High-Capacitance Hybrid Supercapacitor Based on Multi-Colored Fluorescent Carbon-Dots. *Sci. Rep.* **2017**, *7*, 11222. [CrossRef] [PubMed]
32. Gao, W.; Alemany, L.B.; Ci, L.; Ajayan, P.M. New insights into the structure and reduction of graphite oxide. *Nat. Chem.* **2009**, *1*, 403–408. [CrossRef]
33. Jung, I.; Dikin, D.A.; Piner, R.D.; Ruoff, R.S. Tunable Electrical Conductivity of Individual Graphene Oxide Sheets Reduced at "Low" Temperatures. *Nano Lett.* **2008**, *8*, 4283–4287. [CrossRef] [PubMed]
34. Hong, B.J.; An, Z.; Compton, O.C.; Nguyen, S.T. Tunable Biomolecular Interaction and Fluorescence Quenching Ability of Graphene Oxide: Application to "Turn-on" DNA Sensing in Biological Media. *Small* **2012**, *8*, 2469–2476. [CrossRef] [PubMed]
35. Tite, T.; Chiticaru, E.A.; Burns, J.S.; Ioniţă, M. Impact of nano-morphology, lattice defects and conductivity on the performance of graphene based electrochemical biosensors. *J. Nanobiotechnology* **2019**, *17*, 101. [CrossRef] [PubMed]
36. Hwang, M.T.; Heiranian, M.; Kim, Y.; You, S.; Leem, J.; Taqieddin, A.; Faramarzi, V.; Jing, Y.; Park, I.; van der Zande, A.M.; et al. Ultrasensitive detection of nucleic acids using deformed graphene channel field effect biosensors. *Nat. Commun.* **2020**, *11*, 1543. [CrossRef]
37. Bard, A.J.; Faulkner, L.R. *Electrochemical Methods: Fundamentals and Applications*; Wiley: New York, NY, USA, 1980.
38. Li, C.-Z.; Long, Y.-T.; Kraatz, H.-B.; Lee, J.S. Electrochemical Investigations of M-DNA Self-Assembled Monolayers on Gold Electrodes. *J. Phys. Chem. B* **2003**, *107*, 2291–2296. [CrossRef]
39. Xiao, J.; Liu, Z.; Li, C.; Wang, J.; Huang, H.; Yi, Q.; Deng, K.; Li, X. Tunable graphdiyne for DNA surface adsorption: Affinities, displacement, and applications for fluorescence sensing. *Anal. Bioanal. Chem.* **2021**, *413*, 3847–3859. [CrossRef] [PubMed]
40. Asefifeyzabadi, N.; Taki, M.; Funneman, M.; Song, T.; Shamsi, M.H. Unique Sequence-Dependent Properties of Trinucleotide Repeat Monolayers: Electrochemical, Electrical, and Topographic Characterization. *J. Mater. Chem. B* **2020**, *8*, 5225–5233. [CrossRef]
41. Alam, M.N.; Shamsi, M.H.; Kraatz, H.-B. Scanning positional variations in single-nucleotide polymorphism of DNA: An electrochemical study. *Analyst* **2012**, *137*, 4220–4225. [CrossRef] [PubMed]
42. Shamsi, M.H.; Kraatz, H.-B. Probing nucleobase mismatch variations by electrochemical techniques: Exploring the effects of position and nature of the single-nucleotide mismatch. *Analyst* **2010**, *135*, 2280–2285. [CrossRef] [PubMed]
43. Taki, M.; Rohilla, K.J.; Barton, M.; Funneman, M.; Benzabeh, N.; Naphade, S.; Ellerby, L.M.; Gagnon, K.T.; Shamsi, M.H. Novel probes for label-free detection of neurodegenerative GGGGCC repeats associated with amyotrophic lateral sclerosis. *Anal. Bioanal. Chem.* **2019**, *411*, 6995–7003. [CrossRef] [PubMed]
44. Asefifeyzabadi, N.; Alkhaldi, R.; Qamar, A.Z.; Pater, A.A.; Patwardhan, M.; Gagnon, K.T.; Talapatra, S.; Shamsi, M.H. Label-free Electrochemical Detection of CGG Repeats on Inkjet PrinTable 2D Layers of MoS2. *ACS Appl. Mater. Interfaces* **2020**, *12*, 52156–52165. [CrossRef] [PubMed]
45. Xu, S.; Zhan, J.; Man, B.; Jiang, S.; Yue, W.; Gao, S.; Guo, C.; Liu, H.; Li, Z.; Wang, J.; et al. Real-time reliable determination of binding kinetics of DNA hybridization using a multi-channel graphene biosensor. *Nat. Commun.* **2017**, *8*, 14902. [CrossRef]
46. Yang, I.V.; Thorp, H.H. Modification of Indium Tin Oxide Electrodes with Repeat Polynucleotides: Electrochemical Detection of Trinucleotide Repeat Expansion. *Anal. Chem.* **2001**, *73*, 5316–5322. [CrossRef] [PubMed]
47. Ge, Z.; Ma, M.; Chang, G.; Chen, M.; He, H.; Zhang, X.; Wang, S. A novel solution-gated graphene transistor biosensor for ultrasensitive detection of trinucleotide repeats. *Analyst* **2020**, *145*, 4795–4805. [CrossRef]

Article

Rapid Multianalyte Microfluidic Homogeneous Immunoassay on Electrokinetically Driven Beads

Pierre-Emmanuel Thiriet *, Danashi Medagoda, Gloria Porro and Carlotta Guiducci

Laboratory of Life Sciences Electronics, École Polytechnique Fédérale de Lausanne, 1015 Lausanne, Switzerland; danashi.medagoda@epfl.ch (D.M.); gloria.porro@epfl.ch (G.P.); carlotta.guiducci@epfl.ch (C.G.)

* Correspondence: pierre-emmanuel.thiriet@epfl.ch; Tel.: +41-216-931-345

Received: 5 November 2020; Accepted: 17 December 2020; Published: 21 December 2020

Abstract: The simplicity of homogeneous immunoassays makes them suitable for diagnostics of acute conditions. Indeed, the absence of washing steps reduces the binding reaction duration and favors a rapid and compact device, a critical asset for patients experiencing life-threatening diseases. In order to maximize analytical performance, standard systems employed in clinical laboratories rely largely on the use of high surface-to-volume ratio suspended moieties, such as microbeads, which provide at the same time a fast and efficient collection of analytes from the sample and controlled aggregation of collected material for improved readout. Here, we introduce an integrated microfluidic system that can perform analyte detection on antibody-decorated beads and their accumulation in confined regions within 15 min. We employed the system to the concomitant analysis of clinical concentrations of Neutrophil Gelatinase-Associated Lipocalin (NGAL) and Cystatin C in serum, two acute kidney injury (AKI) biomarkers. To this end, high-aspect-ratio, three-dimensional electrodes were integrated within a microfluidic channel to impart a controlled trajectory to antibody-decorated microbeads through the application of dielectrophoretic (DEP) forces. Beads were efficiently retained against the fluid flow of reagents, granting an efficient on-chip analyte-to-bead binding. Electrokinetic forces specific to the beads' size were generated in the same channel, leading differently decorated beads to different readout regions of the chip. Therefore, this microfluidic multianalyte immunoassay was demonstrated as a powerful tool for the rapid detection of acute life-threatening conditions.

Keywords: bead-based immunoassays; dielectrophoresis (DEP); three-dimensional microelectrodes; on-chip incubation; acute kidney injury diagnosis; multimarker analysis; microfluidic-based diagnostics

---

1. Introduction

Acute kidney injury (AKI) is a life-threatening condition characterized by a rapid loss of kidney function [1]. In developed countries, AKI occurs in 20% of hospitalized adult patients and 25% of pediatric patients receiving intensive care [2], and its diagnosis is critical to improve survival. One of the consequences of AKI is the disruption of homeostasis, inducing an accumulation of waste products normally removed by the kidneys, which can lead to severe damages throughout the body. If treated quickly, the effects of AKI are reversible, notably through fluid resuscitation and medication [3] but they can lead to death of the patient without proper intervention.

Currently, AKI is diagnosed through monitoring of the patient's urine output volume and measurement of the level of serum creatinine in blood [3]. Creatinine quantification suffers from diverse limitations, namely, interferences with drugs such as antiretroviral drugs [4], variations in basal creatinine levels between patients, and most importantly a long delay (36 to 48 h) between the occurrence of AKI and a detectable increase in serum creatinine concentration [5]. This delays the diagnostics, with possibly critical consequences. Numerous potential alternative AKI biomarkers are currently investigated by research groups worldwide [6]. Here we focused on the most promising ones,

Cystatin C and Neutrophil Gelatinase-Associated Lipocalin (NGAL). Cystatin C is a molecule present in all tissues and filtered by the kidneys. Its concentration spikes in serum 24 h following injury [7,8]. NGAL is a protein that can be found in neutrophils and some epithelia including renal tubules. As AKI damages the kidneys' epithelium, such disorder induces an increase of NGAL concentration in serum within four hours [9,10]. A combined analysis of both NGAL and Cystatin C would allow for early diagnosis of AKI, reducing the risk of false negatives [9].

The critical requirements of timeliness in the diagnosis of AKI call for fast analytical devices to perform the analysis of the relevant biomarkers directly within intensive care units or emergency facilities. The detection of such markers in clinical settings relies on immunoassays carried out by means of bulky analyzers, such as Abbot Architect i1000SR and c4000, used, respectively, for the quantification of NGAL and Cystatin C. Commercial point-of-care (PoC) systems for the individual detection of these targets are also available on the market, namely, the Triage® system of Alere Inc. for the detection of NGAL through a lateral flow assay [11] and the Cube® of Eurolyser for Cystatin C levels' quantification. We aimed to concomitantly detect Cystatin C and NGAL using an immunoassay approach that could be engineered into a PoC device. Such multiple analysis would be an effective mean to obtain an accurate diagnosis of AKI in the first day following the injury. In order to minimize the number of assay steps, our device performed both the binding of the analyte molecules from the sample and the readout phase. We implemented such detection on antibodies-decorated beads to perform sandwich immunoassays. Indeed, functionalized microbeads suspended in a microfluidic channel provided a fast and effective collection of analytes in-flow. However, the gathering of beads into large clusters was required to obtain a high readout fluorescent signal. The spatial handling of beads is commonly performed either through magnetic forces [12] or by means of mechanical restrictions [13]. These approaches suffer from limitations hindering their use in portable devices, namely, a difficult integration into a portable platform for magnetic devices and clogging issues for mechanical systems. Electrokinetics manipulation of beads through the generation of gradients of electrical fields appears as a potential solution for beads' manipulation in a compact and highly integrated system.

Dielectrophoresis (DEP) is a physical phenomenon appearing when placing a polarizable particle into a non-uniform electric field. It has been successfully employed to sort particles according to their size, shape, and dielectric properties [14,15]. It also proved to be valuable in the diagnostics field through enrichment of targeted protein [16], stream focusing in flow cytometry [17], and spatial confinement of decorated beads [18–20]. Iswardy et al. [18] implemented a DEP-based biosensing platform for the diagnosis of Dengue virus. The analyte is captured on beads functionalized with antibodies specific to one of the virus proteins and is held in place thanks to DEP forces. Ramon-Azcon et al. [19] developed a device detecting pesticide residues in wine. Decorated beads are maintained against the flow with DEP and exposed to the sample of interest. A similar technique was proposed by Park et al. [20], combining the manipulation of beads with DEP forces with a polarization-based preconcentration approach to increase the analyte concentration in the vicinity of beads. However, this solution imposes limitations on the flow rate and, thus, on the amount of analyte accessible for detection. Most importantly, for all the abovementioned technologies, the levels of DEP forces achieved in buffers of regular conductivity are not sufficient to efficiently manipulate the beads; therefore, they require the use of highly diluted solutions (low ionic force) [21], which limits their overall sensing performance.

Here, we introduce high-aspect-ratio vertical electrodes in the microfluidic channel to enable highly effective manipulation of microbeads by electrokinetics in minimally diluted buffers (5x). In our system, microbeads decorated with specific antibodies were incubated with analytes on-chip along DEP trapping regions in presence of flow, drastically reducing mass transport issues. They were later moved and concentrated to detection regions, where the optical fluorescence signal of the markers was amplified by clustering multiple beads. Our device allowed carrying out both analyte binding and signal readout under the continuous flow of the same reagents, avoiding rinsing steps.

In this paper, we present the first microfluidic multianalyte platform for rapid detection of AKI. The simultaneous assessment of Cystatin C and NGAL levels we achieved allows for rapid and accurate diagnosis of AKI within a large temporal window after the injury and with a matching performance compared with a state of the art ELISA kit.

2. Materials and Methods

2.1. Chip Microfabrication

The microfluidic system consists of a 4-mm-wide main channel, spanned by six rows of vertical electrodes. Two inlets allow for sequential injection of beads and reagents, subsequently disposed through a single outlet. The height of both the microfluidic channel and the vertical electrodes is 50 µm.

The platform is fabricated through an additive process on a glass substrate [22]. A detailed illustration of the process flow is reported in Supplementary Information Figure S1. After sputtering of a Ti/Pt/Ti (20/200/20 nm) layer on the entire wafer surface (Pfeiffer Spider 600, Pfeiffer Vacuum, Asslar, Germany), planar metal lines are patterned with photolithography and ion beam etching (Veeco Nexus IBE 350, Veeco, Plainview, TX, USA). Successively, in order to insulate the metal lines from the liquid, a 300-nm layer of oxide is sputtered on the wafer. This layer is opened through dry etching (SPTS Advanced Plasma System (APS), SPTS technologies, Newport, UK) in the electrically active regions of the device. Cylindrical vertical pillars in SU-8 photoresist (Microchem 3025, Microresist Technologies, Berlin, Germany) are then deposited on the exposed metal pads connected to the oxide-passivated lines. A layer of Ti/Pt (20/200 nm) is sputtered onto the entire wafer surface. This layer is subsequently removed through vertical ion beam etching, leaving the metal only on the vertical pillars' sidewalls. Once the fabrication of the electrodes has been carried out, the SU-8 microfluidic channel is patterned at the same height as the electrodes. Finally, the device is sealed with a polydimethylsiloxane (PDMS) coverslip, bonded to the chip by a 3-aminopropyl triethoxysilane (APTES) treatment and a baking step at 150 °C for 2 h. This results in the integration of electrically active 3-D structures within the microfluidic channel, shown in Figure 1.

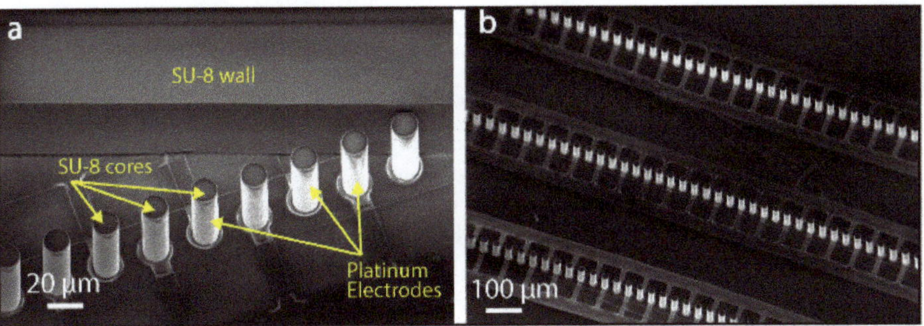

Figure 1. Scanning electron microscope (SEM) images of the concentration lines. (**a**) One of the microbeads' concentration regions. The beads surf along the diagonal line in the space between the line and the SU-8 photoresist wall and stop at their conjunction. (**b**) Three parallel electrodes' lines are patterned in the microfluidic channel to allow for accurate deflection of beads driven by the flow. The height of the vertical electrodes and SU-8 microfluidic channels is 50 µm.

2.2. DEP-Based Manipulation of Beads

Dielectrophoresis (DEP) is a phenomenon that occurs to a polarizable particle placed in a non-uniform electric field. The presence of an electric field triggers the formation of a dipole in a polarizable particle. The dipole experiences a net force if the electric field is spatially non-uniform, which leads the particle to move. The direction of the force induced by DEP on the particle depends on the relative polarizability of the particle and of the surrounding medium. If the particle is more polarizable than the surrounding medium, the particle dipole will be oriented along the electric field, while a particle less polarizable than the surrounding medium results in a dipole oriented against the electric field. The force experienced by a spheroid particle with a radius R placed in an electric field E can be written as:

$$F_{DEP} = 2\pi R^3 \varepsilon_{medium} Re(f_{CM}(\omega)) \nabla E^2 \tag{1}$$

where ε_{medium} is the dielectric permittivity of the medium and $f_{CM}(\omega)$ is the Clausius-Mossotti factor, which is a function of the permittivity and, consequently, of the polarizability of the particle and the surrounding medium [23].

Furthermore, a spherical particle injected into the microfluidic channel is dragged by the flow according to the following equation:

$$F_{drag} = 6\pi\eta R V_0 \tag{2}$$

where η is the fluid viscosity, R the sphere radius, and V_0 the fluid constant velocity at infinity.

In our device, the electric field is generated by linearly arranged vertical electrodes. If the line of electrodes is normal to the flow, the beads flowing through the channel experience a DEP force directed against the drag force (Figure 2a). Provided that the DEP force is sufficient to compensate for the drag force, the beads are immobilized upstream in the vicinity of the electrodes' line as illustrated in Figure 2c. The beads can be kept at this position while experiencing a continuous flow of analyte and reagents, thus allowing their binding to the beads' surface. However, if the row of electrodes is placed at a specific angle with respect to the flow, the DEP force will not be directed against the drag force and we will observe a net force heading the bead in a specific direction (Figure 2b and Video S1 and Video S2). This "surfing" phenomenon, illustrated in Figure 2d, allows for the displacement of beads along diagonal lines and for their accumulation at a dedicated location in the channel, which results in an increase of the readout signal and improves the device sensitivity [24].

Equation (1) establishes the dependence between the size of the beads and the DEP force they will experience. Each line of electrodes can generate a different DEP force and can consequently trap and displace specific beads depending on their size. In this way, beads of different sizes can be driven to distinct regions on the chip (Figure 2f). If beads of different sizes are functionalized with different antibodies, our approach allows a multianalyte analysis with a single fluorescent channel.

2.3. Microfluidic Analytical Device Operation and Experimental Setup

The microfluidic chip for DEP-based immunoassays is shown in Figure 3a. It consists of a single channel with two inlets and a single outlet featuring linear arrangements of three-dimensional electrodes, patterned to obtain incubation lines (horizontal) and concentration lines (diagonal).

Bead-based sandwich immunoassays were performed on our platform as illustrated in Figure 3b,c. The fluorescently labeled detection antibodies (dAbs) were spiked in the sample prior to injection in the chip. Before undergoing injections of reagents solutions, the chip was primed with buffer solution (fetal bovine serum, FBS, diluted five times in deionized water) in order to prevent unspecific bindings to the channel walls. (1) After priming, beads functionalized with capture antibodies (cAbs) were injected in the device and held in suspended small clusters (few tens of beads) upstream of the incubation lines, as illustrated in step 1 in Figure 3c. The number of captured beads was controlled through visual inspection. During this phase, the microbeads–cAbs solution, 10 µL 0.5% (*w/v*) in 1 mL of diluted FBS, was delivered from one inlet at 2 µL/min, maximal flow that could be applied without having beads escaping the incubation region, while the incubation line exerted a holding force (20 Vpp, 1 MHz). The obtained small beads' aggregates were spatially confined by the DEP force, while kept in slight agitation by the flow: This turbulent motion favors the convective transport of target molecules. (2) After a sufficient number of beads was collected (1–2 ms), the beads–cAbs solution flow was stopped and the solution of antigen–dAbs complexes was immediately dispensed from the other inlet at 2 µL/min. The antigen–dAbs solution was flushed in the chip for 15 min. During this incubation step on-chip, the binding between the analyte–dAbs and the cAbs led to the formation of the complete sandwich assay on the beads' surface. (3) After incubation, the beads were released by turning off the electrical signal. Concomitantly, the concentration line was activated (20 Vpp, 1 MHz) and the beads were led to "surf" in-flow along the diagonal concentration lines until they reached a region where they were trapped against the microfluidic wall, as shown in Video S4. (4) The beads carrying the sandwich assays were, hence, accumulated to enhance the total fluorescence signal. We noticed that the so-obtained clusters, imaged in CY5 fluorescence channel at 5000 ms exposure, presented the same intensity per μm^2 provided that their footprint was equal or larger than 50 μm^2. (5) Finally, the beads were released from this concentration region by deactivation of the electrodes' line.

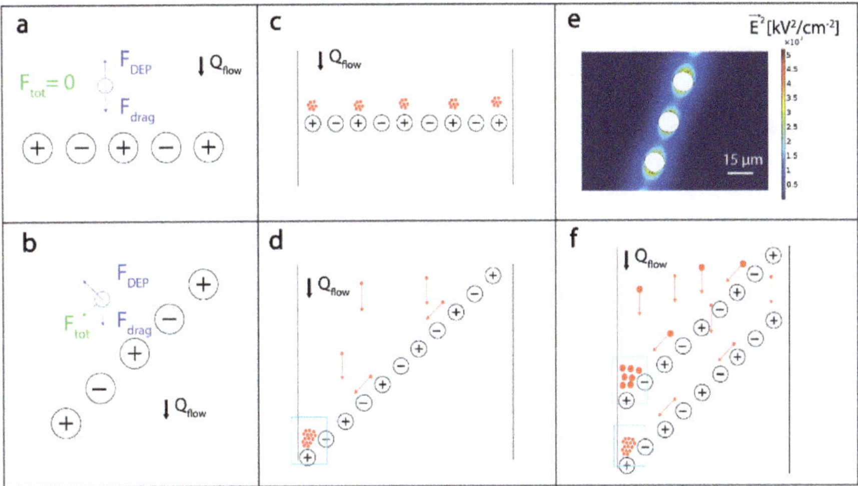

Figure 2. Dielectrophoresis (DEP)-based incubation and accumulation of beads' working principles and simulation. (**a,b**) Illustration of the net force experienced when a bead approaches a horizontal or a diagonal electrodes' line, respectively resulting in the bead immobilization (**a**) or directed "surfing" along the electric field (**b**). (**c,d**) Extending this approach to multiple beads, one can obtain either small clusters for on-chip incubation close to horizontal lines (**c**) or large clusters in the regions where all the surfing beads are immobilized against the microfluidic channel wall (**d**), indicated with a blue rectangle. Beads are depicted in red. (**e**) Finite element simulation of the electrical field generated by electrodes in the diagonal line. The field gradient is higher between electrodes, resulting in a DEP force preventing the beads from crossing the line. Simulations were carried out on Comsol 5.3, using the Electrostatics module and a voltage amplitude difference of 20 V between electrodes. (**f**) Illustration of the multianalyte detection capabilities of our platform. Through the application of different electrical potentials on diagonal lines, we can achieve the clustering of beads of different sizes in separate locations.

A similar protocol can be applied to run two distinct immunoassays in parallel on the same channel. Two different beads' populations (2-µm and 6-µm diameter) were functionalized with cAbs for Cystatin C and NGAL, respectively. The beads–cAbs solution was obtained by adding 10 µL 0.5% (*w/v*) 2-µm beads and 10 µL 0.5% (*w/v*) 6-µm beads in 1 mL 5-fold diluted FBS. The incubation step was performed on two different lines (the line upstream was activated by applying a 14-Vpp, 1-MHz signal and the line downstream by applying a 20-Vpp, 1-MHz signal) in presence of a 0.4 µL/min flow, maximal flow that was applied while holding the small 2-µm beads against the flow. The line upstream was set to exert a weaker electric field so that beads of larger size were trapped by the first line, while the smaller ones could pass through, to be successively trapped by the second line, with a trapping efficiency of 90% and 70% for the 6-µm and the 2-µm beads, respectively. After beads' clustering, incubation was performed flushing a solution containing the analyte–dAbs complexes of both Cystatin C and NGAL, with the dAbs labeled with CY3 and CY5 fluorophores, respectively. Following incubation, the two beads' populations carrying the two distinct sandwich assays were released and concentrated at different locations on-chip. To do so, the two accumulation lines were activated (line upstream electrical stimulus: 14 Vpp, 1 MHz; line downstream electrical stimulus: 20 Vpp, 1 MHz), and the flow was set to 0.8 µL/min, maximal flow allowing accumulation of 2-µm beads. The two distinct clusters were imaged in CY3 and CY5 fluorescent channels with 5000-ms exposure times. On average 20 experiments (incubation, accumulation, and beads' release) could be performed before the appearance of clogging issues preventing the chip from further use.

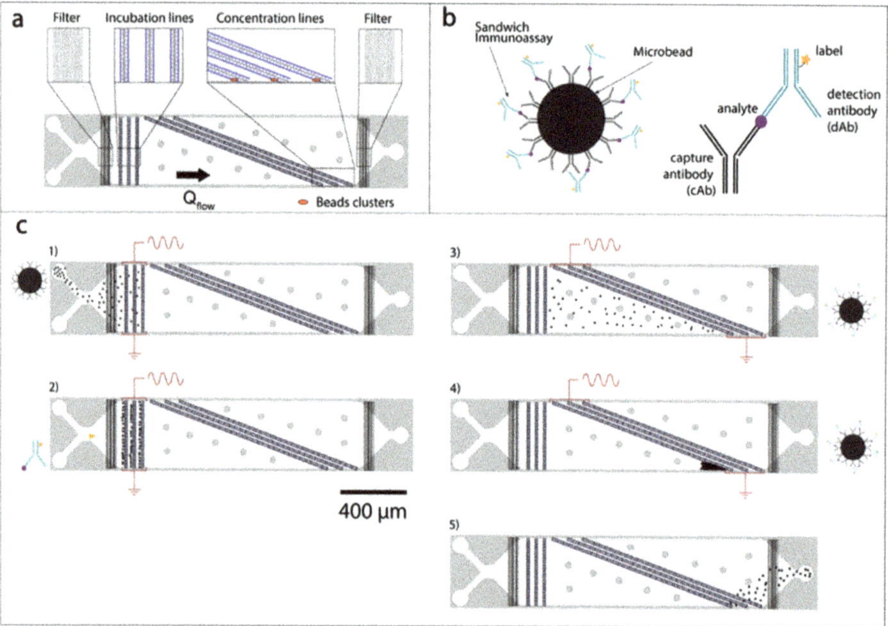

Figure 3. Chip description and operation. (**a**) Presentation of the chip layout. Beads and reagents can be successively injected through the two inlets visible on the left. The device consists of three incubation lines upstream and three concentration lines downstream, at the end of which the beads are accumulated in clusters (shown here in red). (**b**) Illustration of a sandwich immunoassay used for detection of biomarkers. The analyte we aimed to detect was captured by the bead decorated with capture antibody (cAb) and detection was performed thanks to the fluorescently labeled detection antibody (dAb). (**c**) Presentation of the successive steps performed on-chip to operate the platform, namely, (**1**) beads' loading, (**2**) incubation with detection antibodies and (**3**) release from the incubation line, (**4**) clustering in the concentration region, and (**5**) discarding through the outlet. For the sake of clarity, the species bound to the beads and the electrically activated arrays of electrodes are indicated for each step.

The experimental setup employed for on-chip experiments included two PHD ULTRA™ syringe pumps (Harvard Apparatus, Holliston, MA, USA), a TG5012A function generator (Aim-TTi, Huntingdon, UK), an optical fluorescence microscope (Leica DM2500M) with Y3 and Y5 filter cubes (Leica, Wetzlar, Germany), and a camera (ORCA-Flash4.0LT, (Hamamatsu Photonics, Hamamatsu, Japan). The electric signal was provided to the chip through a homemade printed circuit board (PCB), depicted in Figure S2.

2.4. Antibodies' Conjugation to Beads and to Labeling Fluorescent Molecules

Polystyrene beads covalently coated in Streptavidin (Spherotec Inc.) were purchased at sizes of 2-µm diameter (binding capacity = 0.42 nmol/mg) and 6-µm diameter (binding capacity = 0.14 nmol/mg). The 2-µm beads were incubated with biotinylated Cystatin C monoclonal capture antibodies (Cyst13-biotinylated, Novus Biologicals) and the 6 µm-beads were incubated with biotinylated NGAL polyclonal capture antibodies (Human Lipocalin-2/NGAL-biotinylated antibody BAF1757, R&D Systems, UK) for at least four hours. They were then resuspended and incubated for 1–2 h in 1% bovine serum albumin (BSA, Sigma Aldrich, St. Louis, MO, USA) in 0.1 M phosphate buffer (PB, Sigma Aldrich, St. Louis, MO, USA) for blocking. After blocking, beads were washed four times through a procedure of centrifugation, supernatant removal, and resuspension in a solution of 0.05% Tween-20 (Millipore, Burlington, MA, USA) in 0.1 M PB. After the final washing steps, beads were stored in 0.1 M PB at 4 °C. Both antigens for Cystatin C (Human Recombinant Cystatin C, Novus Biologicals, UK) and NGAL (Human Lipocalin-2/NGAL, CF, R&D Systems, UK) were acquired and used as is. Monoclonal detection antibodies for Cystatin C (Cyst24-Dylight 550, Novus Biologicals, UK) were purchased with a Dylight 550 fluorophore, while monoclonal detection antibodies for NGAL (Human Lipocalin-2/NGAL Antibody MAB17571R, R&D Systems, UK) were labeled with a fluorophore using a Lightning-Link Rapid Alexa Fluor 647 antibody labeling kit (Expedeon, San Diego, CA, USA). Prior to experiments, antigen and detection antibodies were incubated together for 15 min in fetal bovine serum (FBS, Sigma Aldrich, St. Louis, MO, USA) diluted five times in MilliQ water.

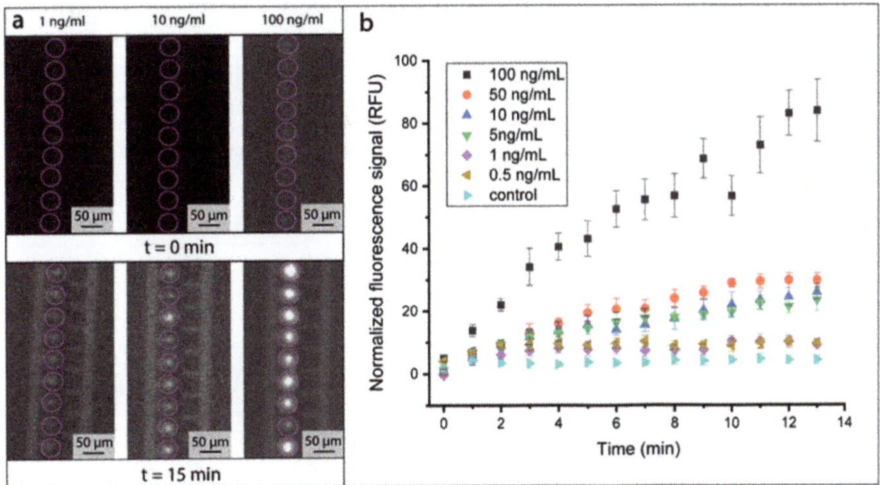

Figure 4. On-chip incubation of Neutrophil Gelatinase-Associated Lipocalin (NGAL) biomarker. (**a**) Observation of the small beads' clusters (circled in pink) before and after 15 min of incubation. The fluorescence signal arose from the binding of dAb–NGAL complex to cAb-decorated beads dielectrically trapped in the regions upstream to the electrode line. Three NGAL concentrations were injected in separate experiments, namely, 1 ng/mL, 10 ng/mL, and 100 ng/mL. (**b**) Fluorescence signal as a function of the incubation time for different NGAL concentrations. After 15 min all concentrations provided a signal greater than the control experiment, consisting of an injection of a solution in absence of NGAL molecules. The error bars were obtained by measuring the fluorescent signal from 10 clusters.

2.5. Data Analysis

All images and videos acquired in the scope of this paper were analyzed using ImageJ software (Fiji). For each cluster, the neighboring background signal was calculated and subtracted from the mean intensity in the regions of interest, to account for possible variations of the background intensity. With this approach, we quantified the normalized fluorescence signals displayed in Figures 4–6 and

Figures S3–S6. In addition to this, in order to plot the calibration curves of Figure 5, Figures S4 and S6, the average signal of the negative control was subtracted from each measurement point. This allowed us to remove artefact effects induced by unspecific binding of the labeled antibody with the beads.

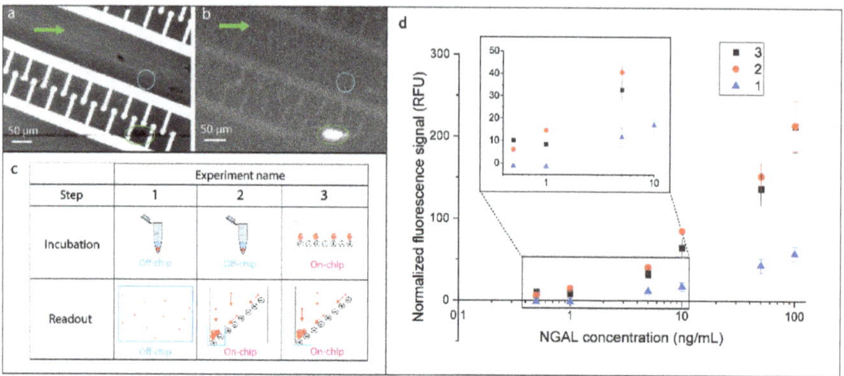

Figure 5. Assessment of the on-chip incubation and accumulation performance. (**a,b**) Brightfield (**a**) and fluorescence (**b**) images of beads clustered after 15 min of incubation (NGAL concentration of 100 ng/mL) and concentration steps. A bright fluorescent signal is clearly visible in the accumulation region circled in green. The blue circle indicates the region chosen as neighboring background for the normalization of the cluster signal. (**c**) Experimental protocol implemented to separately assess the impact on the output signal of our platform of both on-chip incubation and accumulation of beads. (**d**) Calibration curves obtained for the aforementioned experiments. The measured relative fluorescence signal is plotted as a function of the NGAL concentration employed for incubation. Plot 3 presents the dose-response of our system with both incubation and accumulation steps performed on-chip and taken as the reference curve in the following discussion section. Error bars were calculated over three acquisitions.

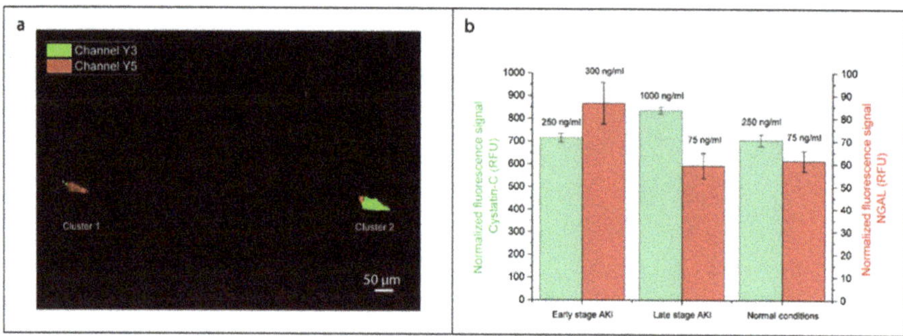

Figure 6. Simultaneous detection of NGAL and Cystatin C for acute kidney injury (AKI) diagnosis. (**a**) Superposition of two clusters acquired, respectively, in CY5 fluorescent channel (NGAL label) and CY3 channel (Cystatin C label). This picture was taken in the case of an "early" patient (NGAL: 300 ng/mL, Cystatin C: 250 ng/mL). (**b**) Fluorescent signal acquired for CY3 channel (Cystatin C detection) and CY5 channel (NGAL detection), respectively, in cluster 2 (downstream, accumulating 2-μm beads capturing Cystatin C) and cluster 1 (upstream, accumulating 6-μm beads capturing NGAL). Three cases were investigated, corresponding to the clinically relevant situations: an "early-stage" patient with high NGAL and normal Cystatin C concentrations, a "late-stage" patient with high Cystatin C and normal NGAL concentrations, and a "healthy" patient with normal NGAL and Cystatin C concentrations. Error bars were calculated over three measurements.

3. Results and Discussion

3.1. Immunoassay Incubation On-Chip

The on-chip incubation of antibody-functionalized beads with antigens and detection antibodies took place on dedicated horizontal incubation lines. We maintained beads decorated with cAbs against the flow and concentrated them in small clusters of about 10 to 20 beads. Significant variability was observed in the size of the growing clusters, reflecting inhomogeneities in the flow lines within the microfluidic channel. Such inhomogeneities were mainly caused by microfabrication defects or debris accumulation at the entrance of the chip. Then, a solution containing the dAb–NGAL complexes was injected at a flow rate of 2 µL/min for 15 min. The binding of the dAb–NGAL complexes to the cAb-decorated beads was observed and quantified in real time by fluorescence measurements of the CY5 channel (Figure 4 and Video S3). Figure 4a illustrates the dependence of measured fluorescence intensity of the small clusters on the concentration of NGAL in the injected solution. Such behavior could be quantified with a minute-range resolution, as shown in Figure 4b. A steady increase of fluorescence signal could be recorded after 15 min upon injection of NGAL–dAbs complexes at concentrations in the range of 0.5–100 ng/mL. Interestingly, none of the plots were reaching a plateau, suggesting the persistence of a transient binding regime. Different concentrations corresponded to similar fluorescent intensities, e.g., 5 ng/mL and 10 ng/mL. Thus, they might have appeared perilous to resolve due to a significant standard deviation. This variability in the clusters' fluorescent signal was caused by the varying size of the small clusters formed near the incubation lines. Nonetheless, the purpose of the horizontal incubation lines was to allow for incubation to happen in the most favorable conditions and not to enhance the readout of the fluorescence intensity. This step took place in the accumulation regions, presented in the next section.

We designed the DEP holding action at the horizontal incubation lines in a way that the size of the clusters would be maintained sufficiently small to keep the accumulated beads in slight agitation. This approach allows for uniform binding of analytes in-flow and access to the whole surface offered by the beads. In fact, in the case of large clusters, the reagents would be depleted at the downstream portion of the aggregate, an issue that can be noticed in previous works [18]. This limitation inspired us for the creation of two separate lines for incubation and accumulation of beads, presented in Figure 3.

Moreover, previous approaches employing DEP force to immobilize beads and expose them to reagents relied on the use of planar electrodes to generate electric fields [20]. This approach relies on a simple fabrication process but leads to the creation of high electrokinetic forces only in close proximity to the chip's surface that features the electrodes. Consequently, as we moved toward the opposite side of the channel, the DEP force experienced by a bead decreased. Vertical electrodes employed in our device, instead, generated a homogeneous electrical field over the entire channel height, ensuring that all the beads entering the incubation region experienced the same DEP force. The height of the channel was then no longer limited and could be substantially increased, to 50 µm in this device and possibly larger, as only the microfabrication process dictates the extension of the microchannel height. Increasing the height of the channel helped us increase the reagent flow rate in the microfluidic chamber by a factor of 10 or more compared to similar detection platforms [18,20], improving the collection of analytes that may be only present at low concentration in serum.

Another advantage deriving from the three-dimensional electrodes employed to generate DEP forces is the use of minimally diluted solutions. Currently, as DEP forces are weakened in high ionic strength solutions [25], most DEP-based microfluidic platforms are forced to operate in extremely diluted (low ionic force) solutions, thus drastically limiting the actual detection capabilities of their systems due to the consequent dilution of the analytes' concentration [20]. Indeed, diluting the sample of interest by a large factor will consequently reduce the output signal. In comparison, our device successfully performed beads' collection and analyte binding in an only 5× diluted serum, a dilution factor commonly found in commercial biomarker assays [26]. A slight dilution has proven to be even suitable for biomarker analysis, as it reduces matrix effect while maintaining favorable binding conditions [27].

Furthermore, the integration of the incubation step on-chip appears as a key milestone in the process of embedding our technology in a point-of-care device (PoC). In fact, it would limit the variance and errors in the concentration readout that might derive from additional manipulations [28].

3.2. Immunoassay Performance

The step directly following on-chip incubation is the accumulation of beads in dedicated areas, to obtain a larger signal and, thus, increase the sensitivity of detection. This section aimed to assess the impact of carrying out beads' incubation and aggregation in our microfluidic system, comparing our approach both to a bulk process of incubation of the reagents and to the detection of the signal from single beads. To do so, the experimental plan described in Figure 5b was designed and implemented. Beads decorated with NGAL cAbs were incubated either on-chip on the horizontal *incubation lines* with the antigen–dAbs complex, as described in the previous section, or off-chip, by placing the Eppendorf containing the beads and the complexes in a rotating mixer. The obtained beads were then either clustered on-chip, as shown in Figure 5a, or observed sparsely on a microscope slide, and the corresponding fluorescent signal was acquired and plotted, as in Figure 5c.

The comparison between plots 1 and 2 of Figure 5d shows the impact of measuring the fluorescence signal of the microbeads in the accumulation regions on-chip vs. measuring the fluorescence of single beads on plates. In fact, in both cases, the prior incubation of beads with the analyte and dAb was performed identically off-chip in vials. We observed a signal increase of 3.5-fold in the case of beads concentrated in specific locations on the chip, demonstrating the validity of our approach to enhance the signal by locally increasing beads' density to get a larger signal. Such amplification allows the detection of 1 ng/mL NGAL concentration in 5-fold diluted FBS, while this concentration could not be resolved by observing the fluorescent intensity of single beads. Common approaches for beads' accumulation involve the application of magnetic forces [29,30] or DEP forces [18]. The increase in signal that we found to be a consequence of the beads' accumulation is in line with what has been previously reported in literature [29].

On the other hand, the influence of incubation conditions on the attained signal could be analyzed comparing plots 2 and 3 of Figure 5d. In this case, the clustering of beads was carried out in the microfluidic platform in both experiments, while the incubation was performed either on-chip or off-chip. On-chip incubated beads reached the same level of antigen binding as the off-chip incubated beads. This demonstrates that the on-chip process matches the performance of a standard incubation in turbulent regime.

The integration of the incubation step within the microfluidic system is a key step in the design of autonomous lab-on-chip platforms. Recent approaches emphasize the need to maintain a certain level of agitation for beads during incubation, in order to maximize the interaction between beads and target analytes in the solution [12,31,32]. Indeed, a local increase of the convection phenomenon in close vicinity of the beads would reduce mass transport issues and speed up the supply of analyte. The incubation step we implemented on the horizontal lines appears to be as efficient as a standard turbulent off-chip incubation, even though it was previously reported that turbulent incubation performs better than exposure to continuous flows [33,34]. Our solution thus succeeded in providing an incubation as efficient as the gold standard off-chip methodology.

The NGAL concentrations that could be detected with our device ranged from 1 to 100 ng/mL, with a limit of detection of 1 ng/mL, calculated using a three-times standard deviation approach. However, as the concentrations expressed in the calibration curve refer to the five times diluted serum, this interval can be translated into a 5–500-ng/mL detection range in non-diluted serum. This interval covers the clinical NGAL concentration values observed in healthy patients (around 80 ng/mL) and patients suffering AKI (above 300 ng/mL) [10,35]. Furthermore, the standard deviation calculated with our platform appears to be small enough to distinguish healthy patients from ill patients. Indeed, the clinical procedure for AKI diagnosis, defined as an increase of more than 100% of the NGAL basal concentration [10,36], is resolvable with our platform, which makes our device suitable for the rapid diagnostics of kidney injury. In order to compare the performance of our platform with a state-of-the-art method, the same samples were tested by ELISA (Figure S6), exhibiting similar performance, with comparable sensitivity over a 1–100-ng/mL concentration range. Notably, the assay time could be shortened down to 15 min with our platform, versus 4 h required to run the ELISA test. Table 1 describes the advantages of our approach over existing solutions for the detection of NGAL.

Table 1. Comparison of advantages of our approach with respect to standard methods used for detection of NGAL.

	DEP Surfing	ELISA	Abbott Architect	Lateral Flow Assay
Limit of detection	Low	Low	Low	Average
Total analysis time	Short	Long	Average	Short
Sample processing	Limited	Extensive	Extensive	Limited
Volume needed	Low	Average	Average	High
Multiple analytes	Easy	Difficult	Easy	Average
Translation to PoC	Easy	Difficult	Difficult	Easy

All experiments carried out within this publication were run in 5 times diluted serum sample (FBS: fetal bovine serum). As none of the previous studies employing DEP to perform immunoassay on beads relies on serum [18,20], we are the first to implement the detection of analytes with DEP in a representative medium.

3.3. Simultaneous On-Chip Analysis of AKI Biomarkers NGAL and Cystatin C

This section aimed to investigate the possibility to concomitantly quantify the concentration of two biomarkers on a single chip. The two chosen analytes, NGAL and Cystatin C, spike into serum at different stages of the kidney injury [9]. Three hypothetical scenarios were defined based on clinical data: a "healthy" patient who presents basal levels for both NGAL (75 ng/mL) and Cystatin C (250 ng/mL), an "early-stage" patient presenting a spike in NGAL (300 ng/mL) and basal Cystatin C levels (250 ng/mL), and finally a "late-stage" patient presenting a spike in Cystatin C (1000 ng/mL) and basal NGAL levels (75 ng/mL). Two antibodies selective to NGAL or Cystatin C were conjugated to 6-μm and 2-μm beads, respectively. Sets of beads with different sizes are necessary to separate and localize beads in different areas on the device (See Materials and Methods section).

In order to describe the efficiency of beads' separation, NGAL and Cystatin C detection antibodies (see Figure 3b) were labeled with fluorophores emitting in different spectral regions, respectively in CY5 and CY3 regions. The spectral overlap between fluorophores was also investigated to ensure negligible spectral overlap due to the fluorophore emission ranges (See Figure S3).

Figure 6a shows the achieved spatial separation of beads based on their size, with the 6-μm and the 2-μm beads, respectively, appearing as in red or green. The performance of this multianalyte approach is quantified in Figure 6b for the three patients described above. As the fluorophores emitting in CY3 and CY5 regions were different in brightness, the comparison between the absolute fluorescent signals was not relevant.

For the CY5 channel (Figure 6b, red), corresponding to the fluorescence of 6-μm beads capturing NGAL collected in cluster 1, the detection of the NGAL spike in the "early" case scenario appears to be clearly resolved. The presence of some signal in cluster 2, shown in Figure S5, for all scenarios suggests that some 6-μm beads could cross the first electrical barrier and get trapped at the downstream location. This contamination, evaluated at 20% of the signal obtained in the 6-μm beads' clusters, was due to defects in the electrical contact between some of the vertical pillars and the planar electrodes, leading to the leaking of beads through the lines and unwanted gathering in the downstream region, which could be reduced by an optimization of the microfabrication process. The signal produced by NGAL beads appears lower than the one observed with a similar concentration in the previous section, which was due to the smaller flow rate used in this experiment for incubation, 0.4 μL/min vs. 2 μL/min.

Regarding the CY3 channel (Figure 6b, green), corresponding to the fluorescence of 2-μm beads capturing Cystatin C accumulated downstream in cluster 2, the detection of the Cystatin C spike in the "late" case scenario can be resolved vs. normal conditions. The corresponding calibration curve for Cystatin C on-chip detection can be found in Figure S4. We observed a limit of detection of 0.5 ng/mL and a detection range covering concentrations from 0.5 to 200 ng/mL, which suits the clinical requirement for detection of AKI in patients [37]. Furthermore, the resolution at 50 ng/mL was calculated and estimated to be of 11 ng/mL. Contaminations can also be noticed (Figure S5) in cluster 1 for all scenarios. We estimated that contaminations were equivalent to 25% of the signal obtained in the 2-μm beads' clusters and arose from a tendency of small beads to stick together and with large beads, therefore, forming clusters while incubating on horizontal lines. Those clustered beads then behaved as larger beads and, thus, accumulated in cluster 1. Such effect could be mitigated through the introduction of a surfactant in the reaction solution. Moreover, assuming that the assay will be calibrated to result in comparable fluorescence signals corresponding to physiological basal levels of NGAL and Cystatin C (healthy patient), a 20% variability due to contaminations would not prevent us from detecting a 100% increase in one of the marker concentrations, as an effect of an AKI condition.

Our platform could achieve the differentiation of healthy, "early AKI" stage and "late AKI" patient by detecting both NGAL and Cystatin C in clinically relevant ranges [10,36–38] within 15 min. As NGAL and Cystatin C spike in the serum at different stages of kidney failure [9], our approach combining detection of both these biomarkers within one test allows for injury detection within a large time window, from immediately after the injury to 48 h, reducing the risk of inaccurate diagnosis and improving survival rate in patients experiencing AKI. Our device could perform the detection of analytes within 15 min requiring a serum volume of only 50 μL, which is compatible with its integration into a point-of-care platform available at the intensive care unit and requiring a small amount of blood to perform the analysis [26].

Previous approaches aiming to simultaneously carry out multiple biomarkers' detection mainly relied on the use of fluorophores embedded in the beads [39,40]. This technology, developed by Luminex, associates a barcode defined as a ratio of fluorophore dye in the bead to each analyte of interest [41], thus allowing for efficient detection of distinct markers in a flow cytometry setup [42]. Despite its performance, the main limitation of this solution is the need to integrate at least two fluorescent filters within the readout platform to ensure barcode reading. This requirement severely hinders a potential integration of this technique in portable devices. Nonetheless, as beads in our DEP-based platform are resolved spatially according to their sizes, the readout can be performed with a single fluorescent channel that could be easily integrated within a PoC device [26]. Another option to optically quantify multiple species without the need for multiple fluorescence channels was proposed by Falconnet et al. [13], who introduced a digitally encoded silicon disk, on the top of which immunoassay would be performed. However, as each barcode has to be observed singularly, the silicon microparticles cannot be accumulated to increase the overall sensitivity of the system. Our method, instead, allows both multimarker analysis and amplification of the outcome signal. Our platform is, thus, the first of its kind, allowing on-chip incubation for optimized analyte collection, multiple biomarkers' detection, signal amplification, and optical readout by means of a single fluorescence channel.

4. Conclusions

In this paper, we presented a fully integrated system for fast and efficient acute kidney injury diagnostics. The introduction of high-aspect-ratio vertical electrodes within the microfluidic channel permitted an accurate manipulation of antibodies-decorated beads through a novel method named "DEP surfing". The incubation step was conducted in a dedicated area under a high flow rate, ensuring an effective collection of analytes on the surface of beads, an important feature for the detection of analytes at low concentration. We introduced the possibility to perform in the same microchamber, while in two separate phases, sample beads' incubation and beads' accumulation, controlled solely by electrical signals. The amplified fluorescent signal acquired with our method proved to be as bright as the one obtained with turbulent mixing. We employed this approach to the concomitant detection of two analytes, thanks to a size-based beads' separation technique: NGAL and Cystatin C could be simultaneously detected within 15 min in a minimally diluted matrix, and the detection performance matches the one of a commercial ELISA kit. The combined detection of both biomarkers allows for the diagnosis of AKI conditions at different stages, which could be greatly beneficial to patients in intensive care units.

Currently, our device is designed for the detection of two biomarkers. Nonetheless, the technology could be easily readjusted for the detection of more analytes through the use of beads of different sizes. Furthermore, since this system relies on the largely established biochemistry of antibody–beads' conjugation and on-sandwich assays, it can be easily translated to the analysis of other acute conditions or infectious diseases.

In order to miniaturize the platform and promote its integration in a one-step, point-of-care device, the active fluidic components would need to be replaced by passive fluidic structures, such as capillary pumps, while the beads could be dried in the channels prior to exposure to the patient sample [43,44]. An automated inspection of the cluster size would contribute to the reduction of the inter-experiments' variability and handling errors.

5. Patents

This work resulted in the deposition of the following patent: Thiriet, P.-E.; Medagoda, D.; Guiducci, C. Dielectrophoresis detection device. European Patent (EP) priority: 5 June 2020 no. 20178445.1

Supplementary Materials: The following are available online at http://www.mdpi.com/2079-6374/10/12/212/s1. Figure S1: Fabrication of vertical electrodes in the microfluidic channel. Figure S2: (**A**) Computer-aided design of the printed circuit board (PCB) layout. (**B**) Picture of the assembled PCB. Figure S3: Spectral overlap of CY3 and CY5 signal from 6-μm NGAL- and 2-μm Cystatin C-decorated beads. Figure S4: Cystatin C dose-response curve. Figure S5: Cross-contamination analysis in the multianalyte experiment. Figure S6: NGAL detection performance comparison between our immunoassay on-chip and a commercial ELISA kit. Video S1: Beads' trapping. Video S2: Beads' release. Video S3: Binding of fluorophores to the beads. Video S4: From incubation to accumulation.

Author Contributions: Conceptualization, P.-E.T., D.M., and C.G.; methodology, P.-E.T., D.M., and G.P.; validation, P.-E.T. and D.M.; formal analysis, P.-E.T.; investigation, P.-E.T. and D.M.; data curation, P.-E.T. and D.M.; writing—original draft preparation, P.-E.T., G.P., and C.G.; writing—review and editing, P.-E.T., D.M., G.P., and C.G.; project administration, C.G. All authors have read and agreed to the published version of the manuscript.

Funding: This work is supported by the Swiss National Foundation (514247).

Acknowledgments: We are thankful for the support of the Center for Nano- and Microtechnology (CMi) at École Polytechnique Fédérale de Lausanne (EPFL) where the microsystems were developed. We would like to thank Lauréanne Putallaz and Fabien Rebeaud from Abionic SA for the valuable discussions and advice and the provision of some reagents. We would like to thank Cécile Plaire and Tsandni Jamal for their valued assistance in experimental work. We would also like to acknowledge the precious support received by the Service de Pharmacologie Clinique at the Centre Hospitalier Universitaire Vaudois (CHUV) and in particular to Thierry Buclin, Christophe Butticaz, Laurent Arthur Decosterd, and Dominique Werner.

Conflicts of Interest: The authors declare no conflict of interest.

References

1. Makris, K.; Spanou, L. Acute Kidney Injury: Definition, Pathophysiology and Clinical Phenotypes. *Clin. Biochem. Rev.* **2016**, *37*, 85–98. [PubMed]
2. Hoste, E.A.J.; Kellum, J.A.; Selby, N.M.; Zarbock, A.; Palevsky, P.M.; Bagshaw, S.M.; Goldstein, S.L.; Cerdá, J.; Chawla, L.S. Global epidemiology and outcomes of acute kidney injury. *Nat. Rev. Nephrol.* **2018**, *14*, 607–625. [CrossRef] [PubMed]
3. Rahman, M.; Shad, F.; Smith, M.C. Acute kidney injury: A guide to diagnosis and management. *Am. Fam. Physician* **2012**, *86*, 631–639. [PubMed]

4. Maggi, P.; Montinaro, V.; Mussini, C.; Di Biagio, A.; Bellagamba, R.; Bonfanti, P.; Calza, L.; Cherubini, C.; Corsi, P.; Gargiulo, M.; et al. Novel antiretroviral drugs and renal function monitoring of HIV patients. *AIDS Rev.* **2014**, *16*, 144–151.

5. Wilson, F.P.; Greenberg, J.H. Acute Kidney Injury in Real Time: Prediction, Alerts, and Clinical Decision Support. *Nephron* **2018**, *140*, 116–119. [CrossRef]

6. Ostermann, M.; Joannidis, M. Acute kidney injury 2016: Diagnosis and diagnostic workup. *Crit. Care* **2016**, *20*, 299. [CrossRef]

7. Murty, M.S.N.; Sharma, U.K.; Pandey, V.B.; Kankare, S.B. Serum cystatin C as a marker of renal function in detection of early acute kidney injury. *Indian J. Nephrol.* **2013**, *23*, 180–183. [CrossRef]

8. Urbschat, A.; Obermüller, N.; Haferkamp, A. Biomarkers of kidney injury. *Biomarkers* **2011**, *16*, S22–S30. [CrossRef]

9. Padhy, M.; Kaushik, S.; Girish, M.P.; Mohapatra, S.; Shah, S.; Koner, B.C. Serum neutrophil gelatinase associated lipocalin (NGAL) and cystatin C as early predictors of contrast-induced acute kidney injury in patients undergoing percutaneous coronary intervention. *Clin. Chim. Acta* **2014**, *435*, 48–52. [CrossRef]

10. Khawaja, S.; Jafri, L.; Siddiqui, I.; Hashmi, M.; Ghani, F. The utility of neutrophil gelatinase-associated Lipocalin (NGAL) as a marker of acute kidney injury (AKI) in critically ill patients. *Biomark. Res.* **2019**, *4*. [CrossRef]

11. Devarajan, P. Neutrophil gelatinase-associated lipocalin (NGAL). *Scand. J. Clin. Lab. Investig. Suppl.* **2008**, *241*, 89–94. [CrossRef] [PubMed]

12. Armbrecht, L.; Dincer, C.; Kling, A.; Horak, J.; Kieninger, J.; Urban, G. Self-assembled magnetic bead chains for sensitivity enhancement of microfluidic electrochemical biosensor platforms. *Lab Chip* **2015**, *15*, 4314–4321. [CrossRef] [PubMed]

13. Falconnet, D.; She, J.; Tornay, R.; Leimgruber, E.; Bernasconi, D.; Lagopoulos, L.; Renaud, P.; Demierre, N.; van den Bogaard, P. Rapid, Sensitive and Real-Time Multiplexing Platform for the Analysis of Protein and Nucleic-Acid Biomarkers. *Anal. Chem.* **2015**, *87*, 1582–1589. [CrossRef] [PubMed]

14. Chen, Q.; Yuan, Y.J. A review of polystyrene bead manipulation by dielectrophoresis. *Rsc. Adv.* **2019**, *9*, 4963–4981. [CrossRef]

15. Salari, A.; Thompson, M. Recent advances in AC electrokinetic sample enrichment techniques for biosensor development. *Sens. Actuators B Chem.* **2018**, *255*, 3601–3615. [CrossRef]

16. Sanghavi, B.J.; Varhue, W.; Rohani, A.; Liao, K.-T.; Bazydlo, L.A.L.; Chou, C.-F.; Swami, N.S. Ultrafast immunoassays by coupling dielectrophoretic biomarker enrichment in nanoslit channel with electrochemical detection on graphene. *Lab Chip* **2015**, *15*, 4563–4570. [CrossRef]

17. Holmes, D.K.; She, J.L.; Roach, P.; Morgan, H. Bead-based immunoassays using a micro-chip flow cytometer. *Lab Chip* **2007**, *7*, 1048–1056. [CrossRef]

18. Iswardy, E.; Tsai, T.-C.; Cheng, I.-F.; Ho, T.-C.; Perng, G.; Chang, H.-C. A bead-based immunofluorescence-assay on a microfluidic dielectrophoresis platform for rapid dengue virus detection. *Biosens. Bioelectron.* **2017**, *95*, 174–180. [CrossRef]

19. Ramón-Azcón, J.; Kunikata, R.; Sanchez, F.-J.; Marco, M.-P.; Shiku, H.; Yasukawa, T.; Matsue, T. Detection of pesticide residues using an immunodevice based on negative dielectrophoresis. *Biosens. Bioelectron.* **2009**, *24*, 1592–1597. [CrossRef]

20. Park, S.; Yossifon, G. Combining dielectrophoresis and concentration polarization-based preconcentration to enhance bead-based immunoassay sensitivity. *Nanoscale* **2019**. [CrossRef]

21. Yasukawa, T.; Suzuki, M.; Sekiya, T.; Shiku, H.; Matsue, T. Flow sandwich-type immunoassay in microfluidic devices based on negative dielectrophoresis. *Biosens. Bioelectron.* **2007**, *22*, 2730–2736. [CrossRef] [PubMed]

22. Thiriet, P.-E.; Pezoldt, J.; Gambardella, G.; Keim, K.; Deplancke, B.; Guiducci, C. Selective Retrieval of Individual Cells from Microfluidic Arrays Combining Dielectrophoretic Force and Directed Hydrodynamic Flow. *Micromachines* **2020**, *11*, 322. [CrossRef] [PubMed]

23. Gascoyne, P.R.C.; Vykoukal, J. Particle separation by dielectrophoresis. *Electrophoresis* **2002**, *23*, 1973–1983. [CrossRef]

24. Thiriet, P.-E.; Medagoda, D.; Guiducci, C. Dielectrophoresis Detection. Device. Patent 20178445.1, 5 June 2020.

25. Nerguizian, V.; Stiharu, I.; Nosayba, A.-A.; Bader, Y.-D.; Alazzam, A. The effect of dielectrophoresis on living cells: Crossover frequencies and deregulation in gene expression. *Analyst* **2019**, *144*, 3853–3860. [CrossRef]

26. Putallaz, L.; van den Bogaard, P.; Laub, P.; Rebeaud, F. Nanofluidics Drives Point-of-care Technology for on the Spot Protein Marker Analysis with Rapid Actionable Results. *J. Nanomed. Nanotechnol.* **2019**, *10*. [CrossRef]

27. Taylor, T.P.; Janech, M.G.; Slate, E.H.; Lewis, E.C.; Arthur, J.M.; Oates, J.C. Overcoming the Effects of Matrix Interference in the Measurement of Urine Protein Analytes. *Biomark Insights* **2012**, *7*, 1–8. [CrossRef]

28. Price, C.P. Point of care testing. *BMJ* **2001**, *322*, 1285–1288. [CrossRef]

29. İçöz, K.; Mzava, O. Detection of Proteins Using Nano Magnetic Particle Accumulation-Based Signal Amplification. *Appl. Sci.* **2016**, *6*, 394. [CrossRef]

30. Jalal, U.M.; Jin, G.J.; Eom, K.S.; Kim, M.H.; Shim, J.S. On-chip signal amplification of magnetic bead-based immunoassay by aviating magnetic bead chains. *Bioelectrochemistry* **2018**, *122*, 221–226. [CrossRef]

31. Yoo, H.; Lee, D.J.; Kim, D.; Park, J.; Chen, X.; Hong, S. Magnetically-focusing biochip structures for high-speed active biosensing with improved selectivity. *Nanotechnology* **2018**, *29*, 265501. [CrossRef]

32. Dittmer, W.U.; de Kievit, P.; Prins, M.W.J.; Vissers, J.L.M.; Mersch, M.E.C.; Martens, M.F.W.C. Sensitive and rapid immunoassay for parathyroid hormone using magnetic particle labels and magnetic actuation. *J. Immunol. Methods* **2008**, *338*, 40–46. [CrossRef]

33. Sreenivasan, K.R. Turbulent mixing: A perspective. *Proc. Natl. Acad. Sci. USA* **2019**, *116*, 18175–18183. [CrossRef] [PubMed]

34. Zimmermann, M.; Delamarche, E.; Wolf, M.; Hunziker, P. Modeling and Optimization of High-Sensitivity, Low-Volume Microfluidic-Based Surface Immunoassays. *Biomed. Microdevices* **2005**, *7*, 99–110. [CrossRef] [PubMed]

35. Min, J.; Nothing, M.; Coble, B.; Zheng, H.; Park, J.; Im, H.; Weber, G.F.; Castro, C.M.; Swirski, F.K.; Weissleder, R.; et al. Integrated Biosensor for Rapid and Point-Of-Care Sepsis Diagnosis. *ACS Nano* **2018**, *12*, 3378–3384. [CrossRef] [PubMed]

36. Hang, C.-C.; Yang, J.; Wang, S.; Li, C.-S.; Tang, Z.-R. Evaluation of serum neutrophil gelatinase-associated lipocalin in predicting acute kidney injury in critically ill patients. *J. Int. Med. Res.* **2017**, *45*, 1231–1244. [CrossRef] [PubMed]

37. Yong, Z.; Pei, X.; Zhu, B.; Yuan, H.; Zhao, W. Predictive value of serum cystatin C for acute kidney injury in adults: A meta-analysis of prospective cohort trials. *Sci. Rep.* **2017**. [CrossRef]

38. Villa, P.; Jiménez, M.; Soriano, M.-C.; Manzanares, J.; Casasnovas, P. Serum cystatin C concentration as a marker of acute renal dysfunction in critically ill patients. *Crit. Care Lond. Engl.* **2005**, *9*, R139–R143. [CrossRef]

39. Ellington, A.A.; Kullo, I.J.; Bailey, K.R.; Klee, G.G. Antibody-Based Protein Multiplex Platforms: Technical and Operational Challenges. *Clin Chem* **2010**, *56*, 186–193. [CrossRef] [PubMed]

40. Kingsmore, S.F. Multiplexed protein measurement: Technologies and applications of protein and antibody arrays. *Nat. Rev. Drug Discov.* **2006**, *5*, 310–320. [CrossRef]

41. Harink, B.; Nguyen, H.; Thorn, K.; Fordyce, P. An open-source software analysis package for Microspheres with Ratiometric Barcode Lanthanide Encoding (MRBLEs). *PLoS ONE* **2019**, *14*, e0203725. [CrossRef]

42. Aira, C.; Ruiz, T.; Dixon, L.; Blome, S.; Rueda, P.; Sastre, P. Bead-Based Multiplex Assay for the Simultaneous Detection of Antibodies to African Swine Fever Virus and Classical Swine Fever Virus. *Front. Vet. Sci.* **2019**, *6*. [CrossRef] [PubMed]

43. Pham, N.M.; Rusch, S.; Temiz, Y.; Lovchik, R.D.; Beck, H.P.; Karlen, W.; Delamarche, E. A bead-based immunogold-silver staining assay on capillary-driven microfluidics. *Biomed. Microdevices* **2018**, *20*, 1–9. [CrossRef] [PubMed]

44. Epifania, R.; Soares, R.R.G.; Pinto, I.F.; Chu, V.; Conde, J.P. Capillary-driven microfluidic device with integrated nanoporous microbeads for ultrarapid biosensing assays. *Sens. Actuators B Chem.* **2018**, *265*, 452–458. [CrossRef]

Publisher's Note: MDPI stays neutral with regard to jurisdictional claims in published maps and institutional affiliations.

 *biosensors*

Article

Development of a Pharmacogenetic Lab-on-Chip Assay Based on the In-Check Technology to Screen for Genetic Variations Associated to Adverse Drug Reactions to Common Chemotherapeutic Agents

Rosario Iemmolo [1], Valentina La Cognata [1], Giovanna Morello [1], Maria Guarnaccia [1], Mariamena Arbitrio [2], Enrico Alessi [3] and Sebastiano Cavallaro [1,*]

[1] Institute for Biomedical Research and Innovation, National Research Council, Via Paolo Gaifami, 18-95126 Catania, Italy; r.iemmolo@isn.cnr.it (R.I.); valentina.lacognata@cnr.it (V.L.C.); g.morello@isn.cnr.it (G.M.); maria.guarnaccia@cnr.it (M.G.)

[2] Institute for Biomedical Research and Innovation, National Research Council, 88100 Catanzaro, Italy; mariamena.arbitrio@cnr.it

[3] Analog, MEMS & Sensor Group Health Care Business Development Unit, STMicroelectronics, Stradale Primosole, 50-95126 Catania, Italy; enrico.alessi@st.com

* Correspondence: sebastiano.cavallaro@cnr.it

Received: 6 November 2020; Accepted: 7 December 2020; Published: 9 December 2020

Abstract: Background: Antineoplastic agents represent the most common class of drugs causing Adverse Drug Reactions (ADRs). Mutant alleles of genes coding for drug-metabolizing enzymes are the best studied individual risk factors for these ADRs. Although the correlation between genetic polymorphisms and ADRs is well-known, pharmacogenetic tests are limited to centralized laboratories with expensive or dedicated instrumentation used by specialized personnel. Nowadays, DNA chips have overcome the major limitations in terms of sensibility, specificity or small molecular detection, allowing the simultaneous detection of several genetic polymorphisms with time and costs-effective advantages. In this work, we describe the design of a novel silicon-based lab-on-chip assay able to perform low-density and high-resolution multi-assay analysis (amplification and hybridization reactions) on the In-Check platform. Methods: The novel lab-on-chip was used to screen 17 allelic variants of three genes associated with adverse reactions to common chemotherapeutic agents: *DPYD* (Dihydropyrimidine dehydrogenase), *MTHFR* (5,10-Methylenetetrahydrofolate reductase) and *TPMT* (Thiopurine S-methyltransferase). Results: Inter- and intra assay variability were performed to assess the specificity and sensibility of the chip. Linear regression was used to assess the optimal hybridization temperature set at 52 °C ($R^2 \approx 0.97$). Limit of detection was 50 nM. Conclusions: The high performance in terms of sensibility and specificity of this lab-on-chip supports its further translation to clinical diagnostics, where it may effectively promote precision medicine.

Keywords: pharmacogenetics; lab-on-chip; microfluidics; biosensors; In-Check platform; adverse drug reaction

---

1. Introduction

Cancer is one of the leading causes of mortality and a major public health problem worldwide [1,2]. Anticancer therapeutic strategies are influenced by tumor clinical characteristics, such as signs, symptoms, histological type, stage and localization. Currently, the most used cytotoxic antineoplastic drugs are pyrimidine analogues (i.e., 5-fluorouracil, capecitabine or tegafur), purine analogues (i.e., mercaptopurine or thioguanine), and platinum compounds (i.e., cisplatin or oxaliplatin). These drugs have a narrow therapeutic index and strictly dose-related effects that are conditioned by inter-individual variability in their metabolism. Therefore, interest in the relationship between genetic variants and cancer treatment outcomes has been growing. Pharmacogenetic (PGx) research in cancer patients has identified specific allelic variants of genes that are related to antineoplastic drug Absorption, Distribution, Metabolism and Excretion (ADME) and can alter pharmacokinetic and pharmacodynamic parameters. This, in turn, results in variable efficacy in tumors of the same histotype [3] or adverse drug reactions (ADRs) [4]. It is now accepted that polymorphic variants in ADME-related genes impact individual patients' drug sensitivity, resistance and toxicity and contribute to 25–50% of inappropriate pharmacological responses, ranking between the fourth to sixth leading causes of death [5,6]. The most common genetic alterations studied are Single Nucleotide Polymorphisms (SNPs), genomic insertions and deletions, and genetic Copy Number Variations (CNVs). SNPs are a single base-pair difference in the DNA sequence of individuals and represent common inherited variations (90%) distributed throughout the genome. When located within a gene coding sequence or in a regulatory region, they may exert a functional role. SNPs are stably inherited within haplotype blocks in linkage disequilibrium (LD) with a specific gene variant and coheredited within the haplotype. They can be considered as biomarkers of a gene variant and are often used in genomic analyses as tags (tagSNPs) to identify a haplotype block in which few or many polymorphic variants are associated with a disease or drug-response phenotype. The analysis of ADME-related gene polymorphisms can be used for drug efficacy prediction, and screening of individuals who should avoid or receive an adjusted dose of the drug is recommended [7–9]. Unfortunately, pharmacogenetic evaluations for individualized antineoplastic drug response are not common in clinical practice for different reasons. The controversial clinical relevance of pharmacogenetic biomarkers, the heterogeneity in efficacy and toxicity of ADRs, the time lag between scientific findings, drug label annotations and clinical practice guidelines are only a few major challenges in the clinical implementation of pharmacogenetics testing [10–12]. Moreover, the genetic variations associated with the response to drugs are manifold and therefore their evaluation undoubtedly takes a long time and prohibitive costs. For these reasons, pharmacogenetic tests are not usually implemented in clinical care but often performed only in centralized laboratories with expensive or dedicated instrumentation used by specialized personnel [7,13,14]. In the last few years, several biosensors have been developed with the purpose of promoting precision medicine with advantages in terms of efficiency, time and costs of analysis. However, only few studies report the development of miniaturized DNA chip performing allelic variant discrimination to avoid ADRs in oncological patients [15–18].

The In-Check platform developed by STMicroelectronics is an innovative technology in the field of microfluidic-based DNA biosensors that combines micro-electro-mechanical-system (MEMS) techniques. The platform consists of three main modules: the Temperature Control System (TCS) that operates and monitors the thermal cycling reactions, allowing a fast temperature ramping with an accuracy of 0.1 °C with improved performance and amplification time; the portable Optical Reader (OR) to perform scanning of the chip; a miniaturized Lab-on-Chip (LoC) device that brings together silicon-based microelectronics with micromachining technology. LoC consists of two separate Polymerase Chain Reaction (PCR) chambers and a microarray area joined together through microfluidic channels allowing a fast, highly sensitive and specific amplification of nucleic acids and competitive hybridization of generated amplicons simultaneously (Figure 1).

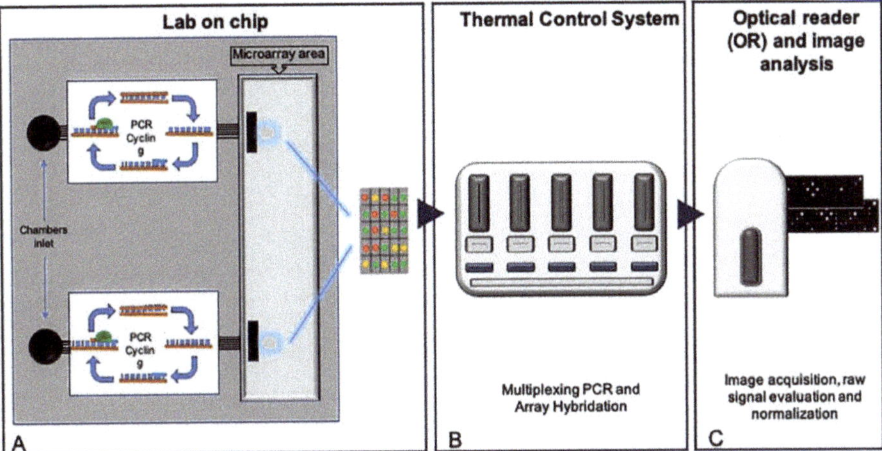

Figure 1. In-Check platform. **(A)** Schematic representation of miniaturized Lab-on-Chip (LoC) device. Two separate PCR chambers and a microarray area are joined together through a microfluidic channel. **(B)** Temperature Control System (TCS) to operate and monitors the reactions thermal cycling. **(C)** Portable Optical Reader (OR) to perform scanning of the chip.

The miniaturized low-density microarray area of the In-Check LoC contains up to 126 spots, offering the opportunity to design and customize the chip for individual applications. The platform is completed with dedicated software that controls the instruments and runs the data analysis [19–21]. The miniaturization of the combined PCR-microarray analytical processes performed into a unique biosensor allows different advantages in terms of high sensitivity, simplicity and high-throughput, with short assay time, low reagent consumption, and easiness to automation [22]. Today, the platform is marketed by Sekisui—Veredus Ltd. and is used with a series of customized panels for infection disease, food safety and biosurveillance (vereduslabs.com). In this work, we describe the development of PGx Lab-on-Chip test based on In-Check technology, which is able to detect multiple allelic variants of three genes (*DPYD*, Dihydropyrimidine dehydrogenase; *MTHFR*, 5,10-Methylenetetrahydrofolate reductase; and *TPMT*, Thiopurine S-methyltransferase) associated to adverse reactions to common chemotherapeutic agents.

2. Materials and Methods

2.1. Development of a Pharmacogenetic Lab-On-Chip

Our custom PGx Lab-on-Chips were developed for the In-Check technology and fabricated as previously described [20]. We designed 31 allele-specific capture probes able to perfect match the selected wild type and 9 different mutant allelic variants of *DPYD*, *MTHFR* and *TPMT* genes (Table 1). An additional artificial oligonucleotide (AT683) was designed and used both as spike-in internal hybridization control and for grid alignment. Purified allele-specific capture probes and hybridization controls were modified with an amino group and a six carbon spacer at their 5′-terminus and spotted onto silicon microarray slide as previously described [19].

Table 1. List of allelic variants detected by the pharmacogenetic Lab-on-Chip and associated with Adverse Drug Reactions (ADRs) to antineoplastic drugs.

Gene	Allelic Variants (hg38)	MAF (1000 Genomes)	Antineoplastic Drug	Level(s)
DPYD (NG_008807.2)	rs2297595 g.226525A > G p.Met166Val	C = 0.0565/283	Cetuximab, fluorouracil, capecitabine, oxaliplatin, bevacizumab	2A
	rs55886062 g.410273T > A g.410273T > G p.Ile560Asn; p.Ile560Ser	C = 0.0002/1	Capecitabine, fluorouracil, Pyrimidine analogues, tegafur	1A
	rs17376848 g.475992T > C p.Phe632Phe	G = 0.0521/261	Leucovorin, fluorouracil, capecitabine, oxaliplatin	3
	rs3918290 g.476002G > A IVS14	T = 0.0030/15	Capecitabine, fluorouracil, Pyrimidine analogues, tegafur	1A
	rs67376798 g.843669A > T p.Asp949Val	A = 0.0022/11	Cetuximab, oxaliplatin, bevacizumab, leucovorin, tegafur, fluorouracil, capecitabine, Pyrimidine analogues	1A
MTHFR (NG_013351)	rs1801131 g.16685A > C p.Glu429Ala	G = 0.2494/1249	Leucovorin, capecitabine, fluorouracil, oxaliplatin, methotrexate, bevacizumab, carboplatin, cisplatin, cyanocobalamin, folic acid, or pemetrexed	3
TPMT (NG_012137.1)	rs1800462 g.16420G > C p.Ala80Pro	G = 0.0022/11	s-adenosylmethionine, purine analogues, mercaptopurine, azathioprine, thioguanine	1A
	rs1800460 g.21147G > A p.Ala154Thr	T = 0.0128/64	s-adenosylmethionine, mercaptopurine, purine analogues, azathioprine, thioguanine, cisplatin	1A/3

2.2. Asymmetric Multiplex PCR and Gene Sequencing

In-Check LoCs can simultaneously perform two asymmetric multiplex PCR reactions in two different PCR chambers. The multiplex PCR design, primer pair sequences and sizes of amplicons are listed in Table 2. PCR mixes were set up as follows: 2.5 U of HotStart Taq (Qiagen, Hilden, Germany), 0.6 mM $MgCl_2$, 0.56 mM of each dNTP, 2.5 µl of 10× PCR buffer, 0.2 µM of forward primer, 2.4 µM of Cy5-labeled reverse primer and 50 ng of standard genomic DNA in a final volume of 12.5 µL. Then, 11.5 µl of each PCR mix was dispensed into the two chamber inlets of the silicon lab-on-chips which were closed with specific clamps and loaded into Temperature Control System (TCS) sets as follow: 95 °C × 300″–35 cycles at 94 °C × 60″, 61 °C × 60″, 72 °C × 60″–72 °C × 600″. To assess the accuracy of multiplex PCRs, amplicons were recovered from the PGx-LoC by brief centrifugation in 50 mL Falcon tubes, and then analyzed on 2% TBE agarose gel and visualized under UV transilluminator. Amplicons were also directly sequenced with unconjugated primer pairs using BigDye terminator v.3.1 kit (Applied Biosystems, Foster City, CA, USA) according to manufacturer's instructions, on ABIPRISM 310 genetic analyzer (Applied Biosystems, Foster City, CA, USA). GeneBank sequences for *DPYD* (#NG_008807.2 RefSeqGene), *MTHFR* (#NG_013351.1 RefSeqGene) and *TPMT* (#NG_012137.2 RefSeqGene) genes were used for multiple alignments using CLC Sequence Viewer 7.6.1 software (CLC bio, Cambridge, MA, USA).

Table 2. Primer pairs used in multiplex PCR. All reverse primers were labeled with Cy5 at 5′ position.

mPCR	Region of Interest	Primer F	GC %	TM	Primer R	GC %	TM	Product Length
#1	MTHFR exon8	TTTGGGGAGCTGAA GGACTAC	52	61.2	CACTCCAGCATCACTCACTTT	48	59.5	177
	DPYD exon23	TGCAGTACCTTGGA ACATTTGG	45	60.1	TGCAGAAGAGCAATATTTGGCA	41	58.4	245
	TPMT exon4	GATCTGCTTTCCTG CATGTTC	48	59.5	TCCAGGAATTTCGGTGATTGG	48	59.5	269
	TPMT exon6	GGACGCTGCTCATC TTCTTA	50	58.4	GACAAAGCTAGTATTGGATTT AGGT	36	60.9	295
#2	DPYD exon7	ACTGAAAATGTACT GCTCATTGCT	38	60.3	CCCCAATCGAGCCAAAAAGG	55	60.5	265
	DPYD exon15	TGTTTCCCCCAG AATCATCCG	52	61.2	TGCATCAGCAAAGCAACTG	47	55	287
	DPYD exon14	AGAAATGGCCGG ATTGAAGT	45	56.4	GACAGAAAGGAAGGAAAGA AACTAA	36	60.9	300

2.3. Microarray Hybridization and Scanning

The next step in the In-Check assay workflow is the microarray hybridization of PCR products. To allow migration of PCR products to the Microarray area, 14.5 µL of hybridization mix (a mixture of 2x hybridization buffer and 500 nM spike-in Cy5 labeled hybridization control) was pumped into PCR chambers. Then, PGx-LoCs were loaded into TCS for 30 min. The hybridization temperature plays a crucial role in hybridization-based genotyping assay. In order to calculate the accuracy of allele-specific capture probes, different hybridization temperatures were tested (50 °C, 52 °C, 54 °C, 56 °C). Three replicates for each hybridization test were performed. After hybridization, PGx-LoCs were centrifuged for 2 min at 3000 rpm into a 50-mL falcon tubes in stringency washing buffer (2× SSC + 0.1% SDS) at room temperature and dried through a second centrifugation for 2 min at 3000 rpm into empty 50 mL falcon tubes. Finally, PGx-LoCs were scanned with a dedicated Optical Reader to acquire fluorescence intensities. Raw signal values were background subtracted and normalized by customized software.

2.4. Statistical Analysis

The custom LoC assay described here was analyzed for its repeatability, sensibility and specificity by studying the Limit of Blank (LoB), Limit of Detection (LoD), and inter- and intra-assay variability as previously reported [22]. For each fluorescent image, the signal intensity for each spot signal was determined by subtracting background intensity from the mean average intensity while the target mean values were used for scaling normalization. The relationship between signal standard deviation and the signal median was evaluated using Pearson correlation coefficient values indicated by R^2. *p*-value ≤ 0.05 was considered significant. A specific software, licensed by STMicroelectronics, was used to perform grid alignment and data analysis.

3. Results

3.1. Design of a Pharmacogenetic LoC Assay for the In-Check Platform

The In-Check platform has immense potentiality in medical diagnostics. It satisfies the cornerstones of biosensors: simplicity in operation, higher sensitivity and the ability to perform multiplex analysis [23]. The functional unit of the In-Check platform is a silicon-based lab-on-chip able to perform low-density multigenic analysis (targets amplification and detection by hybridization) in less than 2 h, with cost-effective and time-consumption advantages. It contains two separate and miniaturized PCR chambers in which two multiplex PCRs can be performed simultaneously (Figure 1).

The microfluidic connection allows samples to naturally migrate from PCR to Microarray chambers. In this area, up to 126 customized allele-specific capture probes can be spotted. Other components of the platform are the Temperature Control System, the Optical Reader and a complete suite of software modules to easily perform image analysis and data interpretation for the management of diagnostic results [20,24].

In this work, we selected 17 different allelic variants, eight of which are nonfunctional variants, for *DPYD* (g.226525A; g.226525G; g.410273T; g.410273G; g.410273A; g.475992T; g.475992C; g.476002G; g.476002A; g.843669A; g.843669T), *MTHFR* (g.16685A; g.16685C) and *TPMT* (g.16420G; g.16420C; g.21147G; g.21147A) genes (Table 1), based on their involvement in drug response according to PharmGKB (http://www.pharmgkb.org), Online Mendelian Inheritance in Man (OMIM—http://www.omim.org) and Single Nucleotide Polymorphism (dbSNP—http://www.ncbi.nlm.nih.gov/SNP) databases. For the analysis of these variants, primer sets were designed to amplify selected regions with a range of melting temperature of 59.4 ± 1.8 °C, guanine-cytosine (GC) content between 36–55%, and target size from 177 bp to 300 bp (Table 2). In addition, 31 allele-specific capture probes were designed to complementary align with wild type (WT) and mutated (M) sequences (Table 3) subdivided as follows: 22 capture probes complementary to *DPYD* hotspots (10 WT and 12 M); 4 capture probes complementary to *MTHFR* hotspots (2 WT and 2 M), 5 capture probes complementary to *TPMT* hotspots (2 WT and 3 M). After in silico testing, allelic capture probes were synthetized and spotted with a duplicated layout on the PGx-LoC microarray area. Additional capture probes complementary to AT683 hybridization control were spotted with a specular pattern for grid alignment (Figure 2).

Table 3. Allelic specific capture probe sequences designed to genotype ADME-related genes.

Probe Name	Probe Sequences (5′-3′)	5′-End Modification	Probe Type
DPYD_ g.226525A	TTTTTTTGGTATTCAAAGCAATGAGTA	5′-C6-NH2	CAPTURE PROBE
DPYD_ g.226525A	TTTTTTTAGGTATTCAAAGCAATGAGT	5′-C6-NH2	CAPTURE PROBE
DPYD_ g.226525G	TTTTTTTGGTATTCAAAGCAGTGAGTA	5′-C6-NH2	CAPTURE PROBE
DPYD_ g.226525G	TTTTTTTAGGTATTCAAAGCAGTGAGT	5′-C6-NH2	CAPTURE PROBE
DPYD_ g.410273T	TTTTTTTCATCAATGATTCGAAGAGCT	5′-C6-NH2	CAPTURE PROBE
DPYD_ g.410273T	TTTTTTTCACATCAATGATTCGAAGAG	5′-C6-NH2	CAPTURE PROBE
DPYD_ g.410273G	TTTTTTTTGAGTCGAAGAGCTTTTGAA	5′-C6-NH2	CAPTURE PROBE
DPYD_ g.410273G	TTTTTTTAATGAGTCGAAGAGCTTTTG	5′-C6-NH2	CAPTURE PROBE
DPYD_ g.410273A	TTTTTTTTGAATCGAAGAGCTTTTGAA	5′-C6-NH2	CAPTURE PROBE
DPYD_ g.410273A	TTTTTTTAATGAATCGAAGAGCTTTTG	5′-C6-NH2	CAPTURE PROBE
DPYD_ g.475992T	TTTTTTTTAAAGGCTGACTTTCCAGAC	5′-C6-NH2	CAPTURE PROBE
DPYD_ g.475992T	TTTTTTTGAACTAAAGGCTGACTTTCC	5′-C6-NH2	CAPTURE PROBE
DPYD_ g.475992C	TTTTTTTTAAAGGCTGACTTCCCAGAC	5′-C6-NH2	CAPTURE PROBE
DPYD_ g.475992C	TTTTTTTGAACTAAAGGCTGACTTCCC	5′-C6-NH2	CAPTURE PROBE
DPYD_ g.476002G	TTTTTTTTTCCAGACAACGTAAGTGTG	5′-C6-NH2	CAPTURE PROBE
DPYD_ g.476002G	TTTTTTTCTTTCCAGACAACGTAAGTG	5′-C6-NH2	CAPTURE PROBE
DPYD_ g.476002A	TTTTTTTTTCCAGACAACATAAGTGTG	5′-C6-NH2	CAPTURE PROBE
DPYD_ g.476002A	TTTTTTTCTTTCCAGACAACATAAGTG	5′-C6-NH2	CAPTURE PROBE
DPYD_ g.843669A	TTTTTTTGGCTATGATTGATGAAGAAAT	5′-C6-NH2	CAPTURE PROBE
DPYD_ g.843669A	TTTTTTTGTGGCTATGATTGATGAAGAA	5′-C6-NH2	CAPTURE PROBE
DPYD_ g.843669T	TTTTTTTGGCTATGATTGTTGAAGAAAT	5′-C6-NH2	CAPTURE PROBE
DPYD_ g.843669T	TTTTTTTGTGGCTATGATTGTTGAAGAA	5′-C6-NH2	CAPTURE PROBE
MTHFR_ g.16685A	TTTTTTTCAGTGAAGAAAGTGTCTTTG	5′-C6-NH2	CAPTURE PROBE
MTHFR_ g.16685A	TTTTTTTCCAGTGAAGAAAGTGTCTTT	5′-C6-NH2	CAPTURE PROBE
MTHFR_ g.16685C	TTTTTTTGTGAAGGAAGTGTCTTTGAA	5′-C6-NH2	CAPTURE PROBE
MTHFR_ g.16685C	TTTTTTTCAGTGAAGGAAGTGTCTTTG	5′-C6-NH2	CAPTURE PROBE
TPMT_ g.16420G	TTTTTTTGTTTGCAGACCGGGGACA	5′-C6-NH2	CAPTURE PROBE
TPMT_ g.16420C	TTTTTTTGTTTCCAGACCGGGGACA	5′-C6-NH2	CAPTURE PROBE
TPMT_ g.21147G	TTTTTTTGGATAGAGGAGCATTAGTTG	5′-C6-NH2	CAPTURE PROBE
TPMT_ g.21147A	TTTTTTTATAGAGGAACATTAGTTGCC	5′-C6-NH2	CAPTURE PROBE
TPMT_ g.21147A	TTTTTTTGGGATAGAGGAACATTAGTT	5′-C6-NH2	CAPTURE PROBE
AT683	AGTGAGGGAGGAGATGGAACCATCT	5′-C6-NH2	hybridization control

	Col 1	Col 2	Col 3	Col 4	Col 5	Col 6	Col 7	Col 8	Col 9	Col 10	Col 11	Col 12	Col 13	Col 14	Col 15	Col 16	Col 17	Col 18	Col 19	Col 20	Col 21
Row 1	AT 683			DPYD ex14 WT	MTHFR ex8 Mut	DPYD IVS14 Mut	TPMT ex6 Mut	DPYD ex7 WT		DPYD ex23 WT		DPYD ex23 WT			DPYD ex14 WT	MTHFR ex8 Mut	DPYD IVS14 Mut	TPMT ex6 Mut	DPYD ex7 WT		AT 683
Row 2		DPYD ex7 Mut		DPYD ex23 Mut			DPYD ex14 Mut	MTHFR ex8 Mut	DPYD ex15 WT	TPMT ex4 Mut	AT 683		DPYD ex7 Mut		DPYD ex23 Mut			DPYD ex14 Mut	MTHFR ex8 Mut	DPYD ex15 WT	TPMT ex4 Mut
Row 3		TPMT ex6 Mut	DPYD ex14 Mut		DPYD IVS14 WT	AT 683		DPYD ex15 Mut	MTHFR ex8 WT	AT 683		AT 683	TPMT ex6 Mut	DPYD ex14 Mut		DPYD IVS14 WT	AT 683		DPYD ex15 Mut	MTHFR ex8 WT	
Row 4	MTHFR ex8 Mut	DPYD ex15 WT	TPMT ex4 WT	DPYD ex7 Mut	TPMT ex8 WT	DPYD ex23 Mut			DPYD ex14 Mut		AT 683	MTHFR ex8 Mut	DPYD ex15 WT	TPMT ex4 WT	DPYD ex7 Mut	TPMT ex6 WT	DPYD ex23 Mut			DPYD ex14 Mut	
Row 5		MTHFR ex8 WT	DPYD ex15 Mut	TPMT ex6 Mut	DPYD ex14 Mut		DPYD IVS14 WT		MTHFR ex5 WT			MTHFR ex8 WT	DPYD ex15 Mut	TPMT ex6 Mut	DPYD ex14 Mut			DPYD IVS14 WT		MTHFR ex5 WT	
Row 6	AT 683	DPYD ex14 WT	MTHFR ex8 Mut	DPYD IVS14 Mut	TPMT ex6 Mut	DPYD ex7 WT		DPYD ex23 WT				DPYD ex14 WT	MTHFR ex8 Mut	DPYD IVS14 Mut	TPMT ex6 Mut	DPYD ex7 WT			DPYD ex23 WT		AT 683

Figure 2. Grid layout of allelic capture and hybridization control probes spotted on the chip array. The overlapping of the virtual grid allows to select each probe cell and test the validity of the call.

3.2. Multiplex PCR Optimization

PGx-LoC was designed for a rapid and multiassay system where few manual operations are required. As previously reported, the first step in the genotyping assay is the optimization of PCR performances in order to amplify and detect specific target site variants in genomic DNA with high fidelity [22]. To implement the protocol on the In-Check platform, primers were tested in combination and two multiplex PCR were performed with an optimal annealing temperature of 61 °C. The specificity of conventional real-time PCR versus PGx-LoC was evaluated through direct sequencing in order to confirm the presence or absence of the target amplicons in the mix, due to the inability to exactly discern them on agarose gel electrophoresis (Figure 3).

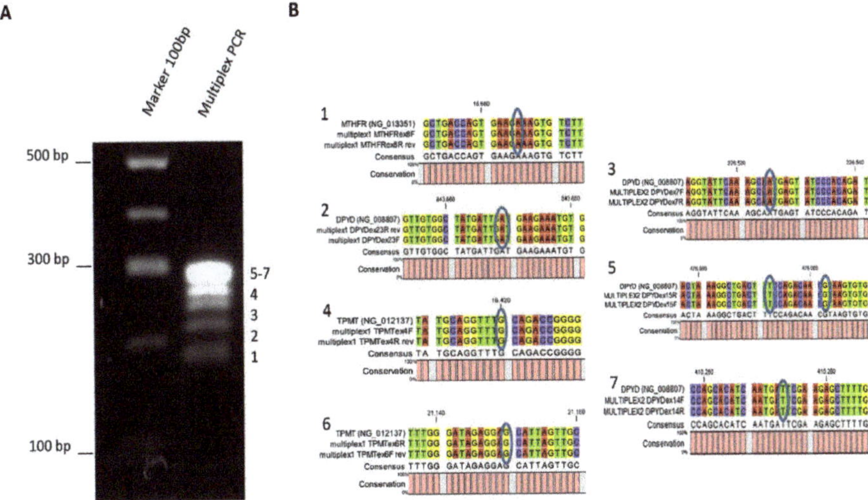

Figure 3. Multiplex PCR performed by the In-Check platform. (**A**) A 2% agarose gel electrophoresis of the multiplex PCR (marker 100bp). On the right, the numerical series identifies the seven amplicons sorted by their lengths. (**B**) Direct sequencing alignments of amplicons. Blue rings indicate selected polymorphisms in MTHFR, DPYD and TPMT genes.

3.3. Analitical Variability of PGx-LoC

In an microarray-based detection assay, different hybridization conditions, such as temperature, denaturation mode, incubation time or the concentration of the analyte, are crucial [25,26]. As reported in our previous study [22], four different hybridization temperatures based upon the temperature melting (Tm) values of allele-specific capture probes, from 50 °C to 56 °C, were tested to evaluate inter- and intra-assay variability. After genomic amplification and hybridization of the selected 17 allelic variants of the wild-type genome shown in Figure 3, fluorescence intensities of allele-specific capture probes were detected and linear regression analysis between median fluorescence intensities and their standard deviations was made. As shown in Figure 4, hybridized wild-type probes were 62% at 50 °C with a R^2 = 0.96, 100% at 52 °C with a R^2 ≈ 0.97, 56% at 54 °C with a R^2 = 0.9147, and 50% at 56 °C with a R^2 = 0.9405. In particular, at 50 °C, 54 °C and 56 °C no fluorescence was detected for the *DPYD* g.226525A, *DPYD* g.410273T, *DPYD* g.843669A and *TPMT* g.21147G related capture probes and no signal was detected for the mutated capture probes. Both fluorescence intensity and linear regression analysis demonstrated that the signal values for wild-type targets were complete and more accurate when hybridization is performed at 52 °C. At this temperature, the hybridization of 100% wild type probes and no hybridization signal in the mutated probes was observed, indicating this as the optimal hybridization temperature (p < 0.0001; R^2 = 0.9659; Figure 4).

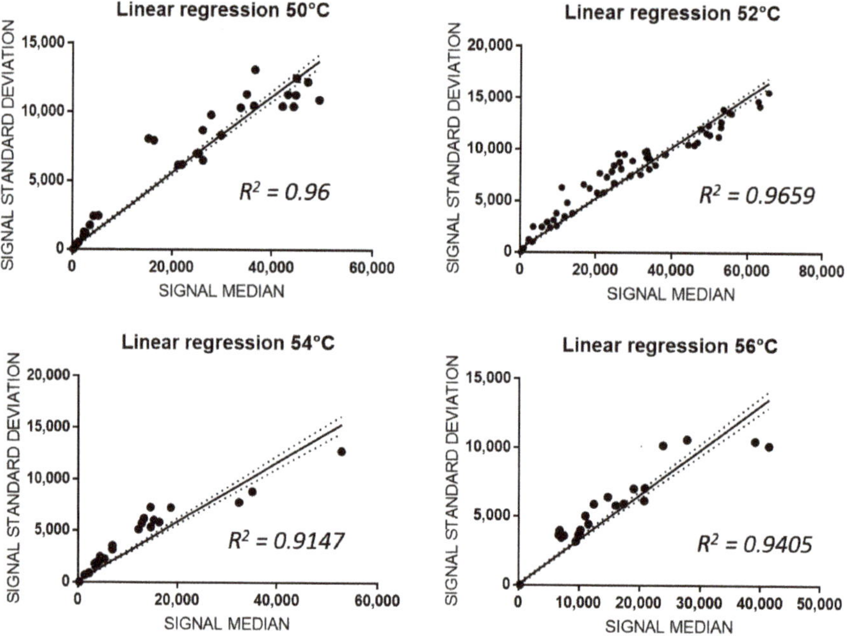

Figure 4. Effects of the hybridization temperature on allele discrimination and SNPs detection. Four different hybridization temperatures were tested and the Linear Regressions of normalized standard deviation and median of fluorescent signals are reported for 50 °C (p < 0.0001; slope = 0.2669 to 0.288; R^2 = 0.96), 52 °C (p < 0.0001; slope = 0.2389 to 0.2559; R^2 = 0.9659), 54 °C (p < 0.0001; slope: 0.273 to 0. 3057; R^2 = 0.9147) and 56 °C (p < 0.0001; slope: 0.311 to 0.3414; R^2 = 0.9405). Best fit in solid lines, 95% intervals of confidence in dashed lines.

Once the optimal hybridization temperature was identified, the PGx-LoC sensibility was measured by testing LoB and LoD. For this purpose, we randomly chose 4 capture probes complementary to two different SNPs of *DPYD* gene (rs17376848_T and rs3918290_G). As reported in other studies with lab-on-chip assays performed by the In-Check platform [22], the measure of LoB in PGx-LoC is a parameter particularly crucial in a test to demonstrate its sensibility in the detection of non-analytical signal when sample without analyte is analyzed. As reported in Figure 5A, the LoB test of PGx-LoC demonstrated its sensibility in blank assay. To evaluate the LoD, *DPYD* exon 15 was preamplified in PGx-LoC to a final concentration of 500 nM in PCR chamber. After the recovery of PCR product, 10-fold serial dilutions were made to 50 nM and 5 nM and hybridized in the array area of the chip. At the concentration of 5 nM, no fluorescence signal was detected, resulting in a microarray image similar to blank (Figure 5B). At the concentration of 50 nM, a fluorescence signal was detected for all the four allelic-capture probes complementary to the selected hotspots (green rings in Figure 5C), revealing it as the lowest quantity of analyte detectable by the capture probes (Figure 5D).

Once the PCR and hybridization conditions were set, complete experiments of pharmacogenetic genotyping through the use of PGx-LoC could be performed (Figure 5E).

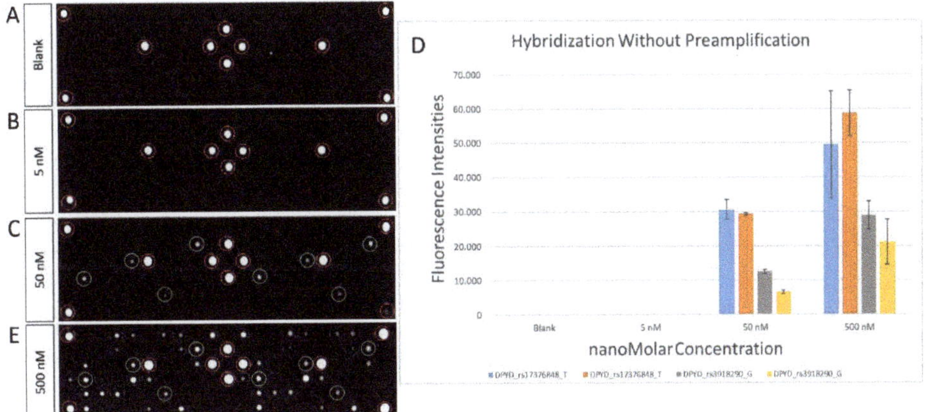

Figure 5. PGx-LoC sensibility measured by Limit of Blank and Limit of Detection. (**A**) Microarray image of Limit of Blank (LoB) with no analyte in the sample. The image reveals only the hybridization control highlighted in red rings. (**B,C**) Microarray images of Limit of Detection (LoD) with 10-fold serial dilution of DPYD exon 15 amplicons preamplified in PGx-LoC: 5 nM of PCR product (**B**); 50 nM of PCR product (**C**). Hybridization control highlighted in red rings. Probes of interest highlighted in green rings. (**D**) Fluorescence intensities of blank (0 nM), 5 nM, 50 nM and 500 nM of PCR products hybridized without preamplification relative to the four allelic capture probes chosen for the sensitivity testing. Data are reported as mean ± SD of three independent experiments. (**E**) Microarray image of a complete experiment of genotyping performed with a PGx-LoC with 500 nM of PCR product. Hybridization control highlighted in red rings. Allelic-capture probes complementary to DPYD exon 15 highlighted in green rings.

4. Discussion

Despite the development of biological agents, chemotherapy still represents a first line pharmacological strategy for different kinds of solid tumors treatments. It is not surprising that patients with the same histological tumor and/or stage disease have different pharmacological responses, sometime toxic. This occurs because in patients two genomes co-exist: the germline genome of the host and the somatic genome of tumor [3,27]. Different germline and somatic polymorphisms in ADME-related genes may significantly affect drug disposition and tolerance, leading to toxicity [28]. To avoid these unpleasant complications, the pharmacogenetic analysis may be conducted on patient's genomes. Unfortunately, pharmacogenetic evaluations for individualized antineoplastic drug response are limited in the clinical practice due to the myriad of PGx variants without unclear functional roles [13].

In this article, we describe the development of an innovative silicon-based lab-on-chip able to perform low-density and high-resolution multi-assay analysis (amplification, hybridization and detection) of three genes (*DPYD*, *MTHFR*, and *TPMT*) associated with adverse reactions to common chemotherapeutic agents.

DPYD converts dihydrouracil and dihydrothymine in uracil and thymine, respectively, through the reduction of NADP. *DPYD* deficiency leads to an autosomal recessive disease associated with a variable clinical phenotype. Pharmacogenetic research defined *DPYD* as a predictive marker of Fluoropyrimidines (FL) response because it is the first and rate-limiting enzyme of FL catabolic pathway [29]. Based on the clinical relevance of *DPYD* polymorphisms, the Clinical Pharmacogenetics Implementation Consortium (CPIC) recommends at least a 50% reduction in starting dose followed by ulterior reduction of dose based on toxicity for heterozygous patients, and to select an alternate antineoplastic drug for homozygous ones [30]. We have selected five different polymorphisms in *DPYD* involved not only in FL (fluorouracil, capecitabine and tegafur) metabolism but also in the response to four additional antineoplastic drugs: oxaliplatin and leucovorin, cetuximab and bevacizumab. Three polymorphisms cause a change in the amino acid sequence (rs2297595, rs55886062, rs67376798), one is a synonymous mutation (rs17376848), and one leads to exon 14 skipping (rs3918290). The resulting products are not functional at a different level and may cause severe toxicity. Fortunately, toxic death after FL administration is extremely uncommon, even if it has been hypothesized that the simultaneous presence of more than one allelic *DPYD* variant could exacerbate the toxicity [31]. Even if associated to likely-benign phenotypes, two mutations were recently recommended to be analyzed: rs56038477 (c.1236G > A) and rs1801160 (c.2194G > A) on exons 12 and 14 respectively [32–34].

The enzyme encoded by *MTHFR* converts irreversibly 5,10-methylenetetrahydrofolate to 5-methyltetrahydrofolate. The latter donates the methyl group to homocysteine to form methionine. In turn, methionine is converted to S-adenosylmethionine (SAM), the methyl donor in DNA methylation [35]. *MTHFR* polymorphisms predispose to serious bone marrow toxicity during treatment with folate synthesis inhibitor [36,37]. We selected the rs1801131 polymorphism that causes the Glu-Ala mutation associated with the reduction of enzyme activity. In tumor tissue, 5-methyltetrahydrofolate is required for optimal fluoropyrimidine efficacy, so that tumors with mutated *MTHFR* should be more sensitive to 5-fluorouracil (5FU) than wild-type tumors. However, *MTHFR*-rs1801131 polymorphism is significantly associated with a worst prognosis in homozygous mutated patients *versus* homozygous wild type patients (2.48 fold relative risk of death) rendering it a significant predictor for 5FU response in colorectal cancer patients [38,39].

TPMT catalyzes the S-methylation of aromatic and heterocyclic sulfhydryl compounds, including the thiopurine drugs 6-mercaptopurine (6MP) and 6-thioguanine (6TG). We selected two SNPs (rs1800462 and rs1800460), present in TPMT 3*A, 3*B and 3*C haplotype. These two SNPs result in lower TPMT protein quantity and also affect TPMT enzymatic activity. Patients carrying these point mutations can develop thiopurine drugs-related toxicity, such as myelosuppression [40]. Based on their genotypes, patients can be separated in slow, intermediate and extensive metabolizers and may suffer from several side effects (reduced enzymatic activities) or have lower therapeutic efficacy (rapid enzymatic activities). CPIC recommends different doses of thioguanine and mercaptopurine ins TPMT haplotypes. In particular, CPIC suggests to start with drastically reduced doses (reducing daily dose by 10-fold and reducing frequency to thrice weekly instead of daily) for poor metabolizers (homozygous mutant variants with two nonfunctional alleles) [41].

The intra-assay variability, sensibility and specificity of the silicon-based lab-on-chip analysis described in this study supports its further translation to clinical diagnostics, where it may promote precision medicine by reducing adverse drug reactions to common chemotherapeutic agents.

5. Conclusions

One of the challenges of modern medicine is to understand the "costs" of a particular therapeutic approach, in terms of the patient's quality of life and benefits against the disease.

Clinical treatment of oncological patients is made difficult by the adverse reactions of antineoplastic drugs, often linked to genetic factors. The acquired knowledge of the human genome has made possible the development of clinical pharmacogenetic tests in oncological patients. In particular, the design of reliable and low-cost miniaturized diagnostic devices, overcoming the limitations of the current technologies, may facilitate the pharmacogenetic screening of oncological patients, improve prevention of ADRs and promote precision medicine. In this work, we report the development of a novel In-Check LoC able to detect multiple allelic variants in *DPYD* (Dihydropyrimidine dehydrogenase), *MTHFR* (5,10-Methylenetetrahydrofolate reductase) and *TPMT* (Thiopurine S-methyltransferase) which are associated with adverse reactions to common chemotherapeutic agents. The inter- and intra-assay variability testing performed to set the sensitivity and specificity of PCR-microarray combined module, and efficiency of multiplex PCRs and hybridization probes, revealed and an optimal hybridization temperature of 52 °C with the 100% of alleli-capture probes signals were a wild-type genome was analyzed ($p < 0.0001$; slope = 0,2389 to 0,2559; R^2 = 0.9659). The high performance of the molecular assay, in terms of sensibility, specificity and efficiency, may enable the platform for diagnostic applications. In addition to technical advances, the miniaturized LoC promises cost-effectiveness, short analysis time, low sample/reagent consumption, and ease of use. In summary, the results reported in this study support the translation of PGx-LoC to clinical diagnostics as point-of-care or first-line pharmacogenetics genotyping array for cancer patients.

Author Contributions: Conceptualization, E.A. and S.C.; Data curation, R.I., V.L.C., G.M., M.G., M.A., E.A. and S.C.; Formal analysis, R.I.; Funding acquisition, E.A. and S.C.; Investigation, R.I., V.L.C. and G.M.; Methodology, R.I., V.L.C., G.M. and S.C.; Project administration, E.A. and S.C.; Resources, E.A. and S.C.; Software, E.A.; Supervision, S.C.; Validation, R.I., V.L.C., G.M. and M.G.; Visualization, R.I., V.L.C., G.M., M.G., M.A., E.A. and S.C.; Writing—original draft, R.I. and M.A.; Writing—review & editing, R.I., V.L.C., G.M., M.G., M.A., E.A. and S.C. All authors have read and agreed to the published version of the manuscript.

Funding: This research was funded by the Italian Ministry of Education, Universities and Research, grant number CTN01_00177_817708.

Acknowledgments: The authors gratefully acknowledge Cristina Calì, Alfia Corsino, Maria Patrizia D'Angelo and Francesco Marino for their administrative and technical assistance.

Conflicts of Interest: The authors declare no conflict of interest.

References

1. Forman, D.; Bray, F.; Brewster, D.; Gombe Mbalawa, C.; Kohler, B.; Piñeros, M.; Steliarova-Foucher, E.; Swaminathan, R.; Ferlay, J. *Cancer Incidence in Five Continents*; Electronic Version; IARC: Lyon, France, 2013; Volume X.
2. Siegel, R.L.; Miller, K.D.; Jemal, A. Cancer statistics, 2017. *CA A Cancer J. Clin.* **2017**, *67*, 7–30. [CrossRef] [PubMed]
3. Danesi, R.; Di Paolo, A.; Bocci, G.; Crea, F.; Del Tacca, M. Pharmacogenetics in oncology. *Eur. J. Cancer Suppl.* **2008**, *6*, 74–78. [CrossRef]
4. Meyer, U.A. Pharmacogenetics and adverse drug reactions. *Lancet* **2000**, *356*, 1667–1671. [CrossRef]
5. Ozdemir, V.; Shear, N.H.; Kalow, W. What will be the role of pharmacogenetics in evaluating drug safety and minimising adverse effects? *Drug Saf.* **2001**, *24*, 75–85. [CrossRef] [PubMed]
6. Spear, B.B.; Heath-Chiozzi, M.; Huff, J. Clinical application of pharmacogenetics. *Trends Mol. Med.* **2001**, *7*, 201–204. [CrossRef]
7. Stingl, J.C.; Brockmoller, J. Personalised pharmacogenetics. Evidence-based guidelines and clinical application of pharmacogenetic diagnostics. *Bundesgesundheitsblatt Gesundh. Gesundh.* **2013**, *56*, 1509–1521. [CrossRef]
8. Bosch, T.M.; Meijerman, I.; Beijnen, J.H.; Schellens, J.H.M. Genetic Polymorphisms of Drug-Metabolising Enzymes and Drug Transporters in the Chemotherapeutic Treatment of Cancer. *Clin. Pharmacokinet.* **2012**, *45*, 253–285. [CrossRef]

9. Roco, A.; Quinones, L.; Agundez, J.A.; Garcia-Martin, E.; Squicciarini, V.; Miranda, C.; Garay, J.; Farfan, N.; Saavedra, I.; Caceres, D.; et al. Frequencies of 23 functionally significant variant alleles related with metabolism of antineoplastic drugs in the chilean population: Comparison with caucasian and asian populations. *Front. Genet.* **2012**, *3*, 229. [CrossRef]

10. Chang, W.-C.; Tanoshima, R.; Ross, C.J.; Carleton, B.C. Challenges and Opportunities in Implementing Pharmacogenetic Testing in Clinical Settings. *Annu. Rev. Pharmacol. Toxicol.* **2020**, *61*. [CrossRef]

11. Eichelbaum, M.; Ingelman-Sundberg, M.; Evans, W.E. Pharmacogenomics and individualized drug therapy. *Annu. Rev. Med.* **2006**, *57*, 119–137. [CrossRef]

12. Lauschke, V.M.; Milani, L.; Ingelman-Sundberg, M. Pharmacogenomic Biomarkers for Improved Drug Therapy-Recent Progress and Future Developments. *AAPS J.* **2017**, *20*, 4. [CrossRef] [PubMed]

13. Arbitrio, M.; Scionti, F.; Di Martino, M.T.; Caracciolo, D.; Pensabene, L.; Tassone, P.; Tagliaferri, P. Pharmacogenomics Biomarker Discovery and Validation for Translation in Clinical Practice. *Clin. Transl. Sci.* **2020**. [CrossRef] [PubMed]

14. Mujica, M.L.; Zhang, Y.; Bédioui, F.; Gutiérrez, F.; Rivas, G. Label-free graphene oxide–based SPR genosensor for the quantification of microRNA21. *Anal. Bioanal. Chem.* **2020**, *412*, 3539–3546. [CrossRef] [PubMed]

15. Grodzinski, P.; Liu, R.H.; Lee, A.P. Integrated DNA Biochips: Past, Present and Future. In *Integrated Biochips for DNA Analysis*; Liu, R.H., Lee, A.P., Eds.; Springer: New York, NY, USA, 2007. [CrossRef]

16. Hardiman, G. Applications of microarrays and biochips in pharmacogenomics. In *Pharmacogenomics in Drug Discovery and Development*; Springer: Berlin/Heidelberg, Germany, 2008; pp. 21–30.

17. Glotov, A.; Nasedkina, T.; Ivaschenko, T.; Urasov, R.; Surzhikov, S.; Pan'kov, S.; Chudinov, A.; Baranov, V.; Zasedatelev, A. Development of a biochip for analyzing polymorphism of the biotransformation genes. *Mol. Biol.* **2005**, *39*, 357–365. [CrossRef]

18. Bürger, J. Biochips-tools of 21st century medicine. *Versicherungsmedizin* **2006**, *58*, 9–13. [PubMed]

19. Petralia, S.; Verardo, R.; Klaric, E.; Cavallaro, S.; Alessi, E.; Schneider, C. In-Check system: A highly integrated silicon Lab-on-Chip for sample preparation, PCR amplification and microarray detection of nucleic acids directly from biological samples. *Sens. Actuators B Chem.* **2013**, *187*, 99–105. [CrossRef]

20. Palmieri, M.; Alessi, E.; Conoci, S.; Marchi, M.; Panvini, G. Developments of the in-check platform for diagnostic applications. In *Microfluidics, BioMEMS, and Medical Microsystems VI*; International Society for Optics and Photonics: Bellingham, WA, USA, 2008; p. 688602.

21. Pernagallo, S.; Ventimiglia, G.; Cavalluzzo, C.; Alessi, E.; Ilyine, H.; Bradley, M.; Diaz-Mochon, J.J. Novel biochip platform for nucleic acid analysis. *Sensors* **2012**, *12*, 8100–8111. [CrossRef]

22. Guarnaccia, M.; Iemmolo, R.; San Biagio, F.; Alessi, E.; Cavallaro, S. Genotyping of KRAS Mutational Status by the In-Check Lab-on-Chip Platform. *Sensors* **2018**, *18*, 131. [CrossRef]

23. Patel, S.; Nanda, R.; Sahoo, S.; Mohapatra, E. Biosensors in Health Care: The Milestones Achieved in Their Development towards Lab-on-Chip-Analysis. *Biochem. Res. Int.* **2016**, *2016*, 3130469. [CrossRef]

24. Guarnaccia, M.; Gentile, G.; Alessi, E.; Schneider, C.; Petralia, S.; Cavallaro, S. Is this the real time for genomics? *Genomics* **2014**, *103*, 177–182. [CrossRef]

25. Wu, L.R.; Fang, J.Z.; Khodakov, D.; Zhang, D.Y. Nucleic Acid Quantitation with Log–Linear Response Hybridization Probe Sets. *ACS Sens.* **2020**, *5*, 1604–1614. [CrossRef] [PubMed]

26. Leimanis, S.; Hernández, M.; Fernández, S.; Boyer, F.; Burns, M.; Bruderer, S.; Glouden, T.; Harris, N.; Kaeppeli, O.; Philipp, P.; et al. A Microarray-based Detection System for Genetically Modified (GM) Food Ingredients. *Plant Mol. Biol.* **2006**, *61*, 123–139. [CrossRef] [PubMed]

27. Wang, L.; McLeod, H.L.; Weinshilboum, R.M. Genomics and drug response. *N. Engl. J. Med.* **2011**, *364*, 1144–1153. [CrossRef] [PubMed]

28. Kummar, S.; Gutierrez, M.; Doroshow, J.H.; Murgo, A.J. Drug development in oncology: Classical cytotoxics and molecularly targeted agents. *Br. J. Clin. Pharmacol.* **2006**, *62*, 15–26. [CrossRef]

29. Van Kuilenburg, A.B.; Vreken, P.; Abeling, N.G.; Bakker, H.D.; Meinsma, R.; Van Lenthe, H.; De Abreu, R.A.; Smeitink, J.A.; Kayserili, H.; Apak, M.Y.; et al. Genotype and phenotype in patients with dihydropyrimidine dehydrogenase deficiency. *Hum. Genet.* **1999**, *104*, 1–9. [CrossRef]

30. Caudle, K.E.; Thorn, C.F.; Klein, T.E.; Swen, J.J.; McLeod, H.L.; Diasio, R.B.; Schwab, M. Clinical Pharmacogenetics Implementation Consortium guidelines for dihydropyrimidine dehydrogenase genotype and fluoropyrimidine dosing. *Clin. Pharmacol. Ther.* **2013**, *94*, 640–645. [CrossRef]

31. Garziera, M.; Bidoli, E.; Cecchin, E.; Mini, E.; Nobili, S.; Lonardi, S.; Buonadonna, A.; Errante, D.; Pella, N.; D'Andrea, M.; et al. HLA-G 3'UTR Polymorphisms Impact the Prognosis of Stage II-III CRC Patients in Fluoropyrimidine-Based Treatment. *PLoS ONE* **2015**, *10*, e0144000. [CrossRef]

32. Henricks, L.M.; Lunenburg, C.A.; de Man, F.M.; Meulendijks, D.; Frederix, G.W.; Kienhuis, E.; Creemers, G.-J.; Baars, A.; Dezentjé, V.O.; Imholz, A.L. DPYD genotype-guided dose individualisation of fluoropyrimidine therapy in patients with cancer: A prospective safety analysis. *Lancet Oncol.* **2018**, *19*, 1459–1467. [CrossRef]

33. Del Re, M.; Cinieri, S.; Michelucci, A.; Salvadori, S.; Loupakis, F.; Schirripa, M.; Cremolini, C.; Crucitta, S.; Barbara, C.; Di Leo, A. DPYD* 6 plays an important role in fluoropyrimidine toxicity in addition to DPYD* 2A and c. 2846A> T: A comprehensive analysis in 1254 patients. *Pharm. J.* **2019**, *19*, 556–563. [CrossRef]

34. Iachetta, F.; Bonelli, C.; Romagnani, A.; Zamponi, R.; Tofani, L.; Farnetti, E.; Nicoli, D.; Damato, A.; Banzi, M.; Casali, B. The clinical relevance of multiple DPYD polymorphisms on patients candidate for fluoropyrimidine based-chemotherapy. An Italian case-control study. *Br. J. Cancer* **2019**, *120*, 834–839. [CrossRef]

35. Varzari, A.; Deyneko, I.V.; Tudor, E.; Turcan, S. Polymorphisms of glutathione S-transferase and methylenetetrahydrofolate reductase genes in Moldavian patients with ulcerative colitis: Genotype-phenotype correlation. *Meta Gene* **2016**, *7*, 76–82. [CrossRef] [PubMed]

36. Chiusolo, P.; Reddiconto, G.; Casorelli, I.; Laurenti, L.; Sora, F.; Mele, L.; Annino, L.; Leone, G.; Sica, S. Preponderance of methylenetetrahydrofolate reductase C677T homozygosity among leukemia patients intolerant to methotrexate. *Ann. Oncol.* **2002**, *13*, 1915–1918. [CrossRef] [PubMed]

37. Yang, L.; Hu, X.; Xu, L. Impact of methylenetetrahydrofolate reductase (MTHFR) polymorphisms on methotrexate-induced toxicities in acute lymphoblastic leukemia: A meta-analysis. *Tumour Biol. J. Int. Soc. Oncodevelopmental Biol. Med.* **2012**, *33*, 1445–1454. [CrossRef] [PubMed]

38. Etienne, M.C.; Formento, J.L.; Chazal, M.; Francoual, M.; Magne, N.; Formento, P.; Bourgeon, A.; Seitz, J.F.; Delpero, J.R.; Letoublon, C.; et al. Methylenetetrahydrofolate reductase gene polymorphisms and response to fluorouracil-based treatment in advanced colorectal cancer patients. *Pharmacogenetics* **2004**, *14*, 785–792. [CrossRef]

39. Ramos-Esquivel, A.; Chinchilla, R.; Valle, M. Association of C677T and A1298C MTHFR Polymorphisms and Fluoropyrimidine-induced Toxicity in Mestizo Patients With Metastatic Colorectal Cancer. *Anticancer Res.* **2020**, *40*, 4263–4270. [CrossRef]

40. Wang, L.; Pelleymounter, L.; Weinshilboum, R.; Johnson, J.A.; Hebert, J.M.; Altman, R.B.; Klein, T.E. Very important pharmacogene summary: Thiopurine S-methyltransferase. *Pharm. Genom.* **2010**, *20*, 401–405. [CrossRef]

41. Relling, M.V.; Gardner, E.E.; Sandborn, W.J.; Schmiegelow, K.; Pui, C.H.; Yee, S.W.; Stein, C.M.; Carrillo, M.; Evans, W.E.; Hicks, J.K.; et al. Clinical pharmacogenetics implementation consortium guidelines for thiopurine methyltransferase genotype and thiopurine dosing: 2013 update. *Clin. Pharmacol. Ther.* **2013**, *93*, 324–325. [CrossRef]

Publisher's Note: MDPI stays neutral with regard to jurisdictional claims in published maps and institutional affiliations.

Article

Gas Crosstalk between PFPE–PEG–PFPE Triblock Copolymer Surfactant-Based Microdroplets and Monitoring Bacterial Gas Metabolism with Droplet-Based Microfluidics

Sunghyun Ki [1] and Dong-Ku Kang [1,2,*]

[1] Department of Chemistry, Natural Science, Incheon National University, Incheon 22012, Korea; shki@inu.ac.kr
[2] Research Institute of Basic Sciences, Incheon National University, Incheon 22012, Korea
* Correspondence: dkkang@inu.ac.kr

Received: 26 September 2020; Accepted: 9 November 2020; Published: 11 November 2020

Abstract: The PFPE–PEG–PFPE (Perfluoropolyether-polyethylene glycol-perfluoropolyether) surfactant has been used in droplet-based microfluidics and is known to provide high droplet stability and biocompatibility. Since this surfactant ensures the stability of droplets, droplet-based microfluidic systems have been widely used to encapsulate and analyze various biological components at the single-molecule scale, including viruses, bacteria, nucleic acids and proteins. In this study, we experimentally confirmed that gas crosstalk occurred between droplets formed by fluorinated oil and the PFPE–PEG–PFPE surfactant. *E. coli* K-12 bacterial cells were encapsulated with Luria–Bertani broth within droplets for the cultivation, and gas crosstalk was identified with neighboring droplets that contain phenol red. Since bacteria produce ammonia gas during its metabolism, penetration of ammonia gas initiates a color change of phenol red-containing droplets. Ammonia gas exchange was also confirmed by reacting ammonium chloride and sodium hydroxide within droplets that encapsulated. Herein, we demonstrate the gas crosstalk issue between droplets when it is formed using the PFPE–PEG–PFPE surfactant and also confirm that the density of droplet barrier has effects on gas crosstalk. Our results also suggest that droplet-based microfluidics can be used for the monitoring of living bacteria by the determination of bacterial metabolites during cultivation.

Keywords: droplet microfluidics; PFPE–PEG–PFPE surfactant; crosstalk between droplets

---

1. Introduction

Droplet-based microfluidics (DMF) has been continuously developed in biomedical engineering and analytical chemistry to monitor various chemical and biological analytes such as viruses, bacteria, nucleic acids and proteins [1–5]. It provides various advantages such as high-throughput, short reaction time, higher efficiency and cost-effectiveness for analysis of interesting biological targets [6–10]. In addition, DMF cannot only generate monodisperse droplets but also control the flow rate to generate droplets of various sizes. That is why it has been widely used in various research fields such as single fundamental biology, diagnostics and high-throughput drug screening [11–15].

For single-molecule analysis using DMF, droplets must be kept stable for various reactions, including thermocycling and long incubation. Oils and surfactants used for the encapsulation must be highly biocompatible and also can provide stability in droplets to avoid breakage and crosstalk between droplets. Recently, various surfactants and oils have been explored in DMF for sample partitioning and also for various applications [16]. One of the most commercially successful application may be the droplet digital PCR (ddPCR) which requires thermocycling and also need to be perfectly separated during reaction [4,7,11,13,17]. Among various surfactants, PFPE-based surfactants have

Biosensors **2020**, *10*, 172; doi:10.3390/bios10110172

been most commonly used in DMF because of their excellent long-term stability and biocompatibility with biological systems [18–22]. Weitz and his colleagues demonstrated the biocompatibility and stability of PFPE–PEG–PFPE surfactants in various applications [18,23]. However, there are no studies on gas crosstalk between droplets.

Therefore, in this study, gas crosstalk was first validated between droplets generated with HFE-7500 and the PFPE–PEG–PFPE surfactant. To summarize the overall experimental process, First, phenol red was encapsulated within droplets as a pH indicator to analyze the crosstalk between droplets. *E. coli* K-12 was also encapsulated within droplets in a single-cell manner to validate the crosstalk with phenol red droplets after mixing and incubation. These results suggest that pH indicators or bacterial metabolites can cross the droplet barrier. Then bacterial metabolite was identified as transferring component between droplets, and it was characterized as an ammonia gas with Nessler's reagent [24–27]. Finally, ammonia gas was artificially produced through a chemical reaction between ammonium chloride and sodium hydroxide within droplets to experimentally validate that the ammonia gas can actually transport into neighboring droplets. As a result, artificially produced ammonia gas can penetrate the droplet barrier, and the phenol red droplet turned red due to pH change. We also confirmed that surfactant density is a major reason that affects the gas permeability between droplets by validation with droplets that were generated with various concentrations of surfactant. Basically, a higher concentration of surfactant reduced gas penetration between droplets and these results suggest that surface density is the major barrier to gas permeability. These results suggest that live bacteria and bacterial growth can be identified in a single-cell manner by monitoring the penetration of ammonia gas from bacterial cell-containing droplets to indicator droplets.

2. Materials and Methods

2.1. Materials

HFE-7100, oxalyl chloride, anhydrous dichloromethane, Jeffamine ED-900, phenol red, trichloro(1H,1H,2H,2H-perfluorooctyl)silane, sodium hydroxide, ammonium chloride were purchased from Sigma-Aldrich (St. Louis, MO, USA). A Krytox 157-FSH was purchased from Chermours (Wilmington, DE, USA). HFE-7500 was purchased from 3 M (Saint Paul, MN, USA). SU-8-negative photoresist and SU-8 developer were purchased from Kayaku advanced materials (Westborough, MA, USA) for the fabrication of master mold. A Sylgard 184 silicone elastomer kit was purchased from Dow Corning (Midland, MI, USA) for the fabrication of microfluidic devices as polydimethylsiloxane (PDMS) material. Luria–Bertani (LB), Luer-Lok syringe and 25G syringe needle were purchased from BD (Franklin Lakes, NJ, USA). Microfluidic devices were punched with a 1 mm biopsy punch (Kay Industries Co., Gifu, Japan) to provide inlets and outlets. A Nessler's reagent (potassium tetraiodomercurate (II)) was purchased from Duksan science (Seoul, Korea). NanoDrop™ 2000c spectrophotometer (Thermo Fisher Scientific, Waltham, MA, USA) was used to determine the concentration of bacterial cells. Plasma cleaner PDC-002 (Harrick Plasma, Ithaca, NY, USA) was used to bond the PDMS layer with the glass slide by O_2 plasma treatment. To analyze droplets under microscope, a countess cell counting chamber slide (chamber slide) was purchased from Invitrogen (Waltham, CA, USA). Aqueous and oil were introduced with syringe pumps (PHD ultra, Harvard Apparatus, Holliston, MA, USA). For droplet generation, droplets were imaged and recorded under a DMi-8 fluorescence microscope (Leica, Wetzlar, Germany) with a DFC-7000T CCD camera (Leica, Wetzlar, Germany) of Leica Microsystems (Wetzlar, Germany). Microscopic images were analyzed using LAS X software (Version 3.4.2) and image J software (Version 1.52t) from Leica Microsystems.

2.2. Process of Synthesis of PFPE–PEG–PFPE Surfactant

The synthetic process of PFPE–PEG–PFPE surfactant was mainly performed by composing in two steps (Figure 1), which are distinguished in (step 1) a substitution from carboxylic acid to acid chloride and (step 2) an amide bonding process. First, Krytox 157-FSH (2.85 mmol, ~7000 g/mol) was dissolved with HFE-7100 and mixed with oxalyl chloride (28.3 mmol, 126.93 g/mol). Then, oxalyl chloride was added 10 times, comparing the volume of Krytox 157-FSH. The mixture was stirred and incubated overnight at 85 °C under a nitrogen atmosphere, and then it could be seen that the product became slightly yellow. Remaining residues of unreacted oxalyl chloride and HFE-7100 were removed under a rotary evaporator. Then, the remaining product was dissolved in HFE-7100 for the reaction with a diamine PEG (1.5 mmol, Jeffamine ED-900) was dissolved in anhydrous dichloromethane for the central hydrophilic block of the surfactant. The mixture was stirred and incubated for 48 h at 65 °C under a nitrogen atmosphere. After 48 h incubation, the mixture was transferred in a 50 mL conical tube for centrifugation at 8000 rpm (or higher) for 10 min to remove residues of white particles. Then, the final product was dried in a vacuum desiccator for 24 h [18,28]. For the structure determination, the final product was analyzed with ATR-FTIR spectroscopy (Spectrum two, Perkin-Elmer, Waltham, MA, USA) (Figure 2), ^{1}H-NMR (Figure 3) and ^{19}F-NMR spectroscopy (400-MR, Agilent Technologies, Santa Clara, CA, USA) (Figure 4) [29].

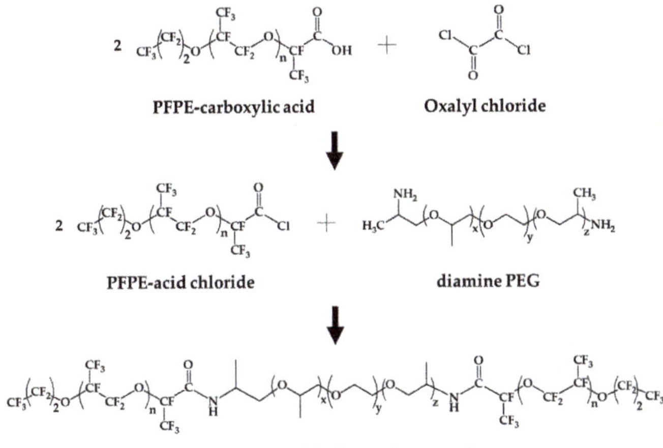

Figure 1. Schematic diagram of the synthesis process for PFPE–PEG–PFPE triblock copolymer surfactant.

2.3. Fabrication of Microfluidic Chip

Silicon wafer-based molds of microfluidic devices were fabricated by conventional photolithography methods [30,31]. Film photomasks were designed for microfluidic devices with AutoCAD software (version 2018), and four-inch silicon wafers were used as a substrate for the microfluidic devices. First, silicon wafers were cleaned using a piranha solution (3:1 concentrated sulfuric acid to 30% hydrogen peroxide) and then 1% hydrogen fluoride to remove debris and oxide layer. Silicon wafers were then patterned with negative photoresist SU-8 50 at 2000 rpm (30 s) using a spin coater. After pre-baking at 65 °C for 6 min, photoresist-coated silicon wafers were soft-baked at 95 °C for 20 min. The photoresist was cured under a wavelength of 400 nm for 23 s at a power of 11 mW/cm^2. Then polymerization was performed at 65 °C for 1 min as a post-exposure baking (PEB) and then baked at 95 °C for 5 min. Unpolymerized photoresist was removed using a SU-8 developer and washed with 100% hexane, 100% methanol and 100% isopropanol. Dried wafers were dried using a Trichloro (1H,1H,2H,2H-perfluorooctyl)silane in a desiccator overnight. In this experiment, microfluidic chips were fabricated with PDMS [32–34]. Briefly, PDMS base and curing agent were mixed in a ratio of 10:1 (*w/w*%), and the mixture was then degassed under desiccator prior to pouring onto silicon wafer-based master mold. It was then polymerized on a 65 °C hot plate overnight. After baking, the PDMS layer was peeled off from the master mold and punched with a 1 mm biopsy punch to provide inlets and outlets. PDMS layer and glass slides were initially treated with an oxygen plasma before there were bonded, and bonded devices were incubated on the hot plate at a temperature of 85 °C for 15 min.

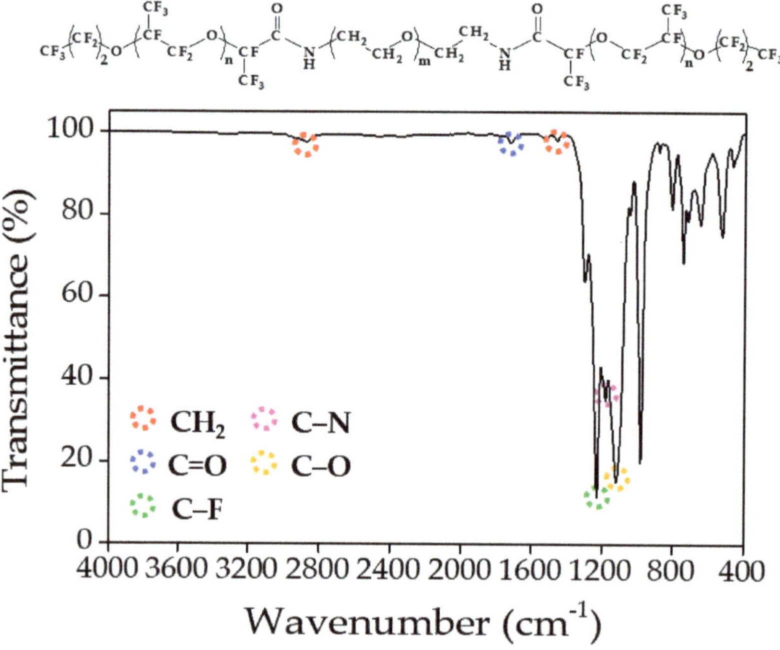

Figure 2. ATR-FTIR spectrum of synthesized surfactant for structural analysis.

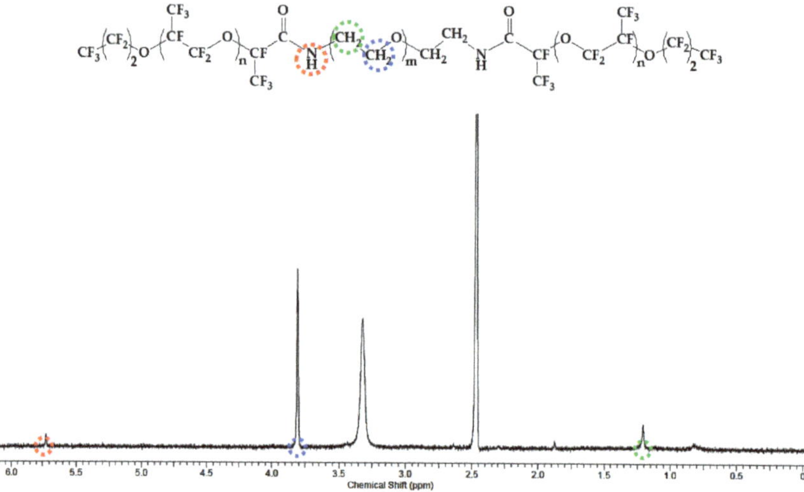

Figure 3. ^{1}H-NMR spectrum of synthesized PFPE–PEG–PFPE surfactant for structural analysis.

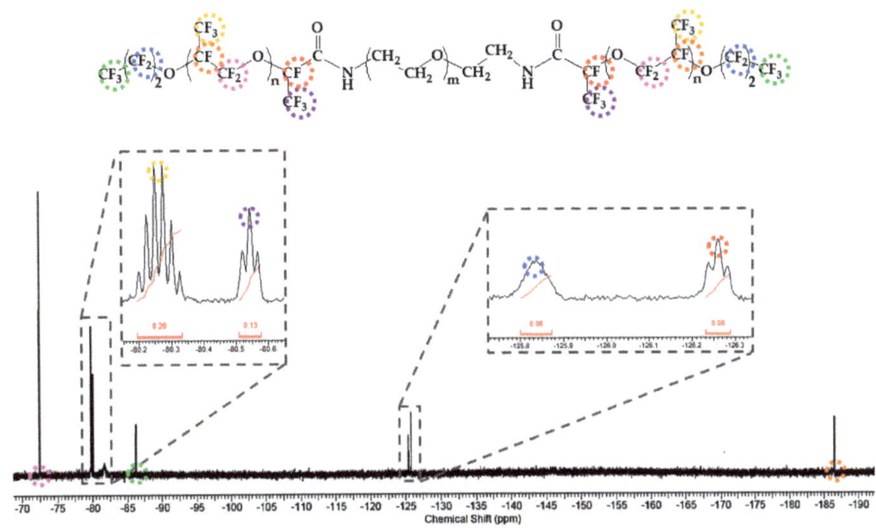

Figure 4. ^{19}F-NMR spectrum of synthesized PFPE–PEG–PFPE surfactant for structural analysis.

2.4. Cell Culture and Droplet Generation

As a model system to prove bacterial metabolism within microdroplets, *Escherichia coli* (*E. coli*) K-12 was used in this study. A single colony was obtained from freshly grown on the agar plate containing LB, and it was then inoculated in 5 mL LB broth at 37 °C overnight. Ten (10) µL of bacterial culture was further inoculated in 5 mL fresh LB broth with shaking at 37 °C for 4 h, and the concentration of bacterial cells was determined by NanoDrop™ 2000c spectrophotometer prior to encapsulation.

A microfluidic device contains an outlet, an oil-phase inlet and two aqueous-phase inlets. To identify ammonia gas crosstalk from bacterial cells, *E. coli* K-12 was encapsulated with HFE-7500 containing 2% (*w*/*w*) surfactant. To mimic ammonia gas production within droplets, 2 M sodium hydroxide and 2 M ammonium chloride were also encapsulated within droplets (ammonia gas generating droplets) using HFE-7500 with 2% surfactant. *E. coli* K-12-containing droplets or ammonia gas generating droplets were then mixed with 1 mg/mL phenol red droplets to identify gas permeability between droplets with chamber slide. More briefly, synthesized surfactant was dissolved in HFE-7500 at a concentration of 2% and introduced through oil inlet. In aqueous phase, *E. coli* K-12 or sodium hydroxide/ammonium chloride was introduced through aqueous inlets and then encapsulated with HFE-7500 at the flow-focusing structure within microfluidic devices (Figure 5) [35]. Droplets were generated to be approximately 80 µm by the optimum flow rate of the oil and aqueous phase. Generated droplets were collected in 1.5 mL tubes for further analysis.

2.5. Detection of E. coli K-12 with Phenol Red, pH Indicator

E. coli K-12 was cultured in fresh LB broth at 37 °C overnight and introduced into the microfluidic devices to be encapsulated within 80 µm droplets in a single-cell manner with 1 mg/mL phenol red using HFE-7500 containing 2% surfactant. More briefly, LB broth was encapsulated into microdroplets with or without 1 mg/mL phenol red. LB broth containing *E. coli* K-12 was also encapsulated with or without 1 mg/mL phenol red to identify bacteria-induced color exchange within droplets (Figure 6). Generated droplets were collected with 1.5 mL tubes and droplets were incubated at 37 °C. After overnight cultivation, droplets were then imaged under a DMi-8 microscope to identify color exchange (Figure 7).

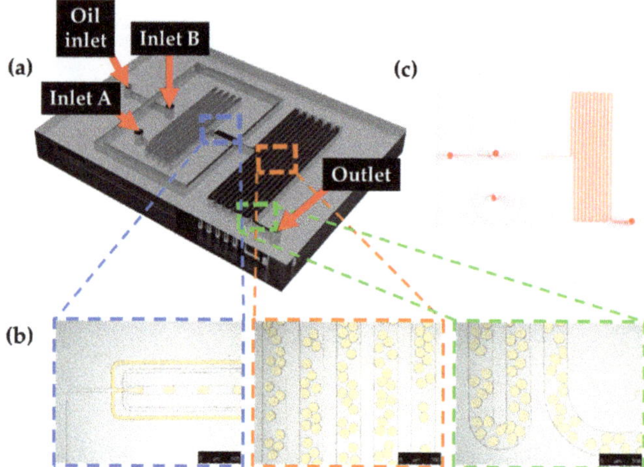

Figure 5. Droplet generation using the microfluidic device. (**a**) Schematic diagram of the microfluidic device for droplet generation. (**b**) Microscopic images of flow-focusing structure (blue box) for droplet generation, incubation channels (orange box) and outlet (green–box) (scale bar = 500 μm). (**c**) A photographic image of the fabricated PDMS microfluidic device. The microfluidic channel was visualized with red food coloring.

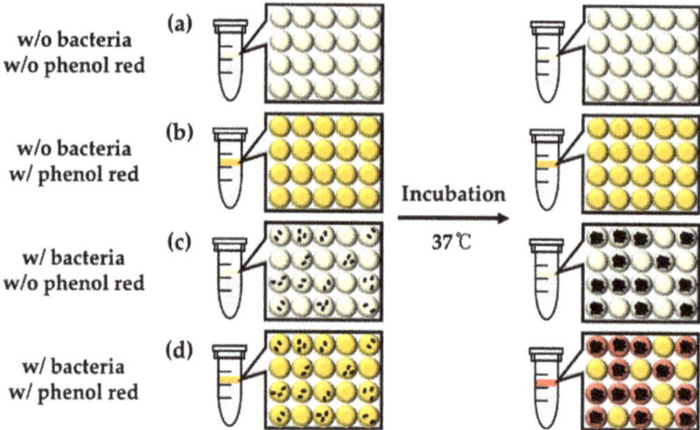

Figure 6. Schematic diagram of the experimental plan to determine bacterial growth by cultivation within droplets using s pH indicator. (**a**) Empty droplets, (**b**) phenol red droplets, (**c**) bacteria droplets, (**d**) phenol red and bacteria droplets for the incubation and detection by the color change of phenol red.

2.6. Identification of Crosstalk between Droplets

E. coli K-12 was freshly cultured in LB broth at 37 °C overnight and encapsulated within 80 μm droplets in a single-cell manner without phenol red. Phenol red was separately encapsulated in 80 μm droplets at a concentration of 1 mg/mL as an indicator of pH change to define crosstalk between droplets. Each droplet was mixed and incubated overnight prior to the identification of metabolite transfer through the droplet barrier (above panels, Figures 8 and 9). To identify phenol red dye permeability between droplets, fresh LB broth was also encapsulated within droplets without *E. coli* K-12 and 1 mg/mL phenol red to be mixed with droplets that contain both *E. coli* K-12 and phenol red. Mixtures of droplets were incubated at 37 °C overnight. Droplets were then imaged under a DMi-8 microscope (below panels, Figures 8 and 9).

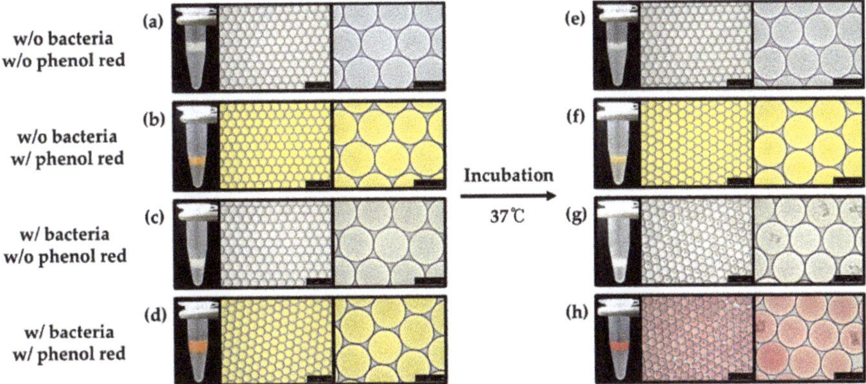

Figure 7. Identification of bacterial growth in droplets using pH indicator. (**a,e**) Empty droplet, (**b,f**) phenol red droplets, (**c,g**) bacteria droplet, (**d,h**) co-encapsulated droplets of both phenol red and bacteria (scale bar of left panels = 250 μm, scale bar of right panels = 75 μm).

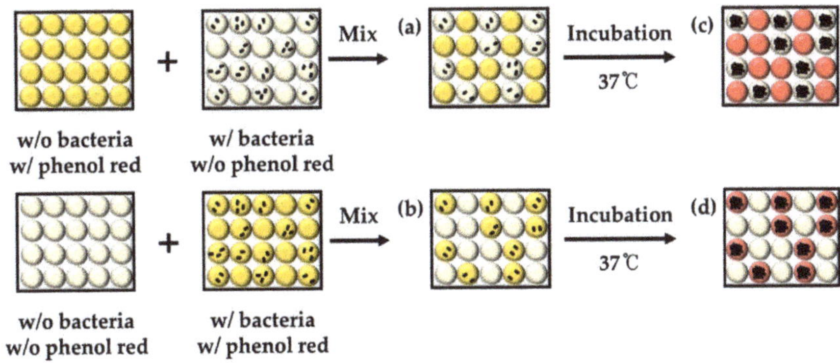

Figure 8. Schematic diagram of crosstalk between mixed droplets. Two types of droplets were mixed and incubated for identifying the gas crosstalk.

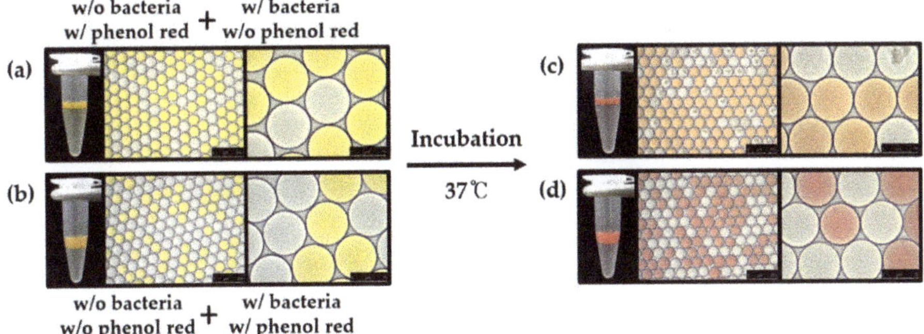

Figure 9. Identification of crosstalk components through droplet barrier. (**a**) Droplet mixture of phenol red-containing droplets with *E. coli* K-12-containing droplets. (**b**) Droplet mixture of empty droplets with co-encapsulated droplets (both phenol red and *E. coli* K-12). Droplet mixtures were imaged before (**a,b**) and after (**c,d**) overnight incubation. (scale bar of left panels = 250 μm, scale bar of right panels = 75 μm).

2.7. Identification of Ammonia Gas Production in Bacterial Cells

In order to confirm ammonia gas production during bacterial cell cultivation, ammonia gas was detected with Nessler's reagent, which is a conventional indicator for ammonia gas, and it was compared with phenol red dye in bulk. *E. coli* K-12 was cultured in LB broth, and 1 mL *E. coli* K-12-containing LB broth was transferred into 1.5 mL tubes. One milliliter of fresh LB broth was also prepared as a negative control as a bacteria-free condition, and 100 μL 100% Nessler's reagent or 100 μL 1 mg/mL phenol red added. The color change was analyzed immediately with a smartphone camera (iPhone 11 pro, Apple, Cupertino, CA, USA) (Figure 10).

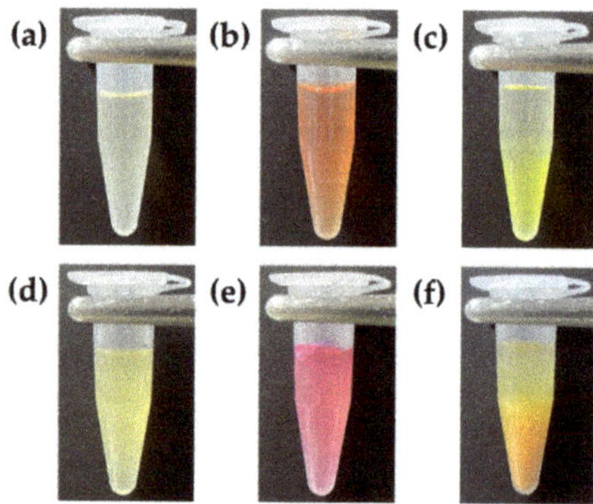

Figure 10. Identification of bacterial metabolite with phenol red and Nessler's reagent. (**a**) Only LB broth, (**b**) LB broth with phenol red, (**c**) LB broth with Nessler's reagent, (**d**) *E. coli* K-12 (**e**) *E. coli* K-12 with phenol red and (**f**) *E. coli* K-12 with Nessler's reagent.

2.8. Validation of Ammonia Gas Crosstalk between Droplets

As a model system of ammonia gas production by chemical reaction within droplets (ammonia gas generating droplets), 2 M sodium hydroxide and 2 M ammonium chloride were co-encapsulated within 80 μm droplets [36]. Phenol red was also solely encapsulated within LB broth at a concentration of 1 mg/mL as indicator droplets to determine the transfer of ammonia gas through the droplet barrier (Figure 11). Ammonia gas generating droplets (10 μL) and indicator droplets (10 μL) were transferred within the same tubes and gently mixed for 10 s prior to imaging under a DMi-8 microscope. A mixture of droplets was then imaged immediately without incubation. To identify gas production through the chemical reaction between sodium hydroxide and ammonium chloride, 2 M sodium hydroxide or 2 M ammonium chloride were separately encapsulated before mixing with indicator droplets (Figure 12). All procedures were performed at room temperature (23 °C) in this experiment.

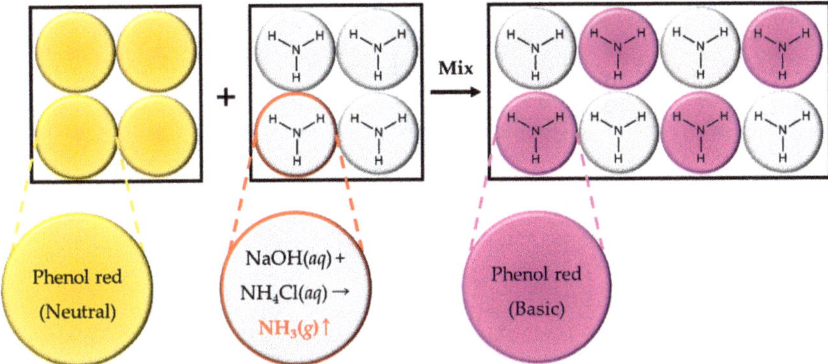

Figure 11. Schematic diagram of artificially produced ammonia gas within droplets and its crosstalk between droplets.

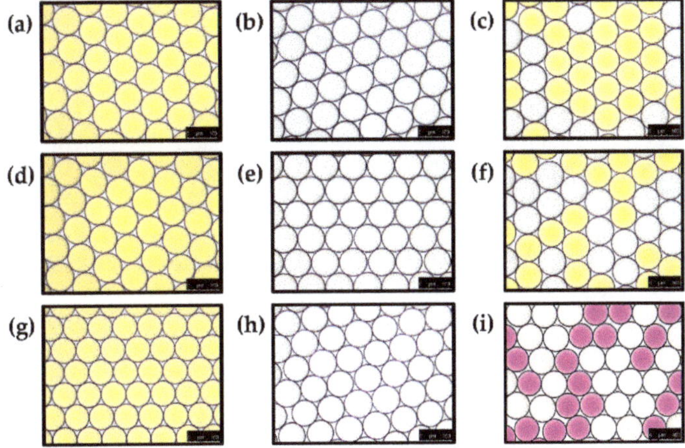

Figure 12. Identification of ammonia gas production within droplets and gas crosstalk between droplets. Phenol red droplets (**a,d,g**) were mixed sodium hydroxide-containing droplets (**b**), ammonium chloride-containing droplets (**e**) or co-encapsulated droplets with sodium hydroxide and ammonium chloride (**h**). Droplet mixtures (**c,f,i**) were then imaged under a DMi-8 microscope (scale bar = 100 μm).

2.9. Effect of Ammonia Gas Concentration on Gas Permeability

To identify the effect of gas concentration on permeability between droplets, sodium hydroxide and ammonium chloride were co-encapsulated at a different concentration from 0.002 M to 2 M. Droplets containing different concentrations of sodium hydroxide and ammonium chloride were separately collected within 1.5 mL tubes. Indicator droplets (10 μL) were also prepared with 1 mg/mL phenol red to be mixed with ammonia gas generating droplets (10 μL). Mixed droplets were immediately transferred into the chamber slide for imaging under a DMi-8 microscope, and the microscopic image was analyzed with the image J software (Figure 13). All procedures were performed at room temperature (23 °C) in this experiment.

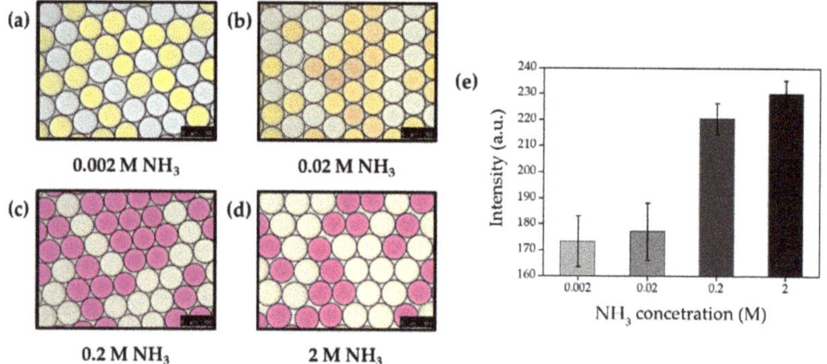

Figure 13. Permeability on ammonia gas concentration between droplets. Ammonia gas generating droplets were generated separately at ammonia gas concentration of 0.002 M (**a**), 0.02 M (**b**), 0.2 M (**c**) and 2 M (**d**) and mixed with indicator droplets, respectively. (**e**) Mean values of color exchange of indicator droplets for ammonia gas concentration (scale bar = 100 μm).

2.10. Effect of Surfactant Concentration on Gas Permeability

To identify the effect of surfactant concentration, 0.02 M sodium hydroxide and 0.02 M ammonium chloride were co-encapsulated with HFE-7500 that contains different concentrations of surfactant at 0.8, 0.2 and 5%. Collected ammonia gas generating droplets (10 μL) were then mixed with indicator droplets (10 μL, 1 mg/mL phenol red), and mixtures of droplets were immediately transferred into chamber slide for imaging under DMi-8 microscope. Microscopic images were analyzed with the image J software, and the mean values of intensity were fitted with OriginPro software (version 9.0, OriginLab, Northampton, MA, USA) (Figure 14). All procedures were performed at room temperature (23 °C) in this experiment.

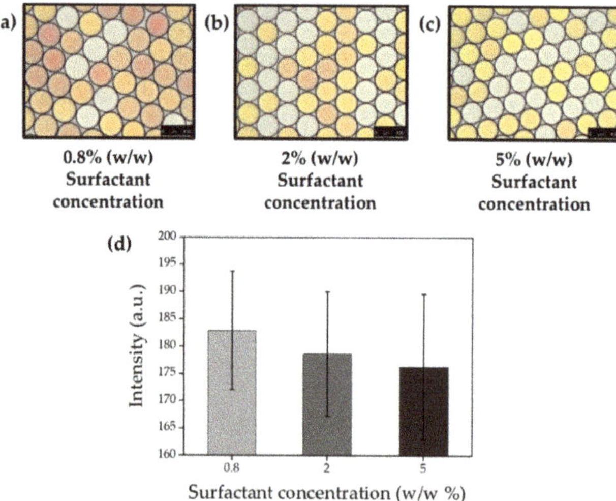

Figure 14. Relationship between surfactant concentration and gas permeability. Microscopic images of droplet mixtures that were generated with HFE-7500 contains 0.8% (**a**), 2% (**b**) or 5% (**c**) surfactant, respectively. (**d**) Mean values of color exchange of indicator droplets for surfactant concentration. (scale bar = 100 μm).

2.11. Time-Course Measurement of Ammonia Gas Crosstalk through Droplet Barrier

To measure the time-course transfer of produced ammonia gas across the droplet barrier, 0.02 M sodium hydroxide and 0.02 M ammonium chloride were co-encapsulated with HFE-7500 containing 2% surfactant. Indicator droplets also were also prepared by encapsulating 1 mg/mL phenol red. Ammonia gas generating droplets, oil and indicator droplets were sequentially introduced into the chamber slide. Once the indicator droplet contacted the ammonia gas generating droplets, a video was recorded with LAS X software under a DMi-8 microscope and the color change was identified with image J software (Figure 15). All procedures were performed at room temperature (23 °C) in this experiment.

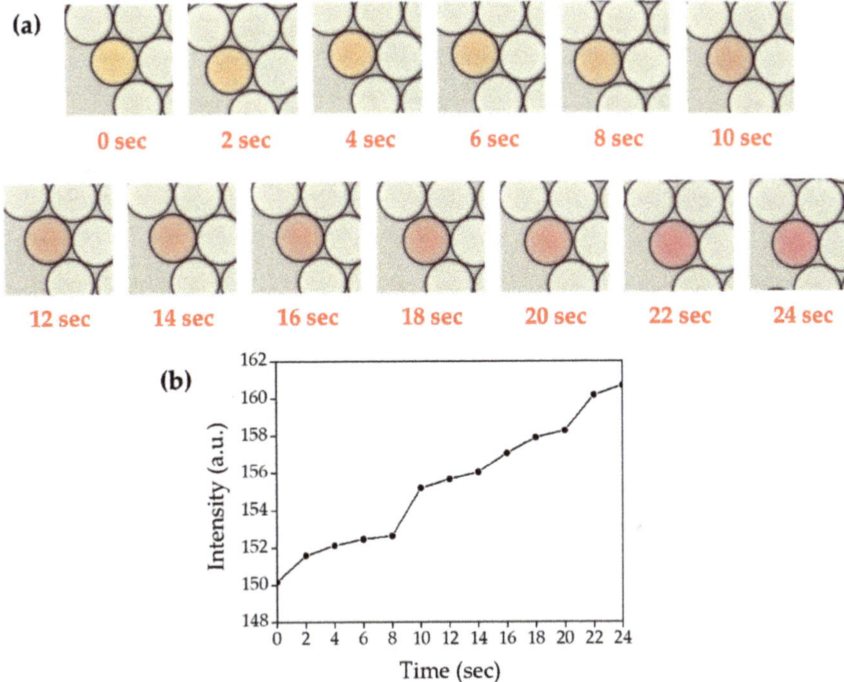

Figure 15. Time-course measurement of ammonia gas crosstalk between droplets. (**a**) Microscopic images of droplet mixture between indicator droplet and ammonia gas generating droplets, the video was recorded for 24 s under DMi-8 microscope. (**b**) Mean values of color change in the indicator droplets.

3. Results

3.1. Leaking of the Droplets during Determination of Bacteria with pH Indicator

Recently, DMF became a prestigious tool for the monitoring of various pathogens such as bacteria, viruses and fungi at a single-cell scale. In this role, the surfactant is one of the important components because it is needed to contribute to providing stability during droplet generation and reactions such as thermocycling. In this study, the PFPE–PEG–PFPE surfactant was synthesized in two steps (Figure 1), and it was identified prior to droplet generation with bacteria and pH indicator because this surfactant is one of the most commonly used in DMF studies. Synthesized surfactant was analyzed with ATR-FTIR spectroscopy (Figure 2), ^{1}H-NMR (Figure 3) and ^{19}F-NMR spectroscopy (Figure 4) to identify its chemical structure. In the ATR-FTIR spectrum (Figure 2), CH_2 stretching, bending bands and C-N stretching bands were respectively indicated at 2876 cm^{-1}, 1460 cm^{-1} and 1180 cm^{-1}, which originated from the Jeffamine ED-900. C–F and C–O stretching bands were indicated at 1228 cm^{-1} and 1120 cm^{-1} that originated from Krytox 157-FSH. The C=O stretching frequency in amide bond has been found around 1724 cm^{-1}. PFPE acid chloride or PFPE carboxylic acid were not observed when the final product was identified with ATR-FTIR. These results demonstrate that initiator or chemical intermediates were well-removed during synthesis and also cleaning procedures.

In the ^{1}H-NMR spectrum (Figure 3), the peak of 5.73 ppm (red circle) is assigned from the amide bond of N-H, and the peak of 3.81 ppm (Blue circle) is assigned from the ether (ROCH$_2$R). The peak of 1.2 ppm (green circle) is indicating methylene (CH$_2$). As shown in Figure 4, the peaks originated from the PFPE group were assigned as follows: −73 ppm (Pink circle), −80.26 ppm (yellow circle), −80.55 ppm (purple circle), −86.71 ppm (green circle), −125.85 ppm (blue circle), −126.27 ppm (red circle) and −186.96 ppm (orange circle) from the ^{19}F-NMR spectrum.

In previous reports, gas permeability was discussed and mentioned, but there was no previous investigation on gas crosstalk between droplets in our understanding. This is why we studied gas permeability through the droplet barrier when droplets were created using HFE-7500 with PFPE–PEG–PFPE surfactant. In Figure 5a, a schematic diagram of the microfluidic device is shown that contains two inlets for the aqueous phase and a single inlet for the oil phase. Droplets were generated through the flow-focusing structure (blue box, Figure 5b), and generated droplets can be incubated within incubation channels (orange box, Figure 5b) prior to the collection through outlet (green–box, Figure 5c). In this study, generated droplets are approximately 80 μm in diameter, which is about 270 pL in volume.

From the beginning of this study, we hypothesized that live bacteria could be determined during cultivation inside droplets with pH indicators because bacterial cells produce ammonia gas as a metabolite, which makes basic pH conditions in cultivation media. In this study, phenol red was co-encapsulated as a pH indicator because it is one of the most commonly used pH indicators for mammalian cell culture, which can turn red in a basic condition. More briefly, single *E. coli* K-12 can be grown within droplets with LB broth, and phenol red can be turned in red due to ammonia gas, which is produced from *E. coli* K-12 during bacterial growth (Figure 6). As shown in Figure 7a,b, there was no color exchange without bacterial growth in droplets. Once single *E. coli* K-12 proliferated within droplets, exponentially grown, *E. coli* K-12 population can be observed under a microscope after overnight incubation (Figure 7g). However, pH indicator (phenol red) was turned in red even droplet does not contain *E. coli* K-12 (Figure 7h). These results demonstrate that pH indicator or bacterial metabolites can penetrate neighber dropletsthrough the droplet barrier. However, it is not clear which components can be transferred between droplets.

3.2. Identification of Crosstalk between Droplets

To determine which components across the droplet barrier among pH indicator or bacterial metabolite, an additional experiment was designed, as shown in Figure 8. To identify which metabolite can be transferred into indicator droplets (phenol red), phenol red or *E. coli* K-12 were encapsulated prior to mix (above panel, Figure 8). The result showed that indicator droplets turned red after overnight incubation that suggests bacterial metabolites could cross through the droplet barrier. However, there was no color change from *E. coli* K-12-containing droplets that suggest phenol red cannot penetrate into *E. coli* K-12-containing droplets (Figure 9a,c). *E. coli* K-12 and phenol red were then co-encapsulated within the same droplets to identify the permeability of phenol red through the droplet barrier into empty droplets (below panel, Figure 8). After incubation, phenol red turned red due to bacterial metabolites, which makes basic condition, but there was no phenol red observed from empty droplets (Figure 9b,d). These results also strongly demonstrate that there is no permeability of phenol red between droplets, and only bacterial metabolites can be transferred through the droplet barrier.

Even if ammonia gas production is reported in previous reports [24] and color change of phenol red indicates pH change in basic condition within droplets, it was not clearly identified that ammonia gas is the metabolite crosstalk between droplets. To validate this, the production of ammonia gas during bacterial growth, bacterial metabolite was determined with Nessler's reagents, which is the conventional indicator of ammonia gas (Figure 10). Phenol red turned red with *E. coli* K-12 culture in LB broth (Figure 10b,e), and Nessler's reagent also immediately reacted with *E. coli* K-12-containing LB broth (Figure 10c,f). These results suggest that pH change was raised due to the production of ammonia gas during bacterial growth and ammonia gas was the metabolite which was reacted with phenol red.

3.3. Validation of Ammonia Gas Crosstalk between Droplets

To identify the permeability of ammonia gas through the droplet barrier, sodium hydroxide and ammonium chloride were co-encapsulated within 80 μm droplets for producing ammonia gas (ammonia gas generating droplets) [24]. As shown in Figure 11, we hypothesized that ammonia gas production could be mimicked by co-encapsulating sodium hydroxide and ammonium chloride. If produced ammonia gas crosstalk into phenol red droplets, the pH indicator will turn red. To validate our hypothesis, droplets were separately generated with 2 M sodium hydroxide (Figure 12b) or 2 M ammonium chloride (Figure 12e). Each droplet was then mixed with phenol red droplets (Figure 12a,d,g), but a color change was not observed (Figure 12c,f). However, phenol red droplets turned red when they were mixed with droplets that co-encapsulated with sodium hydroxide and ammonium chloride to produce ammonia gas within droplets (Figure 12i). These results indicate that ammonia gas was produced from the chemical reaction, and also produced gas was transferred through the droplet barrier.

3.4. Effects of Ammonia Gas Concentration and Surfactant Concentration on Crosstalk between Droplets

To identify the effect of gas concentration on permeability between droplets, sodium hydroxide and ammonium chloride were co-encapsulated at a different concentration from 0.002 M to 2 M. Collected droplets were then mixed with indicator droplets and imaged immediately. More bright color changes were observed from indicator droplets due to a higher production of ammonia gas (Figure 13). This result indicates that gas crosstalk was increased in a higher concentration of gas.

In previous reports on DMF, PFPE–PEG–PFPE surfactants have been used at a concentration of 2% on average, which is why we have identified gas crosstalk with a 2% surfactant in this report. However, the effect of surfactant concentration also identified whether it affects gas permeability between droplets. To validate this hypothesis, 0.02 M sodium hydroxide and 0.02 M ammonium chloride were co-encapsulated with HFE-7500 that contains different concentrations of surfactant at 0.8, 2 and 5%. Collected ammonia gas generating droplets were then mixed with indicator droplets, and gas permeability was decreased with a higher concentration of surfactant (Figure 14). These results suggest that the droplet barrier became more densely when surfactant concentration was increased, and it decreased gas permeability.

3.5. Time-Course Measurement of Ammonia Gas Crosstalk through Droplet Barrier

To measure the time-course transfer of produced ammonia gas through the droplet barrier, 0.02 M sodium hydroxide and 0.02 M ammonium chloride were co-encapsulated with HFE-7500 containing 2% surfactant prior to mix with indicator droplets. Once the indicator droplet was contacted with ammonia gas generating droplets, it started to turn red in a time-dependent manner (Figure 15, Supplementary Video S1). When sodium hydroxide and ammonium chloride were encapsulated at a concentration of 0.2 M for both chemical compounds, the color change was immediately terminated turned red once the indicator droplet was contacted with ammonia gas generating droplets. We also observed that ammonia gas could be transferred through fluorinated oil even without contact between droplets. As shown in Figure 16 (Supplementary Video S2), phenol red-containing droplets turned red slowly due to gas diffusion through HFE-7500 from two white droplets (that contain 0.2 M sodium hydroxide and 0.2 M ammonium chloride). This result suggested that ammonia gas is soluble in HFE-7500 oil.

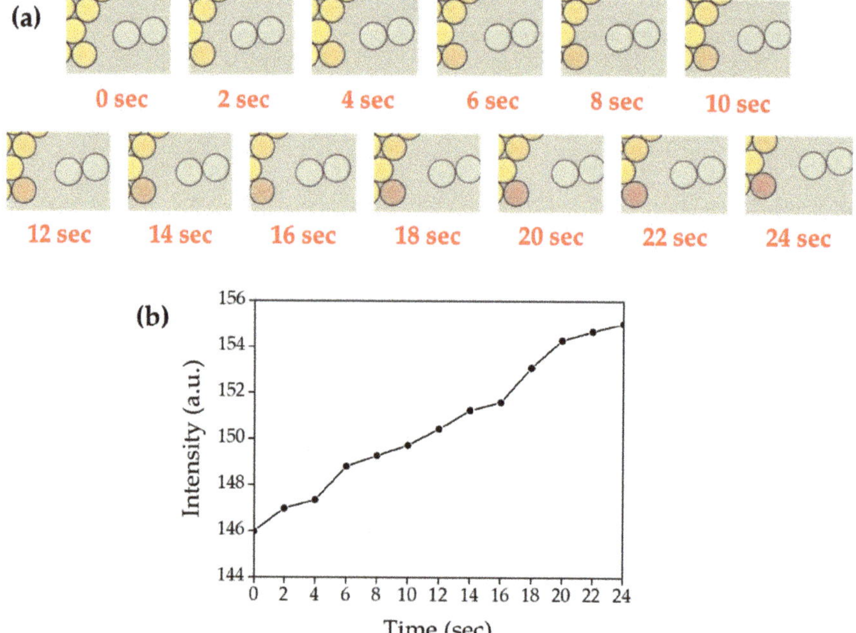

Figure 16. Time-course measurement of ammonia gas crosstalk between droplets. (**a**) Microscopic images of droplet mixture between indicator droplet and ammonia gas generating droplets, the video was recorded for 24 s under a DMi-8 microscope. (**b**) Mean values of color change in the indicator droplets.

4. Discussion

The PFPE–PEG–PFPE surfactant is one of the most commonly used surfactants in DMF, and it was mentioned that fluorinated fluids exhibit high gas solubility [23,35]. However, gas crosstalk between droplets has not been validated yet. In this study, we first demonstrated that bacterial metabolic gas penetrates through barriers between droplets, and it contaminates neighboring droplets. Interestingly phenol red dye cannot penetrate the barrier between droplets due to its relatively higher molar mass (354.38 g/mol), but ammonia gas can across the droplet barrier because of its lower molar mass (17.031 g/mol) in our hypothesis.

It is already known that the weak interactions of fluorinated compounds in relatively high compressible fluids result in high gas solubility, so the interfacial surfactant molecules may be the main barriers that affect gas crosstalk [22]. To validate this hypothesis, the relation of surfactant concentration and gas transport was identified. To provide a higher surface density, surfactant concentration was increased up to 5%, and it decreased gas permeability between droplets and this result indicates surfactant density is the potential reason for the gas crosstalk between droplets.

We also confirmed that ammonia gas was produced during cultivation within droplets, and these results suggest that live bacteria can be grown and monitored using DMF by analyzing the production of ammonia gas. Since bacterial contamination is one of the most serious issues in drinking water, bacterial growth can be monitored by the identification of ammonia gas production using DMF without complex and expensive approaches such as real-time PCR. Even droplets were analyzed with an expensive and large microscope in our study; the microscope can be replaced with inexpensive and portable imaging systems, including smartphones and portable CCD cameras.

Basically, leaking of the droplets is the potential issue on analytical approaches, including basic biological studies and diagnosis because some of the interested biomarkers can be penetrated into neighboring droplets if it is small enough. Since contamination can be a potential reason for false-positive droplets, this issue must be eliminated in case of various diagnosis purposes. Even large-sized, highly hydrophobic and highly hydrophilic components have less chance for crosstalk through the droplet barrier; there is still potential that tiny components can be a useful biomarker for diseases. It turns that a dense droplet barrier must be constructed, which means stronger surfactants must be required for DMF in the future. However, leaking also can be a useful tool that can provide internal components to neighboring droplets. For example, bacteria droplets can be a tiny microcapsule that can supply ammonia gas into the microsized reactor that requires ammonia gas for the reactions such as chemical synthesis, analytical procedures and catalytic reactions.

Supplementary Materials: The following are available online at http://www.mdpi.com/2079-6374/10/11/172/s1, Video S1: Gas crosstalk between droplets when droplets are contected. Video S2: Gas crosstalk through fluorinated oil without contact between droplets.

Author Contributions: Conceptualization, S.K. and D.-K.K.; methodology, S.K.; software, S.K.; validation, S.K.; formal analysis, S.K.; investigation, D.-K.K.; resources, D.-K.K.; data curation, D.-K.K.; writing—original draft preparation, S.K.; writing—review and editing, D.-K.K.; visualization, S.K.; supervision, D.-K.K.; project administration, D.-K.K.; funding acquisition, D.-K.K. Both authors have read and agreed to the published version of the manuscript.

Funding: This work was supported by the Incheon National University Research Grant in 2017 and the National Research Foundation of Korea (NRF) (NRF2016R1A5A1010148).

Conflicts of Interest: The authors declare no conflict of interest.

References

1. Mashaghi, S.; Abbaspourrad, A.; Weitz, D.A.; van Oijen, A.M. Droplet microfluidics: A tool for biology, chemistry and nanotechnology. *TrAC Trends Anal. Chem.* **2016**, *82*, 118–125. [CrossRef]
2. Seemann, R.; Brinkmann, M.; Pfohl, T.; Herminghaus, S. Droplet based microfluidics. *Rep. Prog. Phys.* **2011**, *75*, 016601. [CrossRef] [PubMed]
3. Teh, S.-Y.; Lin, R.; Hung, L.-H.; Lee, A.P. Droplet microfluidics. *Lab Chip* **2008**, *8*, 198–220. [CrossRef] [PubMed]
4. Joensson, H.N.; Andersson Svahn, H. Droplet microfluidics—A tool for single-cell analysis. *Angew. Chem. Int. Ed.* **2012**, *51*, 12176–12192. [CrossRef] [PubMed]
5. Kaminski, T.S.; Scheler, O.; Garstecki, P. Droplet microfluidics for microbiology: Techniques, applications and challenges. *Lab Chip* **2016**, *16*, 2168–2187. [CrossRef] [PubMed]
6. Solvas, X.C.; DeMello, A. Droplet microfluidics: Recent developments and future applications. *Chem. Commun.* **2011**, *47*, 1936–1942. [CrossRef]
7. Rakszewska, A.; Tel, J.; Chokkalingam, V.; Huck, W.T. One drop at a time: Toward droplet microfluidics as a versatile tool for single-cell analysis. *NPG Asia Mater.* **2014**, *6*, e133. [CrossRef]
8. Guo, M.T.; Rotem, A.; Heyman, J.A.; Weitz, D.A. Droplet microfluidics for high-throughput biological assays. *Lab Chip* **2012**, *12*, 2146–2155. [CrossRef]
9. Pekin, D.; Skhiri, Y.; Baret, J.-C.; Le Corre, D.; Mazutis, L.; Salem, C.B.; Millot, F.; El Harrak, A.; Hutchison, J.B.; Larson, J.W. Quantitative and sensitive detection of rare mutations using droplet-based microfluidics. *Lab Chip* **2011**, *11*, 2156–2166. [CrossRef]
10. Zhu, Y.; Fang, Q. Analytical detection techniques for droplet microfluidics—A review. *Anal. Chim. Acta* **2013**, *787*, 24–35. [CrossRef]
11. Brouzes, E.; Medkova, M.; Savenelli, N.; Marran, D.; Twardowski, M.; Hutchison, J.B.; Rothberg, J.M.; Link, D.R.; Perrimon, N.; Samuels, M.L. Droplet microfluidic technology for single-cell high-throughput screening. *Proc. Natl. Acad. Sci. USA* **2009**, *106*, 14195–14200. [CrossRef] [PubMed]
12. Shembekar, N.; Chaipan, C.; Utharala, R.; Merten, C.A. Droplet-based microfluidics in drug discovery, transcriptomics and high-throughput molecular genetics. *Lab Chip* **2016**, *16*, 1314–1331. [CrossRef] [PubMed]
13. Terekhov, S.S.; Smirnov, I.V.; Stepanova, A.V.; Bobik, T.V.; Mokrushina, Y.A.; Ponomarenko, N.A.; Belogurov, A.A.; Rubtsova, M.P.; Kartseva, O.V.; Gomzikova, M.O. Microfluidic droplet platform for ultrahigh-throughput single-cell screening of biodiversity. *Proc. Natl. Acad. Sci. USA* **2017**, *114*, 2550–2555. [CrossRef] [PubMed]

14. Sabhachandani, P.; Sarkar, S.; Zucchi, P.C.; Whitfield, B.A.; Kirby, J.E.; Hirsch, E.B.; Konry, T. Integrated microfluidic platform for rapid antimicrobial susceptibility testing and bacterial growth analysis using bead-based biosensor via fluorescence imaging. *Microchim. Acta* **2017**, *184*, 4619–4628. [CrossRef]

15. Kang, D.-K.; Ali, M.M.; Zhang, K.; Huang, S.S.; Peterson, E.; Digman, M.A.; Gratton, E.; Zhao, W. Rapid detection of single bacteria in unprocessed blood using Integrated Comprehensive Droplet Digital Detection. *Nat. Commun.* **2014**, *5*, 1–10. [CrossRef] [PubMed]

16. Baret, J.-C. Surfactants in droplet-based microfluidics. *Lab Chip* **2012**, *12*, 422–433. [CrossRef] [PubMed]

17. Kang, D.-K.; Ali, M.M.; Zhang, K.; Pone, E.J.; Zhao, W. Droplet microfluidics for single-molecule and single-cell analysis in cancer research, diagnosis and therapy. *TrAC Trends Anal. Chem.* **2014**, *58*, 145–153. [CrossRef]

18. Holtze, C.; Rowat, A.; Agresti, J.; Hutchison, J.; Angile, F.; Schmitz, C.; Köster, S.; Duan, H.; Humphry, K.; Scanga, R. Biocompatible surfactants for water-in-fluorocarbon emulsions. *Lab Chip* **2008**, *8*, 1632–1639. [CrossRef]

19. Chiu, Y.-L.; Chan, H.F.; Phua, K.K.; Zhang, Y.; Juul, S.; Knudsen, B.R.; Ho, Y.-P.; Leong, K.W. Synthesis of fluorosurfactants for emulsion-based biological applications. *ACS Nano* **2014**, *8*, 3913–3920. [CrossRef]

20. Holt, D.J.; Payne, R.J.; Abell, C. Synthesis of novel fluorous surfactants for microdroplet stabilisation in fluorous oil streams. *J. Fluor. Chem.* **2010**, *131*, 398–407. [CrossRef]

21. Platzman, I.; Janiesch, J.-W.; Spatz, J.P. Synthesis of nanostructured and biofunctionalized water-in-oil droplets as tools for homing T cells. *J. Am. Chem. Soc.* **2013**, *135*, 3339–3342. [CrossRef] [PubMed]

22. Gruner, P.; Riechers, B.; Orellana, L.A.C.; Brosseau, Q.; Maes, F.; Beneyton, T.; Pekin, D.; Baret, J.-C. Stabilisers for water-in-fluorinated-oil dispersions: Key properties for microfluidic applications. *Curr. Opin. Colloid Interface Sci.* **2015**, *20*, 183–191. [CrossRef]

23. Wagner, O.; Thiele, J.; Weinhart, M.; Mazutis, L.; Weitz, D.A.; Huck, W.T.; Haag, R. Biocompatible fluorinated polyglycerols for droplet microfluidics as an alternative to PEG-based copolymer surfactants. *Lab Chip* **2016**, *16*, 65–69. [CrossRef] [PubMed]

24. Vince, A.J.; Burridge, S.M. Ammonia production by intestinal bacteria: The effects of lactose, lactulose and glucose. *J. Med. Microbiol.* **1980**, *13*, 177–191. [CrossRef]

25. Hills, G.M. Ammonia production by pathogenic bacteria. *Biochem. J.* **1940**, *34*, 1057–1069. [CrossRef]

26. Cummings, J.H.; Macfarlane, G.T. Role of intestinal bacteria in nutrient metabolism. *Clin. Nutr.* **1997**, *16*, 3–11. [CrossRef]

27. Sezonov, G.; Joseleau-Petit, D.; d'Ari, R. Escherichia coli physiology in Luria-Bertani broth. *J. Bacteriol.* **2007**, *189*, 8746–8749. [CrossRef]

28. Lee, M.; Collins, J.W.; Aubrecht, D.M.; Sperling, R.A.; Solomon, L.; Ha, J.-W.; Yi, G.-R.; Weitz, D.A.; Manoharan, V.N. Synchronized reinjection and coalescence of droplets in microfluidics. *Lab Chip* **2014**, *14*, 509–513. [CrossRef]

29. Scanga, R.; Chrastecka, L.; Mohammad, R.; Meadows, A.; Quan, P.-L.; Brouzes, E. Click chemistry approaches to expand the repertoire of PEG-based fluorinated surfactants for droplet microfluidics. *RSC Adv.* **2018**, *8*, 12960–12974. [CrossRef]

30. Khademhosseini, A.; Suh, K.Y.; Jon, S.; Eng, G.; Yeh, J.; Chen, G.-J.; Langer, R. A soft lithographic approach to fabricate patterned microfluidic channels. *Anal. Chem.* **2004**, *76*, 3675–3681. [CrossRef]

31. Kim, P.; Kwon, K.W.; Park, M.C.; Lee, S.H.; Kim, S.M.; Suh, K.Y. Soft lithography for microfluidics: A review. *Biochip J.* **2008**, *2*, 1–11.

32. McDonald, J.C.; Duffy, D.C.; Anderson, J.R.; Chiu, D.T.; Wu, H.; Schueller, O.J.; Whitesides, G.M. Fabrication of microfluidic systems in poly (dimethylsiloxane). *Electrophor. Int. J.* **2000**, *21*, 27–40. [CrossRef]

33. Ren, K.; Zhou, J.; Wu, H. Materials for microfluidic chip fabrication. *Acc. Chem. Res.* **2013**, *46*, 2396–2406. [CrossRef] [PubMed]

34. Hosokawa, K.; Fujii, T.; Endo, I. Handling of picoliter liquid samples in a poly (dimethylsiloxane)-based microfluidic device. *Anal. Chem.* **1999**, *71*, 4781–4785. [CrossRef]

35. Tan, Y.-C.; Cristini, V.; Lee, A.P. Monodispersed microfluidic droplet generation by shear focusing microfluidic device. *Sens. Actuators B Chem.* **2006**, *114*, 350–356. [CrossRef]
36. Wolfbeis, O.S.; Posch, H.E. Fibre-optic fluorescing sensor for ammonia. *Anal. Chim. Acta* **1986**, *185*, 321–327. [CrossRef]

Publisher's Note: MDPI stays neutral with regard to jurisdictional claims in published maps and institutional affiliations.

 *biosensors*

Article

The Effect of Optically Induced Dielectrophoresis (ODEP)-Based Cell Manipulation in a Microfluidic System on the Properties of Biological Cells

Po-Yu Chu [1,†], Chia-Hsun Hsieh [2,3,4,†], Chien-Ru Lin [5] and Min-Hsien Wu [3,5,6,*]

[1] Ph.D. Program in Biomedical Engineering, Chang Gung University, Taoyuan City 33302, Taiwan; l1a2yyy@gmail.com
[2] Division of Haematology/Oncology, Department of Internal Medicine, New Taipei Municipal Hospital, New Taipei City 23600, Taiwan; wisdom5000@gmail.com
[3] Division of Haematology/Oncology, Department of Internal Medicine, Chang Gung Memorial Hospital at Linkou, Taoyuan City 33302, Taiwan
[4] College of Medicine, Chang Gung University, Taoyuan City 33302, Taiwan
[5] Graduate Institute of Biomedical Engineering, Chang Gung University, Taoyuan City 33302, Taiwan; chianru0608@gmail.com
[6] Department of Chemical Engineering, Ming Chi University of Technology, New Taipei City 24301, Taiwan
[*] Correspondence: mhwu@mail.cgu.edu.tw; Tel.: +886-3-2118-800 (ext. 3599)
[†] Chu and Hsieh contributed equally to this work.

Received: 11 May 2020; Accepted: 14 June 2020; Published: 16 June 2020

Abstract: Cell manipulation using optically induced dielectrophoresis (ODEP) in microfluidic systems has attracted the interest of scientists due to its simplicity. Although this technique has been successfully demonstrated for various applications, one fundamental issue has to be addressed—Whether, the ODEP field affects the native properties of cells. To address this issue, we explored the effect of ODEP electrical conditions on cellular properties. Within the experimental conditions tested, the ODEP-based cell manipulation with the largest velocity occurred at 10 Vpp and 1 MHz, for the two cancer cell types explored. Under this operating condition, however, the cell viability of cancer cells was significantly affected (e.g., 70.5 ± 10.0% and 50.6 ± 9.2% reduction for the PC-3 and SK-BR-3 cancer cells, respectively). Conversely, the exposure of cancer cells to the ODEP electrical conditions of 7–10 Vpp and 3–5 MHz did not significantly alter the cell viability, cell metabolic activity, and the *EpCAM*, *VIM*, and *ABCC1* gene expression of cancer cells. Overall, this study fundamentally investigated the effect of ODEP electrical conditions on the cellular properties of cancer cells. The information obtained is crucially important for the utilization of ODEP-based cell manipulation in a microscale system for various applications.

Keywords: optically induced dielectrophoresis; microfluidic systems; cell manipulation; cell property; cell physiology

---

1. Introduction

The fine manipulation of biological cells has attracted considerable interests in a wide variety of biomedical studies (e.g., cell sorting [1–8], the isolation and purification of cells [1–9], cell patterning [10–14], tissue engineering [10,11,14], or various cell-based bioassays [12,13]). These attempts are technically challenging to realize conventionally, due to the inadequate scale of tools or equipment used for cell manipulation in general biological laboratories. Leveraging the appropriate dimensional feature in a microfluidic system, and various novel approaches for microparticle manipulations (e.g., acoustophoresis [1,13], magnetophoresis [2], thermophoresis [15], and dielectrophoresis (DEP) [3,4]), researchers could achieve the above-mentioned goals in a high performance manner [16,17].

Among the techniques for microparticle manipulation in a microfluidic system, microparticle manipulation using DEP mechanism has been widely adopted for various applications (e.g., microparticle enrichment [18,19], microparticle patterning [10,11], cell-type classification [20], or rare cell isolation [3,4]). DEP-based microparticle manipulation, first presented in the 1950s [21], is well-described elsewhere [22], and can be briefly described as follows. When dielectric microparticles (e.g., biological cells) are suspended in a solution in which an electric field is exerted, charges can be electrically polarized on the microparticles' surface. The interaction between the induced charges on the microparticles and the electric field specifically exerted on them can generate a DEP force on such microparticles [22]. In practice, therefore, the scientists can delicately control the electric field exerted on biological cells via a designed microelectrode array, to manipulate these cells in a manageable manner. Although the DEP technique is proven feasible for the fine manipulation of cells, this technique normally requires a costly, time-consuming, and technically demanding microfabrication process to create a unique metal microelectrode layout that is specific to the application [23,24]. In addition, cell manipulation using the DEP mechanism might be technically challenging for general scientists, due to its complexity. To tackle the technical issues, the optically induced dielectrophoresis (ODEP)-based technique could open up a new horizon for the manipulation of microparticles in a more efficient, flexible, and user-friendly manner [23,24].

Microparticle manipulation based on the ODEP mechanism was first proposed in 2005 [23]. Similar to the DEP mechanism described earlier, an electric voltage is applied between the top and bottom substrates of an ODEP system to generate a uniform electric field in the solution layer sandwiched between the two substrates. Under this circumstance, the dielectric microparticles suspended in the solution layer are electrically polarized. In contrast to the DEP mechanism, however, the key technical merit of the ODEP mechanism is that it can easily and quickly create or modify an electrode layout through the control of optical patterns, by acting as a virtual electrode [23,24]. When the photoconductive layer of an ODEP system is briefly projected with light, it can lead to a significant reduction in electrical impedance in the light-illuminated area. This phenomenon can cause the exerted voltage to decrease across the solution layer inside the light-projected region. This phenomenon, therefore, generates a locally nonuniform electric field within an ODEP system. For microparticle manipulation, scientists can simply utilize a commercial projector to project optical images on an ODEP system, by which the interaction between the generated nonuniform electric field and an electrically polarized microparticle is used to manipulate the microparticle. In terms of operation, one can control the light image projected onto the ODEP system and thus the nonuniform electric field created to manipulate dielectric microparticles within the ODEP system.

The use of ODEP mechanism in microfluidic systems has been successfully applied for the high-accuracy positioning and assembling of metallic beads [25], dynamic analysis of cancer-immune microenvironment [26], or the isolation and purification of rare cell species in clinical samples (e.g., Raji cells [7], circulating tumor cells (CTCs) [5,9], and CTC clusters [6]) in a higher performance manner than their conventional counterparts. Although the incorporation of the ODEP mechanism in a microscale system has provided a powerful tool for the biological cell-relevant studies or applications, one fundamental question has to be answered before its widespread applications—Whether the cell manipulation using the ODEP mechanism influences the cellular properties. If this is the case, the use of such a mechanism for cell manipulation could complicate the subsequent biological studies or applications. To the best of our knowledge, however, this fundamental issue has not yet been well explored.

To investigate the effect of ODEP-based cell manipulation (the operating conditions explored—(1) the magnitude of AC voltage: 5–10 V; (2) frequency of AC voltage: 1–5 MHz, (3) operating time: 3 min) on the properties (e.g., cell viability, metabolic activity of cells, or gene expression of cells) of biological cells (e.g., PC-3 and SK-BR-3 cancer cells), we performed various experiments. Within the experimental conditions investigated, the ODEP-based cell manipulation with the best performance (e.g., highest cell manipulation velocity) occurred at 10 Vpp and 1 MHz, for the two cancer cell types explored. Under this ODEP electrical condition, however, the cell viability of the cancer cells tested was significantly affected (e.g., 70.5 ± 10.0% and 50.6 ± 9.2% reduction for the PC-3 and SK-BR-3 cancer cells, respectively), possibly due to the electrical lysis of cells. Conversely, the exposure of cancer cells to the ODEP field with particular conditions (e.g., 7–10 Vpp, 3–5 MHz, and 3 min exposure time) might not significantly alter the cell viability, cell metabolic activity, and gene expression of cancer cells. Within this ODEP operating condition, the highest maximum velocity (e.g., 106.7 ± 30.6 μm s^{-1} and 100.0 ± 20.0 μm s^{-1} for the PC-3 and SK-BR-3 cancer cells, respectively) of a light image that can manipulate cells occurred at the voltage magnitude and frequency of 10 Vpp and 3 MHz, respectively. The above-mentioned cell manipulation velocities are technically sufficient for various applications. As a whole, this study fundamentally investigated the effect of ODEP electrical conditions on the cellular properties of cancer cells. The information obtained is crucially important for the utilization of ODEP-based cell manipulation in a micro-scale system for various applications.

2. Materials and Methods

2.1. Microfluidic Chip and Experimental Setup

To explore the effect of ODEP on the properties of biological cells, we designed a simple microfluidic chip. Figure 1a schematically presents the top-view layout of the microfluidic chip, mainly encompassing a microchamber (L = 6.0 mm, W = 4.0 mm, and H = 50.0 μm) connecting two microchannels (L = 2.0 mm, W = 1.0 mm, and H = 50.0 μm). In this work, an ODEP field was applied to the microchamber (Figure 1a). The structure of the microfluidic chip is illustrated in Figure 1b. Briefly, the microfluidic chip consists of two custom-made polydimethylsiloxane (PDMS) (Sylgard® 184, Dow Corning, Midland, MI, USA) adapters for tubing connection (A), an indium–tin oxide (ITO) glass (10 Ω, 0.7 mm; Innolux Corp., Miaoli County, Taiwan) (B), a processed adhesive tape (L298, Sun-yieh, New Taipei City, Taiwan; thickness: 50 μm) containing hollow microchamber and microchannel structures (C), and a bottom ITO glass (D) with a coating layer of photoconductive material (a 1.0 μm-thick hydrogenated amorphous silicon (a-Si:H) layer). The fabrication processes (e.g., PDMS replica mold, metal mold-punching fabrication, plasma-enhanced chemical vapor deposition (PECVD)-based thin-film technology, and plasma oxidation-aided bonding) were well-described previously [8,27,28]. After each substrate was fabricated (Figure 1b), substrate A was bonded with substrate B through the surface treatment of plasma oxidation. This step was followed by the assembly with the substrate D with the aid of the processed double-sided adhesive tape (substrate C) (Figure 1b). In operation, the prepared cell suspension was manually loaded into the microchamber, using a pipette and a tip. For cell manipulation using the ODEP mechanism, the commercially available ODEP-based cell isolation equipment (Celnostics, Ace Medical Technology Co., Ltd., Taipei City, Taiwan) was used to achieve the ODEP-based cell manipulation in the proposed microfluidic chip. Within this equipment, briefly, a function generator (AFG-2125, Good Will Instrument Co., Ltd., New Taipei City, Taiwan) was utilized to apply an AC bias between the two ITO glasses. A computer-interfaced digital projector (EB-X05, Epson, Nagano, Japan) was used to illuminate light images onto the photoconductive material (i.e., the Substrate D) of microfluidic chip to generate the ODEP force on the cells. An illustration of the important operating modules of the equipment is schematically presented in Figure 1c (a photograph of the overall experimental setup is provided in a Supplementary Figure S1).

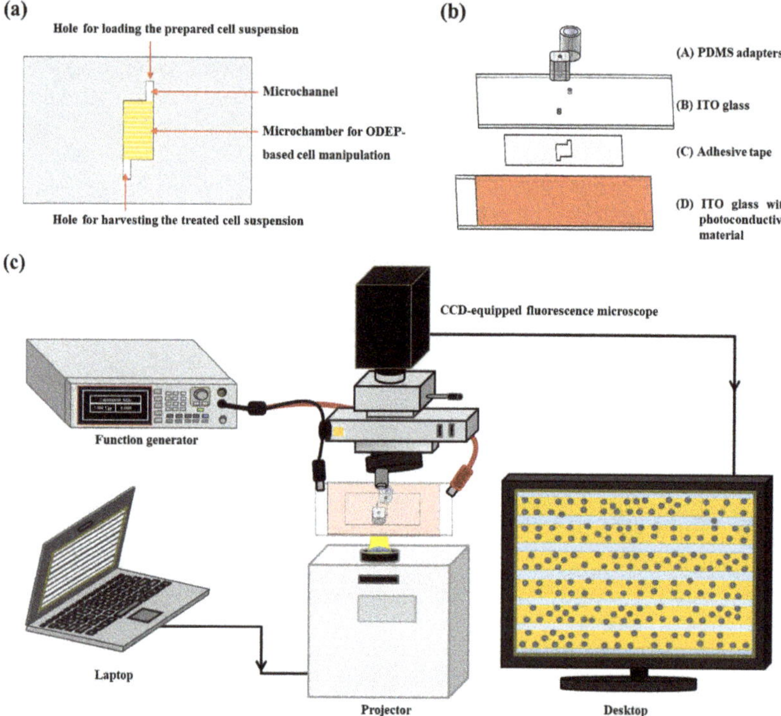

Figure 1. Schematic presentation of (**a**) the design of the microfluidic chip (top view), (**b**) the structure of the microfluidic chip ((A) polydimethylsiloxane (PDMS) adapters for tubing connection, (B) indium-tin oxide (ITO) glass, (C) processed adhesive tape, and (D) ITO glass with a coating layer of photoconductive material), and (**c**) the overall experimental setup.

2.2. The Assessment of the ODEP Manipulation Force Generated on Cells

The working principle of ODEP for cell manipulation is described in the Introduction. The ODEP force generated on a cell can theoretically be described by Equation (1) [24,27,28]:

$$F_{DEP} = 2\pi r^3 \varepsilon_0 \varepsilon_m \mathrm{Re}[f_{CM}]\nabla|E|^2 \tag{1}$$

In Equation (1), r (cellular radius), ε_0 (vacuum permittivity), ε_m (relative permittivity of the surrounding solution), $\nabla|E|^2$ (gradient of electric field squared), and $\mathrm{Re}[f_{CM}]$ (real part of the Clausius–Mossotti factor (f_{CM})) are the important parameters [24,28]. The f_{CM} is described by Equation (2) [29–31]:

$$f_{CM} = \frac{\varepsilon_{cell}^* - \varepsilon_m^*}{\varepsilon_{cell}^* + 2\varepsilon_m^*} \tag{2}$$

where ε_{cell}^* and ε_m^* represent the complex permittivity of the cell and the surrounding solution, respectively. For a single-cell model, the complex permittivity of the cell and the surrounding solution can be further described by Equations (3) and (4):

$$\varepsilon_{cell}^* = C_{mem}^* \frac{3R\varepsilon_{int}^*}{3\varepsilon_{int}^* + 3C_{mem}^*R}, C_{mem}^* = \frac{\varepsilon_{mem}}{d} - \frac{j\sigma_{mem}}{d}, \varepsilon_{int}^* = \varepsilon_0\varepsilon_{int} - \frac{j\sigma_{int}}{\omega} \tag{3}$$

$$\varepsilon_m^* = \varepsilon_0\varepsilon_m - \frac{j\sigma_m}{\omega} \tag{4}$$

In Equations (3) and (4), C^*_{mem} represents the complex cell membrane capacitance, ε^*_{int} represents the complex permittivity of the cellular interior (i.e., cell cytoplasm), R represents the radius of the cellular interior, d represents the thickness of cell membrane, ε represents the relative permittivity of the cell membrane, cellular interior, or surrounding solution (denoted by the subscript *mem*, *int*, or *m*, respectively), σ represents the conductivity of the cell membrane, cell interior, or surrounding solution (denoted by the subscript *mem*, *int*, or *m*, respectively), j represents the imaginary vector ($\sqrt{-1}$), and ω represents the angular frequency (i.e., $\omega = 2\pi f$) of the applied AC field, respectively [29–31]. Under a given solution condition, overall, the magnitude and frequency of the electric voltage applied could play important roles in the ODEP force generated on a particular cell [24,28]. In this work, the effect of electric conditions (i.e., the magnitude and frequency of the electric voltage exerted) on the ODEP force generated on cells was experimentally evaluated. In the evaluation, the ODEP manipulation force, a net force between the ODEP and the friction force, acting on the manipulated cells was then experimentally assessed, based on the method described previously [8,27,28]. In a steady state, the ODEP manipulation force acting on a cell was balanced by the viscous drag of fluid acting on such a cell under a continuous flow condition. As a result, the hydrodynamic drag force of a moving cell was used to evaluate the net ODEP manipulation force of a cell according to Stokes' law (Equation (4)):

$$F = 6\pi r \eta v \tag{5}$$

In Equation (5), r (cellular radius), η (fluidic viscosity), and v (the velocity of a moving cell) are the important parameters. Under the given solution and cellular size conditions, overall, the ODEP manipulation force of the manipulated cell could then be experimentally assessed through the measurement of the maximum velocity of a moving optical image that can manipulate such a cell [8,27,28]. In practice, briefly, a light bar image with different moving velocities (e.g., from low to high velocities) was used to manipulate a cell (e.g., attracted and pulled a cell). Through this process, the maximum velocity of a moving optical image that can manipulate such a cell was then determined. In this work, therefore, the above-mentioned velocity was utilized as an index for the evaluation of the ODEP manipulation force generated on a specific cell under a particular electric condition. Based on this, the effect of electric conditions (e.g., magnitude of AC electric voltage: 7–10 Vpp and frequency of AC electric voltage: 1–5 MHz) on the ODEP manipulation of the cells tested (e.g., PC-3 and SK-BR-3 cancer cells) was evaluated. Briefly, the cell sample tested was prepared in a cell suspension (cell density: 10^6 cells mL^{-1}), followed by loading into the microchamber of the microfluidic chip (Figure 1a). The maximum velocity of a moving light bar (L: 1.3 mm W: 100.0 μm) that could manipulate these cells was then assessed [27,28].

2.3. Evaluation of the Properties of Cancer Cells Treated with Varied ODEP Operating Conditions

For the analysis of the ODEP effect on the cellular properties, the cancer cells tested (e.g., PC-3 and SK-BR-3 cancer cell lines, two of the commonly-used cancer cell lines in cancer-related studies [32,33]) were first treated with the ODEP fields under different conditions for 3 min, followed by assaying their cellular properties, including cellular viability, cellular metabolism activity, and gene expression. In this study, the biological assays were carried out at 1.5 ± 0.2 h after the ODEP exposure treatment. In brief, the background medium of the prepared cancer cell suspension (cell density: 5×10^6 cells mL^{-1} for PC-3 cancer cells, and 3×10^6 cells mL^{-1} for SK-BR-3 cancer cells) was first replaced by a 9.5% (*w/v*) sucrose buffer solution (relative permittivity: ~76.19 [34], fluid viscosity: ~1.0389×10^{-2} g s^{-1} cm^{-1} [35], osmolality: 270–290 mOsmol kg^{-1}, and conductivity: 1–5 μS cm^{-1}), the commonly used low conductivity buffer solution for ODEP-based cell manipulation [5,8,9,27,28]. The processed cell suspension sample was then loaded into the microchamber of a microfluidic chip (Figure 1a). In this process, however, a small portion of cells might retain in the microchannel area. In this work, the cell manipulation using ODEP was first performed to quickly (operation time—within 5 s) transport these cells to the microchamber area before ODEP treatment. At the microchamber zone, 40 rectangular light bar images (L: 4.0 mm, W: 100.0 μm, the interval between light bars: 50.0 μm) were then illuminated to provide an ODEP field on the cancer cells tested. In this study, ODEP fields with different electric conditions (magnitude of AC electric voltage: 7–10 Vpp and frequency of AC electric voltage: 1–5 MHz) were used to treat the cancer cells tested. After exposure to the ODEP field for 3 min, the treated cells were then harvested for the subsequent bioassays. In the process, briefly, a suction-type syringe pump was utilized to collect the treated cancer cells through the hole for harvesting cells (Figure 1a). Before the following bioassays, the collected cells were kept in the form of cell suspension in the original sucrose buffer solution under the thermal condition of 25 °C.

In this work, the commonly-used Cell Counting Kit-8 (CCK-8) (CK04–05, Dojindo, Kumamoto, Japan) and ATP Colorimetric/Fluorometric assays (K354-100, BioVision Inc., Milpitas, CA, USA) were utilized to assay the cell viability and cell metabolic activity of cancer cells, respectively. The bioassays were carried out according to the manufacturer's instructions. For further analysis of whether cell manipulation using ODEP could affect cellular gene expression, the expression levels of genes in cancer cells treated with different ODEP fields were experimentally evaluated. As the ODEP-based cell manipulation was commonly used for the isolation and purification of circulating tumor cells (CTCs) [5,9], the important CTC-related gene expressions were thus explored in this work. For the cancer cells (PC-3 and SK-BR-3 cancer cells) tested, the mRNA levels of epithelial-to-mesenchymal transition (EMT)-associated genes [*EpCAM* (Hs00158980_m1) and *VIM* (Hs00958111_m1)], the multidrug resistance-associated protein 1 (MRP1) gene [*ABCC1* (Hs01561502_m1)], and the housekeeping gene [*GAPDH* (Hs02758991_g1)] were experimentally quantified. The bioassay was based on a method described previously [8,9,27]. In brief, RNA was extracted from the cancer cells tested using a bromochloropropane (BCP)-based TRI Reagent procedure (Thermo Fisher Scientific, San Jose, CA, USA [36]). This process was followed by the reverse transcription using a SuperScript® IV Reverse Transcriptase Kit (Thermo Fisher Scientific, San Jose, CA, USA). The mRNA level was subsequently quantified using a StepOne™ Real-Time PCR System (Thermo Fisher Scientific, San Jose, CA, USA).

2.4. Statistical Analysis

In this study, data were obtained from three separate experiments, and are presented as the mean ± standard deviation (n = 9). To compare the results from different operating conditions, we used one-way ANOVA and Tukey's honestly significant difference (HSD) post-hoc test for the statistical analysis.

3. Results and Discussion

3.1. Effect of the Electric Conditions on ODEP-Based Cell Manipulation

In this study, the effect of the ODEP electrical conditions (e.g., magnitude of AC electric voltage: 7–10 Vpp, and frequency of AC electric voltage: 1–5 MHz) on the cell manipulation of the cancer cells (e.g., PC-3 and SK-BR-3 cancer cells) tested were experimentally evaluated. Figure 2 reveals the quantitative relationship between the maximum velocity of a moving light bar that can manipulate the cancer cells tested (thus, the ODEP manipulation force of the cancer cells; Equation (5)) and the frequency of electric voltage applied under different voltage magnitude conditions. The results (Figure 2a,b) showed that the highest values of the maximum velocity of a light image that can manipulate the cancer cells, both occurred at 10 Vpp and 1 MHz for the two cancer cell types explored (i.e., 475.0 ± 25.0 μm s^{-1} and 458.3 ± 38.2 μm s^{-1} for the PC-3 and SK-BR-3 cancer cells, respectively). When the voltage frequency was lower than 1 MHz, some undesirable phenomena such as cell adhesion (10 Vpp and 500 kHz) (Figure 2c), cell aggregation (10 Vpp and 750 kHz) (Figure 2d), and cell lysis (10 Vpp and 100 kHz) (Figure 2e), would occur that could significantly affect the cellular properties [28]. Within the experimental conditions explored, moreover, the appearance rate of the above-mentioned phenomena was near 100% when the operating conditions were set at the particular electric conditions as indicated. These resulting phenomena might be due to the enhancement of electrically induced cell deformability, mutual DEP, and electrical lysis of cells at a voltage frequency near or below 1 MHz [26]. When the electric condition was set at 7 Vpp and 5 MHz (Figure 2a,b), conversely, we observed that the maximum velocities of a light image that can manipulate the cancer cells tested both reached their lowest levels for the two cancer cell types tested (i.e., 3.3 ± 5.8 µm s^{-1} and 10.0 ± 10.0 µm s^{-1} for the PC-3 and SK-BR-3 cancer cells, respectively) within the experimental conditions investigated. When the voltage magnitude and frequency were lower and higher than 7 Vpp and 5 MHz, respectively, the above-mentioned velocities were further decreased, which might significantly affect cell manipulation using the ODEP mechanism. Based on the facts described above, the optimal window of electric conditions for effective cell manipulation using ODEP was set at 7–10 Vpp and 1–5 MHz for the voltage magnitude and frequency, respectively. The ODEP operating conditions within such ranges were also reported for the ODEP-based cell manipulation for different applications [7,8,27,28]. Within this electric condition range, moreover, the maximum velocity of a light image that could manipulate the cancer cells tested decreased significantly with an increase of voltage frequency (Figure 2a,b), which was also consistent with previous findings [28]. Furthermore, under a given voltage frequency condition, the increase in voltage magnitude was found to increase the maximum velocity of a light image that could manipulate the cancer cells tested. This phenomenon could be explained by Equation (1), in which the ODEP force generated on a cell is proportional to the electric field squared. In addition to the values of the maximum velocity of a light image (and thus the ODEP force) investigated in the above-mentioned studies, it might also be valuable to explore the values of electric-field distribution in ODEP-based cell manipulation.

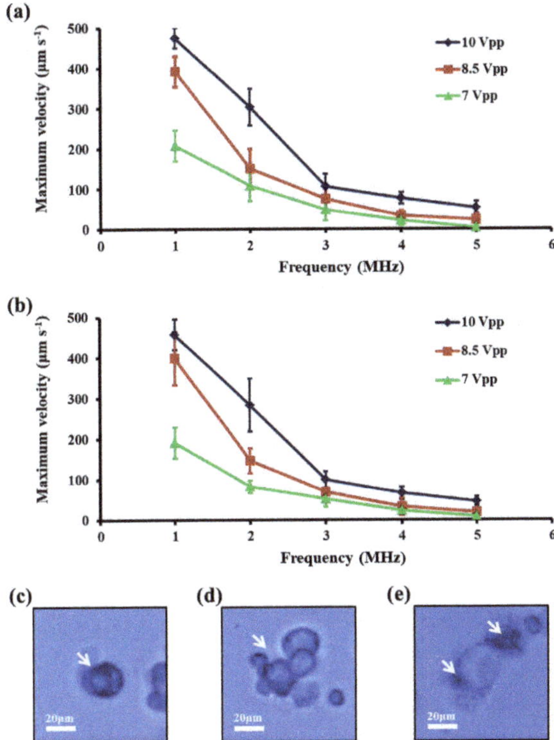

Figure 2. The quantitative relationship between the maximum velocity of a moving light bar that could manipulate the (**a**) PC-3 and (**b**) SK-BR-3 cancer cells tested and the frequency (1, 2, 3, 4, and 5 MHz) of the electric voltage applied under different voltage magnitudes (7, 8.5, and 10 Vpp) conditions and bright-field microscopic images showing the phenomena of (**c**) cell adhesion (10 Vpp and 500 kHz), (**d**) cell aggregation (10 Vpp and 750 kHz), and (**e**) cell lysis (10 Vpp and 100 kHz), during the exposure of cancer cells to the ODEP field with a voltage magnitude and frequency, as indicated.

3.2. Effect of the ODEP Field on the Cellular Viability of Cancer Cells

In this study, PC-3 and SK-BR-3 cancer cells, two of the commonly used cancer cell lines in the cancer-related studies, were used as model cells to investigate the effect of the ODEP-based cell manipulation on cellular properties. In the operations, the prepared cancer cell suspension was loaded into the microchamber of the ODEP microfluidic chip (Figure 1a), followed by exposure to the ODEP field with varied electric conditions (i.e., 7–10 Vpp and 1–5 MHz for the voltage magnitude and frequency, respectively), as determined previously (Figure 2). In our preliminary tests, the impact of ODEP exposure time on the cell viability of PC-3 cancer cells was investigated. Results (Supplementary Figure S2) showed that the cell viability of such cells had no significant difference ($p > 0.05$) compared to the control (i.e., the cancer cells without exposure to the ODEP field) within the 5 min exposure time, when the voltage magnitude and frequency were set at 10 Vpp and 5 MHz, respectively. When the ODEP exposure time was as long as 10 and 15 min, conversely, the cell viability of the PC-3 cancer cells might be significantly affected ($p < 0.05$) (e.g., 43.7 ± 5.5% and 51.3 ± 10.3% reduction for the ODEP exposure time of 10 and 15 min, respectively) in comparison to the control. Based on the preliminary results, therefore, the ODEP exposure time was set at 3 min, and within the time period, general cell manipulation using the ODEP mechanism was normally carried out (e.g., the rapid separation of Raji cells [7] or isolation and purification of CTCs [5,9]). In practice, 40 rectangular light bar images were illuminated on the microchamber zone. During the period of ODEP exposure, the cancer cells tested were attracted within the light bars, as shown in Figure 3(aII,aV). After the exposure to the ODEP field, we observed microscopically that the treated cancer cells within the microchamber were aligned in accordance with the original light bar images, as exhibited in Figure 3 (aIII,aVI). Then, the ODEP-treated cancer cells were collected for the following bioassays.

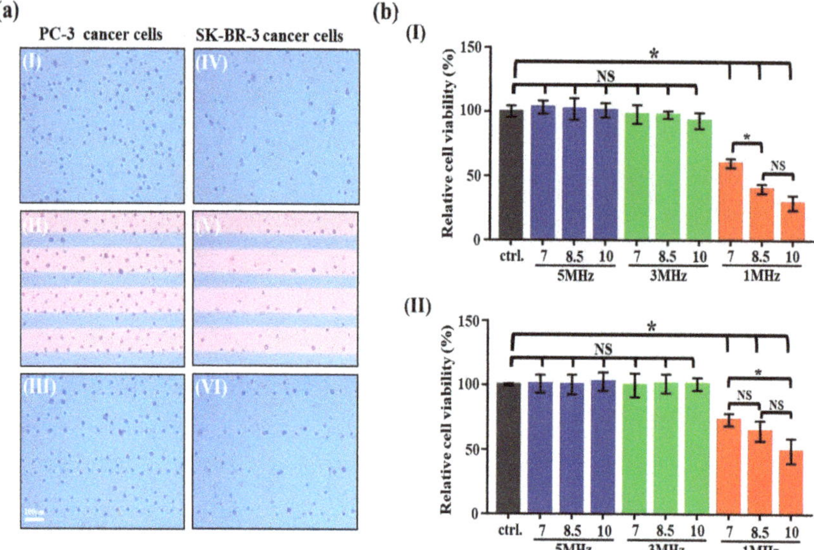

Figure 3. (a) Microscopic observation (close-up view) of the cancer cells in the microchamber of the microfluidic chip before (the upper row), during (the middle row), and after (the lower row) treatment with ODEP exposure (the left column—PC-3 cancer cells; the right column—SK-BR-3 cancer cells), (b) the comparison of relative cell viability (%) of (I) the PC-3 cancer cells and (II) the SK-BR-3 cancer cells treated with the ODEP field with varied electric conditions (voltage magnitude and frequency: 7, 8.5, and 10 Vpp and 1, 3, and 5 MHz, respectively) (NS—No significant difference ($p > 0.05$), * significant difference ($p < 0.05$)).

In this study, the Cell Counting Kit-8 (CCK-8) was utilized to investigate the viability of cancer cells treated with various ODEP fields. The results (Figure 3(bI)) revealed that the cell viability of PC-3 cancer cells had no significant difference ($p > 0.05$) compared to that of the control cells (i.e., the cancer cells without exposure to the ODEP field), when the voltage magnitude and frequency were set at 7–10 Vpp and 3–5 MHz, respectively. A similar result was also found for the SK-BR-3 cancer cells tested (Figure 3(bII)). However, once the voltage frequency was set at 1 MHz, the cell viability of the PC-3 and SK-BR-3 cancer cells decreased significantly ($p < 0.05$) compared to that of the control, irrespective of a voltage magnitude of 7, 8.5, or 10 Vpp. This finding could be explained as follows. If the equivalent circuit of a biological cell is assumed to be a lumped circuit, the electrical property of the cell membrane will be similar to the electric capacitance [28]. When the voltage frequency is altered from high to low frequency (e.g., from 10 MHz to 100 kHz), the electrical impedance of a cell membrane might increase. This phenomenon could, therefore, result in the shift of the exerted electric voltage from the cell cytoplasm to the cell membrane of a cell [28]. This fact might in turn lead to an increase in the transmembrane potential in a cell membrane when the voltage frequency is changed from high to low frequency. The above-mentioned phenomena could explain the low cell viability caused by cell electrical lysis (Figure 2e) when the voltage frequency is set at low conditions (e.g., 1 MHz) (Figure 3b).

Under the low voltage frequency condition of 1 MHz (Figure 3(bI)), moreover, the cell viability of the PC-3 cancer cells significantly decreased ($p < 0.05$) when the magnitude of voltage was higher than 8.5 Vpp (observed cell viability: 40.4 ± 3.6%). A similar result was also found for the SK-BR-3 cancer cells tested (Figure 3(bII)), in which the cell viability of the SK-BR-3 cancer cells was significantly downregulated ($p < 0.05$) (the observed cell viability: 49.4 ± 9.5%) when the magnitude of the voltage was higher than 10 Vpp. Overall, the lower cell viability of cancer cells found at the higher voltage condition could be explained by the fact that the electrical lysis of cells is prone to occur when a cell is exposed to a high-voltage magnitude [37,38]. Taken together, the increase of voltage magnitude or decrease of voltage frequency, respectively, could raise the maximum velocity of a light image that can manipulate cells (i.e., the ODEP manipulation force of cells; Equation (5), Figure 2). Nevertheless, the increase in voltage magnitude or decrease in voltage frequency, respectively, could accordingly increase the tendency of electrical lysis of cells, which could affect cellular viability. Within the experimental conditions explored (i.e., 7–10 Vpp and 1–5 MHz), the determinant electrical condition that could significantly affect cell viability was a low-voltage frequency of 1 MHz, when the voltage magnitude range was 7–10 Vpp.

3.3. Effect of the ODEP Field on the Cellular Metabolic Activity and Gene Expression of the Cancer Cells Tested

As suggested by the previous results (Figure 3b), cell manipulation using the ODEP mechanism at a low voltage frequency of 1 MHz could be lethal to the cancer cells tested. Such an ODEP electrical condition was ruled out in subsequent studies. In the following evaluations, the effect of the ODEP fields with varied electrical conditions (i.e., voltage magnitude and frequency: 7–10 Vpp and 3–5 MHz, respectively) on the metabolic activity and gene expression of cancer cells were explored. For the former, the ATP synthesis of cells was used as an indicator based on the previous studies [39]. The results (Figure 4a) showed that the ATP level of the ODEP-treated PC-3 cancer cells showed no significant difference ($p > 0.05$) compared to that of the control (i.e., the cancer cells without exposure to the ODEP field), within the experimental conditions explored. A similar result was also observed for the SK-BR-3 cancer cells (Figure 4b). The findings in this study could indicate that the ODEP field could not affect the ATP synthesis of a cell within the electric conditions tested. This result was, to some extent, similar to that found in a previous DEP-based study [40].

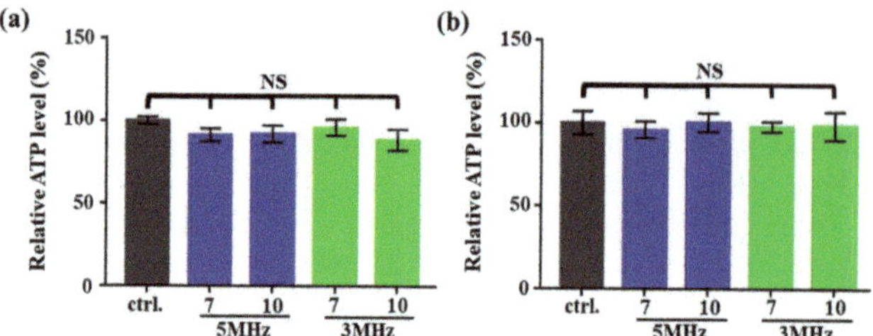

Figure 4. Comparison of the relative ATP levels (%) of (**a**) the PC-3 cancer cells and (**b**) the SK-BR-3 cancer cells treated with the optically induced dielectrophoresis (ODEP) field under varied electric conditions (voltage magnitude and frequency—7 and 10 Vpp and 3 and 5 MHz, respectively) [NS—No significant difference ($p > 0.05$)].

In addition to the effect on cellular metabolic activity, the ODEP effect on the gene expression of cancer cells was investigated. In this study, the gene expression of EMT-associated genes (e.g., *EpCAM* and *VIM*) [8] and the MRP1 gene (e.g., *ABCC1*) [8] in the PC-3 and SK-BR-3 cancer cells was quantified and then normalized to the gene expression of *GAPDH*. The results (Figure 5a) showed that the gene expression of the above-mentioned genes in the PC-3 cancer cells was not significantly different ($p > 0.05$) from that of the control cells (i.e., the cancer cells without exposure to the ODEP field) within the ODEP operating conditions (i.e., 7–10 Vpp, 3–5 MHz, and 3 min exposure time) studied. A similar outcome (Figure 5b) was also found for the SK-BR-3 cancer cells.

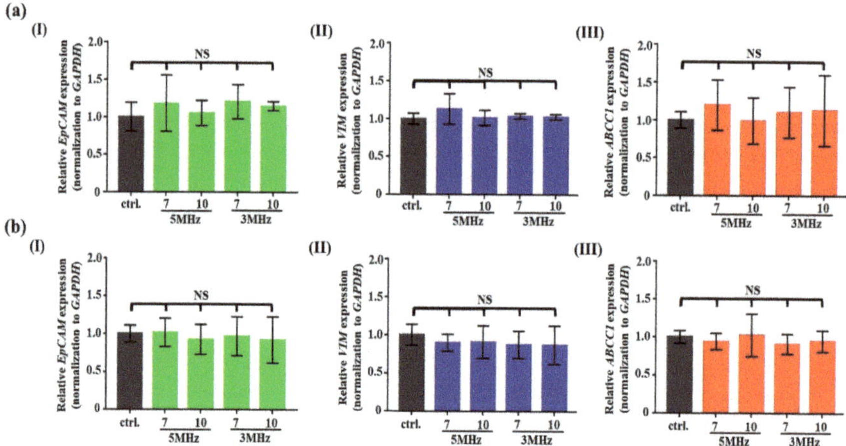

Figure 5. The comparison of relative gene expression ((I) *EpCAM*, (II) *VIM*, and (III) *ABCC1* genes) of (**a**) the PC-3 cancer cells and (**b**) the SK-BR-3 cancer cells treated with the ODEP field with varied electrical conditions (voltage magnitude and frequency: 7 and 10 Vpp and 3 and 5 MHz, respectively) (NS—No significant difference ($p > 0.05$).

Previous studies similar to this work revealed that the exposure to a DEP field (e.g., 20 Vpp, 250 kHz, and 1 h exposure time) could downregulate the expression of cell differentiation-related genes (e.g., *ADGRE5/CD97*, *RUNX2*, and *NES*) in mesenchymal stem cells (e.g., UE7T-13 cells) [41]. It was also reported that the exposure to a DEP field (e.g., 21 Vpp, 5 MHz, and 15 min exposure time) might not affect the cell morphology, cell oxidative respiration rate, and cell cycle dynamics of fibroblast-like cells (e.g., BHK-21/C13 cells) [40]. After exposure to the DEP field, however, the gene expression of the c-fos protein (commonly activated by environmental stress [42]) was observed to be upregulated [40]. The reasons behind the discrepancies found in the previous and current studies might be complicated and could be related to the cell species, electrical conditions, and the genes explored. Further experiments will be required to systematically investigate this issue. In addition to the gene expressions investigated in this study, it might also be valuable to explore the expression profiles of the genes regulating the signal pathways downstream of voltage-gated ion channels.

4. Conclusions

The ODEP-based microsystem provides a powerful tool for biological cell-relevant studies or applications. Before its widespread application, one fundamental issue has to be addressed—Whether cell manipulation using the ODEP mechanism influences the native properties of cells. To address this issue, we carried out various experiments. In this study, the impact of the ODEP electrical conditions ((1) the magnitude of AC voltage: 5–10 V; (2) frequency of AC voltage: 1–5 MHz), (3) operating time: 3 min) on the properties (e.g., cell viability, metabolic activity of cells, or gene expression of cells) of biological cells (e.g., PC-3 and SK-BR-3 cancer cells) was explored. Within the experimental conditions studied, the ODEP-based cell manipulation with the highest performance (e.g., highest cell manipulation velocity) occurred at 10 Vpp and 1 MHz for the two cancer cell types explored. Under this ODEP electrical condition, however, the cell viability of the cancer cells tested was significantly affected (e.g., 70.5 ± 10.0% and 50.6 ± 9.2% reduction for the PC-3 and SK-BR-3 cancer cells, respectively) possibly due to cell electrical lysis. Conversely, the exposure of cancer cells to the ODEP field with a particular condition range (i.e., 7–10 Vpp, 3–5 MHz, and 3 min exposure time) might not significantly alter the cell viability, cell metabolic activity, and the gene expression of cancer cells. Within this ODEP operating condition, the highest maximum velocity (i.e., 106.7 ± 30.6 μm s^{-1} and 100.0 ± 20.0 μm s^{-1}, for the PC-3 and SK-BR-3 cancer cells, respectively) of a light image that could manipulate cells occurred at the voltage magnitude and frequency of 10 Vpp and 3 MHz, respectively. The above-mentioned cell manipulation velocity ranges were technically sufficient for various applications. Overall, this study fundamentally investigated the effect of ODEP electrical conditions on the cellular properties of cancer cells. The information obtained is crucially important for the utilization of ODEP-based cell manipulation in a microscale system for various applications.

Supplementary Materials: The following are available online at http://www.mdpi.com/2079-6374/10/6/65/s1, Figure S1: Photograph of the overall experimental setup, Figure S2: Comparison of relative cell viability (%) of the PC-3 cancer cells treated with varied ODEP exposed times as indicated (electrical condition: 10 Vpp and 5 MHz; ODEP exposed time: 1~15 minutes) [NS: No significant difference ($p > 0.05$), * significant difference ($p < 0.05$)].

Author Contributions: Conceptualization, P.-Y.C., C.-H.H., and M.-H.W.; methodology, P.-Y.C., C.-H.H., and C.-R.L.; validation, P.-Y.C., C.-H.H., and C.-R.L.; writing—original draft preparation, P.-Y.C.; writing—review and editing, P.-Y.C., C.-H.H., and M.-H.W. All authors have read and agreed to the published version of the manuscript.

Funding: This work was funded by the Ministry of Science and Technology, R.O.C. (MOST 107–2221-E-182–033-MY3) and Chang Gung Memorial Hospital (CMRPD2G0061–62, CMRPD2H0121–23, and CMRPD2J0031–32).

Conflicts of Interest: The authors declare no conflict of interest.

References

1. Augustsson, P.; Magnusson, C.; Nordin, M.; Lilja, H.; Laurell, T. Microfluidic, label-free enrichment of prostate cancer cells in blood based on acoustophoresis. *Anal. Chem.* **2012**, *84*, 7954–7962. [CrossRef]
2. Cho, H.; Kim, J.; Jeon, C.W.; Han, K.H. A disposable microfluidic device with a reusable magnetophoretic functional substrate for isolation of circulating tumor cells. *Lab Chip* **2017**, *17*, 4113–4123. [CrossRef]
3. Gupta, V.; Jafferji, I.; Garza, M.; Melnikova, V.O.; Hasegawa, D.K.; Pethig, R.; Davis, D.W. ApoStream™, a new dielectrophoretic device for antibody independent isolation and recovery of viable cancer cells from blood. *Biomicrofluidics* **2012**, *6*, 024133. [CrossRef] [PubMed]
4. Waheed, W.; Alazzam, A.; Mathew, B.; Christoforou, N.; Abu-Nada, E. Lateral fluid flow fractionation using dielectrophoresis (LFFF-DEP) for size-independent, label-free isolation of circulating tumor cells. *J. Chromatogr. B Anal. Technol. Biomed. Life Sci.* **2018**, *1087*, 133–137. [CrossRef] [PubMed]

5. Chou, W.P.; Wang, H.M.; Chang, J.H.; Chiu, T.K.; Hsieh, C.H.; Liao, C.J.; Wu, M.H. The utilization of optically-induced-dielectrophoresis (ODEP)-based virtual cell filters in a microfluidic system for continuous isolation and purification of circulating tumour cells (CTCs) based on their size characteristics. *Sens. Actuators B* **2017**, *241*, 245–254. [CrossRef]

6. Chiu, T.K.; Chao, A.C.; Chou, W.P.; Liao, C.J.; Wang, H.M.; Chang, J.H.; Chen, P.H.; Wu, M.H. Optically-induced-dielectrophoresis (ODEP)-based cell manipulation in a microfluidic system for high-purity isolation of integral circulating tumor cell (CTC) clusters based on their size characteristics. *Sens. Actuators B* **2018**, *258*, 1161–1173. [CrossRef]

7. Liang, W.F.; Zhao, Y.L.; Liu, L.Q.; Wang, Y.C.; Dong, Z.L.; Li, W.J.; Lee, G.B.; Xiao, X.B.; Zhang, W.J. Rapid and label-free separation of burkitt's lymphoma cells from red blood cells by optically-induced electrokinetics. *PLoS ONE* **2014**, *9*, e90827. [CrossRef] [PubMed]

8. Liao, C.J.; Hsieh, C.H.; Chiu, T.K.; Zhu, Y.X.; Wang, H.M.; Hung, F.C.; Chou, W.P.; Wu, M.H. An optically induced dielectrophoresis (ODEP)-based microfluidic system for the isolation of high-purity CD45(neg)/EpCAM(neg) cells from the blood samples of cancer patients-demonstration and initial exploration of the clinical significance of these cells. *Micromachines* **2018**, *9*, 563. [CrossRef]

9. Chiu, T.K.; Chou, W.P.; Huang, S.B.; Wang, H.M.; Lin, Y.C.; Hsieh, C.H.; Wu, M.H. Application of optically-induced-dielectrophoresis in microfluidic system for purification of circulating tumour cells for gene expression analysis-Cancer cell line model. *Sci. Rep.* **2016**, *6*, 32851. [CrossRef]

10. Huan, Z.J.; Chu, H.K.; Liu, H.B.; Yang, J.; Sun, D. Engineered bone scaffolds with Dielectrophoresis-based patterning using 3D printing. *Biomed. Microdevices* **2017**, *19*, 102. [CrossRef]

11. Huang, K.C.; Lu, B.; Lai, J.W.; Chu, H.K.H. Microchip System for Patterning Cells on Different Substrates via Negative Dielectrophoresis. *IEEE Trans. Biomed. Circuits Syst.* **2019**, *13*, 1063–1074. [CrossRef] [PubMed]

12. Kwon, J.S.; Ravindranath, S.P.; Kumar, A.; Irudayaraj, J.; Wereley, S.T. Opto-electrokinetic manipulation for high-performance on-chip bioassays. *Lab Chip* **2012**, *12*, 4955–4959. [CrossRef]

13. Evander, M.; Johansson, L.; Lilliehorn, T.; Piskur, J.; Lindvall, M.; Johansson, S.; Almqvist, M.; Laurell, T.; Nilsson, J. Noninvasive acoustic cell trapping in a microfluidic perfusion system for online bioassays. *Anal. Chem.* **2007**, *79*, 2984–2991. [CrossRef]

14. Lin, Y.H.; Yang, Y.W.; Chen, Y.D.; Wang, S.S.; Chang, Y.H.; Wu, M.H. The application of an optically switched dielectrophoretic (ODEP) force for the manipulation and assembly of cell-encapsulating alginate microbeads in a microfluidic perfusion cell culture system for bottom-up tissue engineering. *Lab Chip* **2012**, *12*, 1164–1173. [CrossRef] [PubMed]

15. Magnez, R.; Thiroux, B.; Taront, S.; Segaoula, Z.; Quesnel, B.; Thuru, X. PD-1/PD-L1 binding studies using microscale thermophoresis. *Sci. Rep.* **2017**, *7*, 1–8. [CrossRef] [PubMed]

16. Gossett, D.R.; Weaver, W.M.; Mach, A.J.; Hur, S.C.; Tse, H.T.K.; Lee, W.; Amini, H.; Di Carlo, D. Label-free cell separation and sorting in microfluidic systems. *Anal. Bioanal. Chem.* **2010**, *397*, 3249–3267. [CrossRef] [PubMed]

17. Bhagat, A.A.S.; Bow, H.; Hou, H.W.; Tan, S.J.; Han, J.; Lim, C.T. Microfluidics for cell separation. *Med. Biol. Eng. Comput.* **2010**, *48*, 999–1014. [CrossRef] [PubMed]

18. Cemazar, J.; Douglas, T.A.; Schmelz, E.M.; Davalos, R.V. Enhanced contactless dielectrophoresis enrichment and isolation platform via cell-scale microstructures. *Biomicrofluidics* **2016**, *10*, 014109. [CrossRef]

19. Kale, A.; Patel, S.; Xuan, X.C. Three-Dimensional Reservoir-Based Dielectrophoresis (rDEP) for Enhanced Particle Enrichment. *Micromachines* **2018**, *9*, 123. [CrossRef]

20. Zhang, Z.Z.; Zheng, T.Y.; Zhu, R. Characterization of single-cell biophysical properties and cell type classification using dielectrophoresis model reduction method. *Sens. Actuators B* **2020**, *304*, 127326. [CrossRef]

21. Pohl, H.A. The motion and precipitation of suspensoids in divergent electric fields. *J. Appl. Phys.* **1950**, *22*, 869–871. [CrossRef]

22. Abd Rahman, N.; Ibrahim, F.; Yafouz, B. Dielectrophoresis for biomedical sciences applications: A review. *Sensors (Basel)* **2017**, *17*, 449. [CrossRef] [PubMed]

23. Chiou, P.Y.; Ohta, A.T.; Wu, M.C. Massively parallel manipulation of single cells and microparticles using optical images. *Nature* **2005**, *436*, 370–372. [CrossRef] [PubMed]

24. Hwang, H.; Park, J.K. Optoelectrofluidic platforms for chemistry and biology. *Lab Chip* **2011**, *11*, 33–47. [CrossRef] [PubMed]

25. Zhang, S.L.; Juvert, J.; Cooper, J.M.; Neale, S.L. Manipulating and assembling metallic beads with Optoelectronic Tweezers. *Sci. Rep.* **2016**, *6*, 32840. [CrossRef]
26. Ke, L.Y.; Kuo, Z.K.; Chen, Y.S.; Yen, T.Y.; Dong, M.X.; Tseng, H.W.; Liu, C.H. Cancer immunotherapy mu-environment LabChip: Taking advantage of optoelectronic tweezers. *Lab Chip* **2018**, *18*, 106–114. [CrossRef]
27. Chu, P.Y.; Liao, C.J.; Hsieh, C.H.; Wang, H.M.; Chou, W.P.; Chen, P.H.; Wu, M.H. Utilization of optically induced dielectrophoresis in a microfluidic system for sorting and isolation of cells with varied degree of viability: Demonstration of the sorting and isolation of drug-treated cancer cells with various degrees of anti-cancer drug resistance gene expression. *Sens. Actuators B* **2019**, *283*, 621–631.
28. Chu, P.Y.; Liao, C.J.; Wang, H.M.; Wu, M.H. The influence of electric parameters on the manipulation of biological cells in a microfluidic system using optically induced dielectrophoresis. *Int. J. Electrochem. Sci.* **2019**, *14*, 905–918. [CrossRef]
29. Valley, J.K.; Jamshidi, A.; Ohta, A.T.; Hsu, H.Y.; Wu, M.C. Operational regimes and physics present in optoelectronic tweezers. *J. Microelectromech. Syst.* **2008**, *17*, 342–350. [CrossRef]
30. Pethig, R. Review Article-Dielectrophoresis: Status of the theory, technology, and applications. *Biomicrofluidics* **2010**, *4*, 022811. [CrossRef]
31. Ohta, A.T.; Chiou, P.Y.; Phan, H.L.; Sherwood, S.W.; Yang, J.M.; Lau, A.N.K.; Hsu, H.Y.; Jamshidi, A.; Wu, M.C. Optically controlled cell discrimination and trapping using optoelectronic tweezers. *IEEE J. Sel. Top. Quant.* **2007**, *13*, 235–243. [CrossRef]
32. Zhang, K.X.; Waxman, D.J. PC3 prostate tumor-initiating cells with molecular profile FAM65B(high)/MFI2(low)/LEF1(low) increase tumor angiogenesis. *Mol. Cancer* **2010**, *9*, 319. [CrossRef]
33. Cynthia, M.; Rodriguez, P.V. Natzidielly lerma, giulio francia, preclinical models to study breast cancer. *Clin. Cancer Drugs* **2014**, *1*, 90–99.
34. Cyrus, G.; Malmberg, A.A.M. Dielectric constants of aqueous solutions of dextrose and sucrose. *J. Res. Natl. Bur. Stand.* **1950**, *45*, 299–303.
35. Hidayanto, E.; Tanabe, T.; Kawai, J. Measurement of viscosity and sucrose concentration in aqueous solution using portable Brix meter. *Berk. Fis.* **2010**, *13*, A23–A28.
36. Chomczynski, P.; Mackey, K. Substitution of Chloroform by Bromochloropropane in the Single-Step Method of RNA Isolation. *Anal. Biochem.* **1995**, *225*, 163–164. [CrossRef] [PubMed]
37. Li, H.; Denzi, A.; Ma, X.; Du, X.T.; Ning, Y.Q.; Cheng, X.H.; Apollonio, F.; Liberti, M.; Hwang, J.C.M. Distributed Effect in High-Frequency Electroporation of Biological Cells. *IEEE Trans. Microw. Theory* **2017**, *65*, 3503–3511. [CrossRef]
38. Islam, M.S.; Shahid, A.; Kuryllo, K.; Li, Y.F.; Deen, M.J.; Selvaganapathy, P.R. Electrophoretic concentration and electrical lysis of bacteria in a microfluidic device using a nanoporous membrane. *Micromachines* **2017**, *8*, 45. [CrossRef]
39. Herrick, J.R.; Brad, A.M.; Krisher, R.L.; Pope, W.F. Intracellular adenosine triphosphate and glutathione concentrations in oocytes from first estrous, multi-estrous, and testosterone-treated gilts. *Anim. Reprod. Sci.* **2003**, *78*, 123–131. [CrossRef]
40. Archer, S.; Li, T.T.; Evans, A.T.; Britland, S.T.; Morgan, H. Cell reactions to dielectrophoretic manipulation. *Biochem. Biophys. Res. Commun.* **1999**, *257*, 687–698. [CrossRef]
41. Yoshioka, J.; Yoshitomi, T.; Yasukawa, T.; Yoshimoto, K. Alternation of gene expression levels in mesenchymal stem cells by applying positive dielectrophoresis. *Anal. Sci.* **2016**, *32*, 1213–1216. [CrossRef] [PubMed]
42. Kovacs, K.J. c-Fos as a transcription factor: A stressful (re)view from a functional map. *Neurochem. Int.* **1998**, *33*, 287–297. [CrossRef]

Review

Role of Paper-Based Sensors in Fight Against Cancer for the Developing World

Amey Dukle [1], Arputharaj Joseph Nathanael [1,*], Balaji Panchapakesan [2] and Tae-Hwan Oh [3]

1. Centre for Biomaterials, Cellular and Molecular Theranostics (CBCMT), Vellore Institute of Technology (VIT), Vellore 632014, Tamil Nadu, India
2. Small Systems Laboratory, Department of Mechanical Engineering, Worcester Polytechnic Institute, Worcester, MA 01609, USA
3. School of Chemical Engineering, Yeungnam University, Gyeongsan 38541, Korea
* Correspondence: joseph.nathanael@vit.ac.in

Abstract: Cancer is one of the major killers across the globe. According to the WHO, more than 10 million people succumbed to cancer in the year 2020 alone. The early detection of cancer is key to reducing the mortality rate. In low- and medium-income countries, the screening facilities are limited due to a scarcity of resources and equipment. Paper-based microfluidics provide a platform for a low-cost, biodegradable micro-total analysis system (µTAS) that can be used for the detection of critical biomarkers for cancer screening. This work aims to review and provide a perspective on various available paper-based methods for cancer screening. The work includes an overview of paper-based sensors, the analytes that can be detected and the detection, and readout methods used.

Keywords: paper-based sensors; cancer screening; disposable sensors; sensors; paper fluidics; microfluidics

Citation: Dukle, A.; Nathanael, A.J.; Panchapakesan, B.; Oh, T.-H. Role of Paper-Based Sensors in Fight Against Cancer for the Developing World. *Biosensors* **2022**, *12*, 737. https://doi.org/10.3390/bios12090737

Received: 22 July 2022
Accepted: 31 August 2022
Published: 7 September 2022

Publisher's Note: MDPI stays neutral with regard to jurisdictional claims in published maps and institutional affiliations.

1. Introduction

Cancer is a major cause of death worldwide [1,2]. It is estimated to be the cause of every 1 in 6 deaths [3,4]. According to the World Health Organisation (WHO), in the year 2020 more than 10 million people lost their life to cancer [5]. Worldwide, an estimated 19.3 million new cancer cases and almost 10.0 million cancer deaths occurred in 2020 [6]. Although the causes of cancer may vary depending on the type, it has been observed that the incidence rate of disease is on the rise [7,8].Worldwide, by 2040, 28.7 million new cases of cancer are projected to occur, a 47% rise compared to 19.3 million in 2020. An increase in the global cancer burden in the next fifty years will come from low- and middle-income countries (400% in low-income countries, 168% in middle-income countries, and 53% in high-income countries) [9,10]. Though there are different treatment strategies that have been developed and the disease is no longer 'incurable' [11–14], the success rate of the treatment depends on the stage of disease progression [15–17]. An individual undergoing treatment in the early stages of cancer has a many times higher chance of survival than in the later stages [18,19]. Estimates suggest that approximately 30–50% of cancer deaths can be prevented by early detection and treatment [20–22].

Although the incidence rate of cancer is higher in wealthy nations compared to low- and middle-income countries, low- and middle-income countries have a lower survival rate, partly due to the late presentation of the disease [23]. The barriers of cancer care in developing countries are due to late-stage presentation, quality of care, affordability, and a lack of access to advanced clinical resources. Late-stage presentation puts tremendous burden on clinicians [10]. The premature death and loss of productive life in the working population results in a significant economic impact on these countries.

In low- and middle-income countries, access to health care facilities is not readily available [24–28]. The number of physicians in low- to middle-income countries can be as low as 0.1 to 2 per 1000 people. In many cases, they are heavily burdened, leading to long waiting times [29]. In many low- and middle-income countries, cancer screening is not covered under insurance, thus discouraging patients from undergoing cancer screening.

To address the issue of cancer screening in these countries, various strategies have been developed [30]. These screening devices need to be easy to manufacture, low-cost, portable, and should not require any special training. Paper-based sensors are gaining significance in this field as they possess all the features of an ideal screening device [31–50].

Paper was invented by the Egyptians in the fourth century BC. It is one material that has existed continuously since the beginning of Egyptian civilization. Paper-based products are the most sold products in the world. For example, the Bible is the most sold book in the world and is printed on paper. The question then becomes: can we print cancer screening devices on paper to enable their use as low-cost sensors for cancer screening? Paper-based sensors for cancer screening are akin to an at-home pregnancy test kit. These kits will indicate whether a specific cancer biomarker is present in the person's body. In many cases, the result will be qualitative, i.e., yes/no type. However, novel sensors have been developed that provide a quantitative output [51].

In case the test returns a positive result, the patient can consult medical professionals so that exhaustive testing can be performed and the stage of the disease identified.

The most common sensing principle used in paper-based sensors for cancer screening is ELISA, wherein the analyte is labelled and detected using a sandwich assay [52]. The results are then readout using a plate reader to specify analyte concentration against a standard curve [53–57]. Various other methods are also being explored for the development of sensors.

This work reviews the various paper-based sensors that have been developed for cancer screening. In the next sections, a brief introduction about paper-based sensors, the analytes used for cancer screening, and the detection and read-out techniques used are discussed.

2. Paper-Based Sensors: Low-Cost Screening Devices for the Developing World

Advancements in microfluidics have led to their widespread usage in sensing applications. Microfluidics, as the name suggests, uses small fluidic channels where liquids, proteins and cells can be manipulated through flow control [58]. The channels are designed in such a way that various operations such as separation of phases, biological cells, the size-based separation of particles, to name a few, can be performed. However, fabrication of microfluidic devices may not be possible in low-income countries due to the lack of availability of materials. Even poly dimethyl siloxane (PDMS)-based soft lithography techniques require a master that can be expensive to make in low-income countries and set-ups such as mask makers and lithography may not be readily available. Paper is a low-cost alternative for sensors in the fight against cancer that may not need lithographic processing. Droplet-based paper devices could be highly useful for biomarker detection. Even microfluidic devices using screen-printing techniques can be printed on paper.

Paper-based sensors are devices that are fabricated on a paper substrate. These devices are printed on a cellulose-based paper substrate using readily available printers, making them easily accessible.

A typical paper-based sensor is based on exploiting the capability of paper to wick liquid, leading to capillary action in the paper. A typical paper-based sensor is developed on an hydrophilic paper substrate with wicking capabilities. Using surface treatment methods, a hydrophobic barrier is created in order to guide the flow of liquid through the specific path or channels.

The most commonly used paper in the fabrication of paper-based sensors is filter paper. Due to its pores, it has sufficient wicking capabilities, providing a moderate flow rate. Whatmann-branded filter paper manufactured by General Electric Health Care is the most widely used filter paper due to its uniform pore size and distribution. In applications where filter paper is not suitable, nitro-cellulose paper is used as a substrate. The main advantage of nitrocellulose substrate is its easy and efficient binding of proteins [59].

For example, nitrocellulose film is used for protein immobilization and filter paper is used for its water absorption. In one particular study for detection of bladder cancer, a glass-cellulose film was used for sampling, a nitro-cellulose film was used for protein immobilization, and filter paper was used for sample transfer due to its adsorption capabilities [60].

Table 1 highlights the different types of paper substrates that have been developed for paper-based sensing applications.

Table 1. Paper substrates of interest for paper-based microfluidics.

Paper Type	Properties	Sensing Methods	Applications/Notes/ References
Whatman Filter Paper Grade 1	Size: 26 × 31 mm to 600 × 600 mm sheets or 10 mm to 150 cm circles. Porosity: 11 µm Nominal thickness: 180 µm Medium retention and flow rate	Colorimetric, Surface Plasmon Resonance SERS, Electrochemical, Chemiluminescence, Phosphorescence, Photometric, Chromogenic sensing, Fluorescence, Dye based sensing, Spectrometry	Analytical separation [61] Electrophoretic separation [62] Soil analysis [63–65] Food testing [66] Point of care testing [67] Protein [68] Atmospheric dust [69] Gas detection [70] HIV detection [71] Explosive Sensing [72] Automated DNA extraction and amplification [73]
Whatman Filter Paper Grade 2	Size: 460 × 570 mm to 580 × 680 mm or 42.5 mm to 500 cm circles. Porosity: 8 µm Nominal thickness: 190 µm More retention than Grade 1 and slower flow rate	Same sensing methods are applicable as in Grade 1	Same applications as Grade 1 except slower flow rate and higher retention due to smaller pore size.
Whatman Filter Paper Grade 3	Size: 26 × 31 mm to 600 × 600 mm or 23 mm to 320 mm circles. Porosity: 6 µm Nominal thickness: 390 µm More retention than Grade 1, 2 and slower flow rate	Poor colorimetric sensing due to slower flow rates	Same applications as Grade 1 except slower flow rate. Poor for colorimetric sensing due to lower color contrast
Whatman Filter Paper Grade 4, 5, 6	Main difference is porosity; Grade 4: 25 µm Grade 5: 2.5 µm and Grade 6: 3 µm.	Poor colorimetric sensing of Grade 5 and 6 is expected due to slower flow rates.	Same applications as Grade 1 except slower flow rate of Grade 5 and 6. Grade 4 suitable for large particles monitoring in air. Soil Suction Testing [74]
Whatman® Grade 903	W × L = 450 mm × 450 mm, 140 µm thickness, porosity: 4–7 µm	Compatible with most sensing methods. Super refined cellulose	Whole-blood collection [75], HIV load, and drug-resistance testing [76]. Element detection in neonatal blood spots (NBSs) using sector-field inductively coupled plasma-mass spectrometry [77].
Whatman® FTA filter paper cards	N/A	Highly sensitive for rapid nucleic acid extractions and storage.	Nucleic acid extraction from cells [78]; fine needle aspirates for cancer testing [79]; tissue analysis [80]; and virus and bacterial RNA detection and preservation [81].
Nitrocellolose membrane	pore size : 0.2 µm	Same sensing methods are applicable as in Grade 1	Western Blotting [82] Fabrication of Lateral Flow Assay [83]
Nanocellulose paper	Nanofibrillated cellulose (NFC) coated with layer of reactive nanoporous silicone nanofilament	Mainly restricted to applications requiring hydrophobic substrate	Paper-based electronics [84]
Microcrystaline Cellulose/ Polyvinyl Alcohol Paper	Porosity: 90%, pore size (between 23 and 46 µm), thickness (from 315 to 436 µm), and high light transmission under water (>95%)	Similar to nanocellulose paper	low-cost cell culture platform [85]
Omniphobic RF paper	"fluoroalkylated paper" ("RF paper") by vapor-phase silanization of paper with fluoroalkyl trichlorosilanes	Resist wetting by liquids with a wide range of surface tensions correlates with the length and degree of fluorination of the organosilane and with the roughness of the paper	Same as nanocellulose paper [48]
Photo paper	Commercially sold by Epson, Canon etc.	Same applications as Grade 1	Pumpless paper-based analytical devices [86]

For patterning of the microfluidic channels on paper substrate, various printing techniques such as wax-printing, ink-jet printing, screen printing, lithography, plasma processing, and manual pattern drawing have been explored. In a typical printing-based fabrication process, a CAD model of the microfluidic channels is printed using wax/ink-jet–printer on the substrate. Since the model is printed on only one side of the substrate, the substrate is heated to cause the reflow of the hydrophobic ink/wax barrier over the complete cross-section of the substrate. These barriers act as microfluidic channels guiding the flow of the fluid (Figure 1).

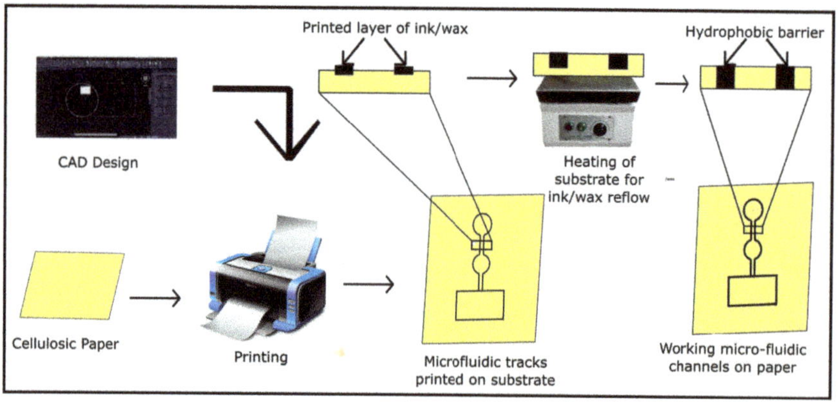

Figure 1. Workflow of the fabrication process for printing microfluidic channels on a paper substrate.

Although paper-based microfluidic channels can be used for most of applications as conventional microfluidic chips, there may be concerns regarding the utility of paper substrate for use in microfluidic applications requiring multiple layers of fluidic channels for phase separation applications. For such applications Japanese paper, folding techniques such as origami and Kirigami are being explored [87].

The low cost of substrate, the ability to print continuously, the ability to make arrays of devices, and minimal capital requirements for the fabrication setup make paper-based sensors an interesting candidate for use in mass screening process. Paper being a naturally derived substance is biodegradable in nature. This will also reduce the environmental impact due mass production and usage of screening kits.

3. Design and Working of a Typical Paper-Based Sensor

The paper-based sensor used for screening must be fast, accurate, reliable, and must have a low limit of detection. A low limit of detection will allow for successful detection even during the early onset of cancer. A typical sensing setup comprises the following elements:

- Analyte: It can be simply be defined as the chemical substance to be measured. In the case of cancer screening, cancer specific biomarkers, tumour markers, antigen, and proteins are essential analytes . More about the different types of analytes for cancer screening in Section 4
- Labeling: In most of the biosensors, labeling plays an important role. For the detection of the analyte, labels that attach to the molecule are used. The selection of label depends on the detection method used.
- Recognition: The recognition element is used to convert the biological information into signals. The most common detection method used in cancer screeing is enzyme-linked immunoassay (ELISA). In Section 5, various recognition methods that have been used for paper-based cancer screening are discussed.
- Readout: The readout method is used to obtain the outcome of the test. Some common readout methods are electrochemical, optical, and colorimetric. Depending on the detection technique used, the results obtained can either be qualitative (yes/no) type or quantitative (numerical values).

The paper-based microfluidic platform is divided into different zones, with each zone having a specific functionality, as shown in Figure 2. In the sampling zone, the sample is placed on the paper substrate. While passing through the microfluidic channels, labeling elements get attached to the analyte. In the detection zone, the analyte is detected and a signal is generated for the readout. In the case of colorimetry-based devices, there is a special zone termed as 'control zone'. In case a test yields a positive or negative result, there is a color change observed in the control zone.

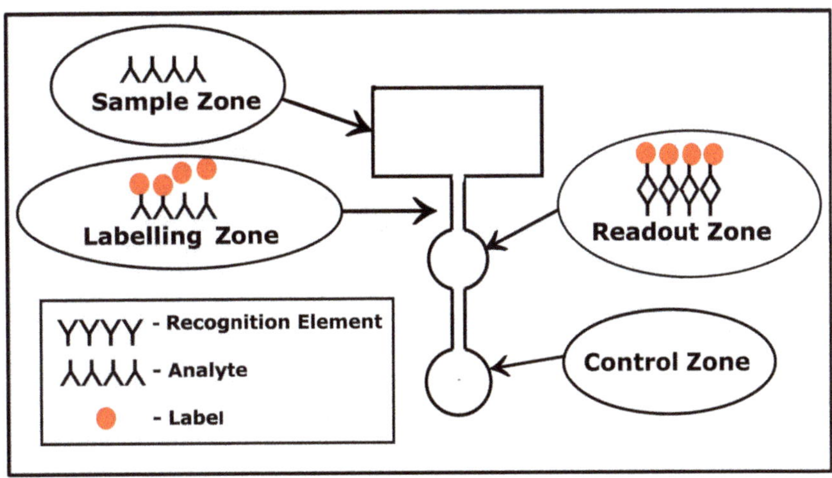

Figure 2. Typical layout of a paper-based sensor with the different zones. Each zone performs a specific function.

4. Analytes for Cancer Screening

For cancer detection, biomarkers play an important role. According to the National Cancer Institute (NCI), biomarkers are defined as "a biological molecule found in blood, other body fluids, or tissues that is a sign of a normal or abnormal process, or of a condition or disease" [88]. Though biomarkers may be generated due to various factors such as somatic mutations, transcriptional changes, or post-translational modifications, they an important differentiators for an affected individual compared to a healthy individual [88]. A handful of biomarkers have been approved for cancer detection. For example, high levels of carcino embryogenic antigen may mean the presence of cancer. Similarly, CA 125 is a protein that is detected in blood for ovarian cancer. Lysophosphatidic acid, leptin, osteopotin, and insulin-like growth factor receptor 2 are used as a biomarker for ovarian carcinoma. Early prostate cancer antigen 2 is used as a novel biomarker for prostate cancer [89]. Biomarkers are not only useful for the detection of screening of cancer, they are also used in monitoring the effectiveness on any treatment or therapy. Various biomarkers are present in blood, saliva, urine, stools, etc., making it possible to obtain samples for analysis in a non-invasive or minimally invasive manner. There are various biomarkers that are used for cancer screening, including antigens, micro-RNAs, proteins, antibodies, and tumor cells. For example, the antibody–antigen interaction creates a signal that is measured qualitatively or quantitatively on a paper substrate.

5. Recognition Element

Recognition elements are responsible for the recognition of target analytes (ex: receptors) and their conversion into a signal, which can be qualitative, semi-quantitative, or quantitative [90]. An ideal recognition element has a highly specific binding affinity towards the analyte of interest [90]. Recognition elements can either be natural or artificial bio-molecules that are synthetically obtained [90–94]. When the target analyte molecule attaches to the recognition element, it undergoes a biochemical reaction producing a signal [90]. In some cases, for the detection of analyte, labeling agents such as nanoparticles are attached to the analyte molecule [90].

5.1. Antibodies

Antibodies are a type of natural bio-receptor that can be derived from living organisms [90]. Monoclonal antibodies are widely preferred in cancer screening applications due to their high specificity to the target antigen [95,96]. Through the development of hybridoma technology, it is possible to obtain a reliable and uniform supply of monoclonal antibodies [96]. This has also led to better reliability and accuracy of sensors using monoclonal antibodies. Covalent binding of antibodies to cellulose paper discs has been developed for colorimetric immunoassays. The antibodies were coated on the amine-functionalized cellulose paper discs. Through a glutaraldehyde cross-linking agent, the antibodies showed enhanced binding activity to the target when compared to the periodate oxidation method [97]. Other methods for antibody immobilization on a paper with shelf-life up to 12 months have been described [98,99]. Polyclonal antibodies have multiple binding sites, each specific to a particular antigen. Although they are cheaper to produce compared to monoclonal antibodies, they are not suitable for sensing applications due to the multiple binding sites [100].

For the detection of analytes, the principle of enzyme linked immunosorbent assay (ELISA) is applied in sensor design. Nanoparticles such as gold are used as antibody carriers and signal enhancers for ELISA. In the range between 0 and 60 U/mL, the ELISA assay adopting gold nanoparticles as an optical signal enhancer resulted in higher sensitivity and shorter assay time when compared to classical ELISA procedures. This was used to detect breast cancer biomarkers [101].

A sensor for the detection of prostate-specific antigens (PSA) used multiwalled carbon nanotubes (MWCNT) activated with anti-PSA antibody for the detection [102] . Due to the site-selective interaction between the antigen and the antibody there was a change in the resistance. This change in resistance could be measured using a benchtop multi-meter [102]. This method was found to be cheaper and faster than the ELISA method used in cancer diagnosis [102]. Similarly, single-wall carbon nanotube based biosensors have been used in the identification of cancer antibody–antigen interactions in blood samples using electrical conductance measurements. Following the measurements, a classification algorithm was implemented to differentiate between cancer and controls with 90% accuracy [103–105].

In a colorimetric sensor for the detection of pancreatic cancer biomarker (PEAK1), gold nanoparticles that were used as a labeling element acted as a color dye catalyst to produce colorimetric signals [106]. A photothermal-effect-based sensor used a graphene oxide (GO)-gold–anti-EpCAM antibody composite as the recognition element for the detection of MCF-7 cancer cells specific antigen [107]. After laser irradiation at the test zone, the temperature contrast was recorded for the detection of cell numbers (Figure 3) [107].

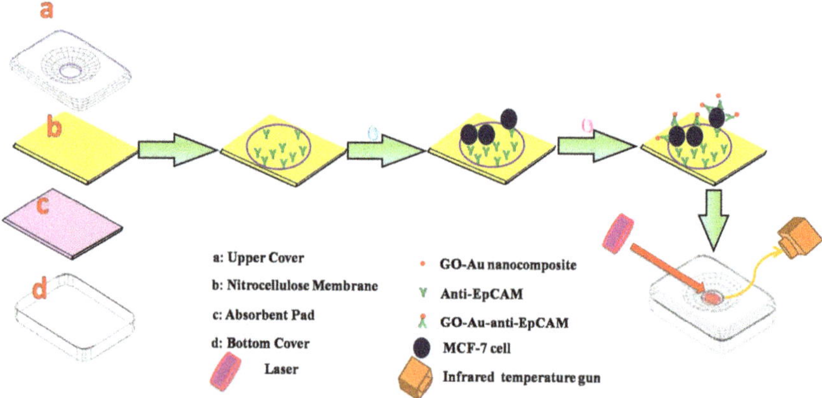

Figure 3. Working of photothermal effect based sensor for MCF-7 cancer cell detection. (Reproduced with permission from [107] Copyright 2016, Elsevier).

5.2. Aptamers

Aptamers are single-stranded nucleic acids that are folded into a specific architecture [108,109]. Due to their specific binding of the target proteins, they are used for sensing applications [108,110]. Their size and chemical stability make them widely preferred for the detection of proteins and small molecules [110,111]. Their low cost makes them a preferred alternative to antibodies in sensing applications [111]. For example, carbon-nanotube-based RNA apatamer sensors were developed for detecting IL6 in blood samples. Apatamer sensors based on field effect transistor arrays suggested a shift in drain current versus gate voltage for 1 pg and 1 ng of IL-6 exposure. The concentration of 1 pg falls below the diagnostic gray zone for cancer (2.3 pg–4 ng/mL), which is an indicator of early-stage cancer [112].

For the design of aptamer-based sensors, various strategies such as sandwich, target-induced structure switching, or competitive replacement modes have been used for biosensor design [111]. Electrochemical sensing is the most preferred sensing method with aptamer as the recognition element; however, other methods such as optical sensing have also been explored [110].

In a fluorescence-based paper-based sensor designed for the detection of multiple types of cancer cells, graphene oxide-coated with mesoporus-silica-labelled high-specificity aptamers was used as a labeling element [113]. Using the excitation wavelength of 350 nm, a color change was produced that could be observed through naked eye [113].

6. Sensing and Readout Methods

For achieving higher utility of paper-based sensors, the readout method used should be cost-effective and portable. The method provides fast and accurate results without a requirement for extensive handling by experts.

In most of the cases, qualitative readout methods should suffice. However, with focus on providing health professionals with important data at the point-of-care, qualitative readout methods are also gaining significance. Although there are various readout methods for sensing applications, electrochemical and optical are the most widely used readout methods for paper-based sensing applications. With advancements in smartphone and machine learning technologies, there have been works that use smartphones for signal interpretation and readout.

6.1. Modified Electrodes

For electrochemical sensing, the potential difference between the electrodes is proportional to the concentration of the analyte. In paper-based sensors, the working electrode is modified such that the binding of the analyte produces an electrical signal through a change in resistance, current, capacitance, or impedance.

Various nanocomposites have been used for the fabrication of modified electrodes. These nanocomposites perform a dual function: recognition and amplification. Amino functional graphene (NH2-G)/thionine (Thi)/gold nanoparticles (AuNPs) nanocomposites are coated with recognition elements such as üimmobilized anti-CEA [114] and anti-NSE [115] for the detection of specific analytes. The sensor could provide fast results with a low limit of detection of 10 pg/mL [114]. In a more recent work, an aptasensor with two working electrodes capable of the simultaneous detection of CEA and NSE has been developed. Along with the NH2-G/Thi/Au nanocomposite, Prussian blue (PB)- poly (3,4- ethylenedioxythiophene) (PEDOT)- AuNPs nanocomposite was used for the fabrication of the second electrode, which was coated with immobilized CEA and NSE aptamers [116]. The device worked on the principle of electrochemluminescnce and could achieve fast and accurate detection of CEA and NSE with a limit of detection of 2 pg/mL and 10 pg/mL, respectively (Figure 4) [116].

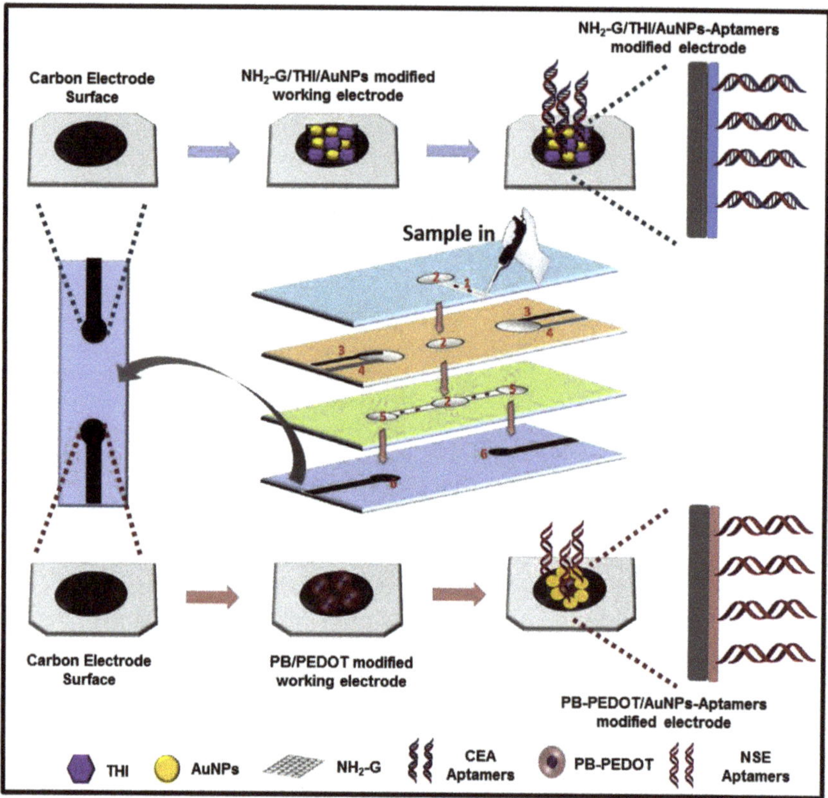

Figure 4. Mechanics of modified electrode paper-based apta sensor. (Reproduced with permission from [116] Copyright 2019, Elsevier).

6.2. Electrochemical

In electrochemical sensing method, the analyte generates an electrical signal proportional to its concentration [90]. The signals may be generated through a biorecognition event, modified electrodes, or enzyme mediated electrodes [90,117]. For electrochemical detection, the sensor should have three electrode systems with reference, working, and counter electrodes. For measuring the signals, electrochemical devices such as electrochemical workstations or bench-top multimeters are used.

Various routes such as the use of labeling agents or modified electrodes may be used for generating electrical output from biological signals. In a sensor developed for the detection of cancer antigens, a marker for ovarian cancer, a reduced graphene oxide/gold nanoparticle/thionine nanocomposite was used as working electrode [118].

For the reliable detection of signals, signal amplification techniques are used. In paper-based sensors for the detection of CEA using horseradish peroxidase (HRP)–O-phenylenediamine–H_2O_2 as a detection element, graphene was coated on the substrate for accelerating the electron transfer and amplifying the signals [119].

6.3. Optical

For optical sensing, signals are generated through a recognition process by the formation of an antigen–antibody complex [90]. The optical signals could be fluorosence, chemiluminescence, or color change [90]. Other than the signals that display a direct color change, a photo-detector is used for measuring the signals [90].

Surface-enhanced Raman scattering is a popularly used method for signal detection in paper-based sensors. Gold nanostar@Raman reporter@silica-sandwiched nanoparticles have been developed as surface-enhanced Raman scattering (SERS) probes for the paper-based lateral flow strip (PLFS) assay [120]. A sensor for the detection of CEA used a portable raman sensor for measurement (Figure 5) [120]. Using a paper-based lateral flow strip capable of plasma separation and using silica nanoparticles for labeling the sensor displayed a limit of detection of 1 ng/mL [120].

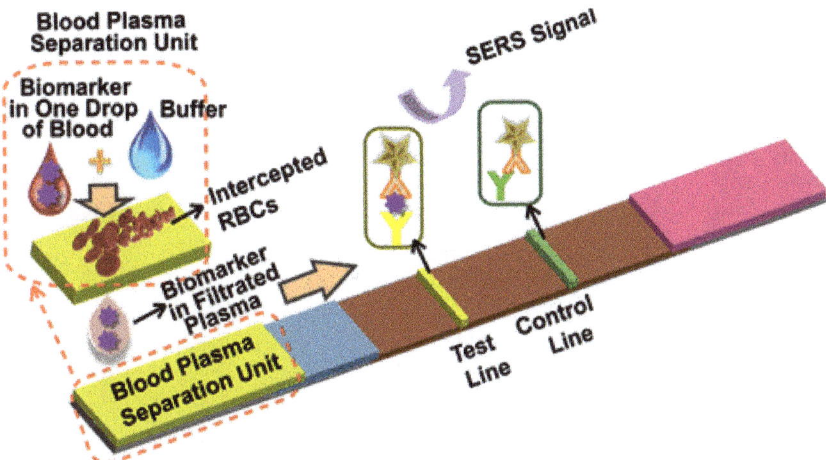

Figure 5. Working of a paper-based sensor for CEA concentration detection using surface-enhanced Raman scattering (SERS). (Reproduced with permission from [120] Copyright 2021, American Chemical Society).

For naked eye detection, luminiscent reporters are used as labeling elements These can be nanoparticles [51], conjugated polyelectrolytes [121], or multi walled carbon nanotubes [102].

6.4. Smartphone/Machine-Learning-Based

Smartphones are devices that are readily available, even in low- and middle-income countries. Mobile health is becoming increasingly popular in developing countries [122]. It is widely explored as a tool for the efficient delivery of services, including in healthcare. Smartphones have been explored as a readout method for both optical and electrical signals [123–125]. For optical signals, a smartphone camera is used for data acquisition [126–128]. Using a custom application, the acquired image is compared with reference values and the result is calculated [126–130].

Smart phone-based imaging was used for calculating and displaying results in a multi-layered paper-based sensor for cancer screening [131]. The movable layers allowed one to control the flow of the solution. Using the special design and smartphone-based readout, it was possible to achieve a low detection limit of 0.015 ng/mL[131].

Smartphones are also used for coupling with an electrochemical sensing device for the readout of signals [132]. A screen-printed sensor with multi-walled carbon nanotubes (MwCNT)/thionine (Thi)/gold nanoparticles (AuNPs) electrodes is capable of detecting cancer antigen (CA125) with a limit of detection of 2mU/mL [132]. The sensor uses a electrochemical detector powered using a smartphone, and it transfers data to the smartphone, where it is readout using a custom app (Figure 6) [132].

Table 2 summarizes recent works using paper-based sensors for cancer screening. It provides the breakdown of the sensor in terms of the biomarker(s) detected, recognition element, readout method used, and the types of cancer detected.

Table 2. Paper-based sensors for cancer screening.

Biomarker Detected	Recognition Element	Readout Method	Types of Cancer	Reference
MCF-7 Cells	Graphene Oxide- Gold nanoparticle nanocomposite with anti-EpCAM antibody.	Protothermal contrasting and visual readout	Breast cancer	[107]
AFP, CEA, CA125, and CA153.	Horse radish peroxidase (HRP)- O phenylene diamine H_2O_2	Electrochemical Immunodevice	Multiple	[119]
PSA	Bipolar electrode	electrochemiluminescence	Prostate cancer	[133]
microRNA-141 (miR-141) and microRNA-21 (miR-21)	Metal–organic framework (MOF) conjugated bio-probe, methylene blue (MB) and ferrocene (Fc) with distinguishable electrochemical signal,	Elctrochemical	Early detection of cancer	[134]
CEA	NH2-G/Thi/AuNPs nanocomposites modified electrode	Electrochemical	Multiple	[114]
miRNA-21	Positively charged conjugated polyelectrolyte (CPEs) "poly(3-alkoxy-4-methylthiophene)" (PT)	Colometric Through Naked Eye	Lung Cancer	[121]
NMP22 and BTA	Antibodies	Colometric With Naked Eye	Bladder Cancer	[60]
miRNA-21	DNA-templated Ag/Pt nanoclusters (DNA-Ag/Pt NCs),	Colometric Through Naked Eye	Lung Cancer	[135]
miRNA-21 and miRNA-31	Duplex-specific nuclease (DSN)	Laser-induced fluorescence (LIF)	miRNAs in cancer cells	[136]
blood cancer cells and skin cancer cell	photonic crystal fiber (PCF)	optical	blood and skin	[137]
Neuron-specific enolase (NSE)	NH2-G/Thi/AuNPs nanocomposites modified electrode	electrochemical detector and Android's smartphone	Lung Cancer	[115]
cancer antigen 125 (CA125)	reduced graphene oxide/thionine/gold nanoparticles (rGO/Thi/AuNPs) nanocomposites coated working electrode	electrochemical	ovarian cancer,lung cancer, endometrial cancer and breast cancer	[118]
CEA	plasma separation	optical :raman scattering readout	Multiple	[120]
free hydrogen sulfide in prostate cancer cells	polyvinylpyrrolidone (PVP) membrane containing silver/Nafion	Colorimetric	Prostate cancer	[138]
PSA	multi wall carbon nanotubes MWCNTs activated PSA antibody (monoclonal antibody of the prostate specific antigen)	Electrochemical: Bench top multimeter	Prostate cancer	[102]
CEA	Anti CEA	Colorimetric	Multiple	[139]
PEAK1	Anti PEAK1	Colorimetric using gold nps	pancreatic cancer	[106]
PEAK1	nanomaterial graphene oxide coated electrode immobilized with anti-PEAK1	Electrochemical	pancreatic cancer	[140]
CEA PSA	[Ru(bpy)3]2+-labeled signal antibody CEA and PSA	Electrochemiluminescence	Multiple	[141]
Cytochrome c (Cyt c)	Cyt c aptamer and Raman reporter Cy5-labeled complementary DNA	optical :raman scattering	Lung Cancer	[142]
CEA and NSE	DNA aptamer	Electrochemical	Multiple	[116]
EGFR	anti-EGFR aptamers	Electrochemical	gastric, breast, ovarian, and colorectal cancers	[87]
MCF-7 cells	Aptamer-modified electrode	Electrochemiluminescence	Breast cancer	[143]
VEGF-C	NMB/NH2-SWCNT/AuNps modified Working electrode	Electrochemical	Cancer progression	[144]
urokinase plasminogen activator	graphene-AuNP platform and fluorescence of quantum dots	Colorimetric	Cancer progression	[145]
Micro RNA MiR-17	"light-switch" molecule [Ru(phen)2dppz]2+ modified electrode.	Electrochemiluminescence	Breast cancer	[146]
Osteopontin	Biotinylated aptamer for precapture and antibody for detection	Optical through naked eye	Cancer prognosis	[147]
Diphenylthiocarbazone	CuO NPs-labeled seconday Anibodies captured by antibodies	Fluorescence resonance energy transfer (FRET)	Prostate cancer	[148]
AFP and MUC16	AuNP labeling and anti-AFP and anti-MUC16 antibodies	colorimetric spot test	Multiple types	[149]
Perilipin-2	Gold nanorattles with PLIN-2 assay	Plasmonic biosensor	Renal cancer	[144]
CA 125	Ag/rGO nano-ink based electrodes with anti-CA	Electrochemical	Ovarian Cancer	[150]
CEA	Graphene-PEDOT:PSS modified electrode	Electrochemical	Multiple	[151]

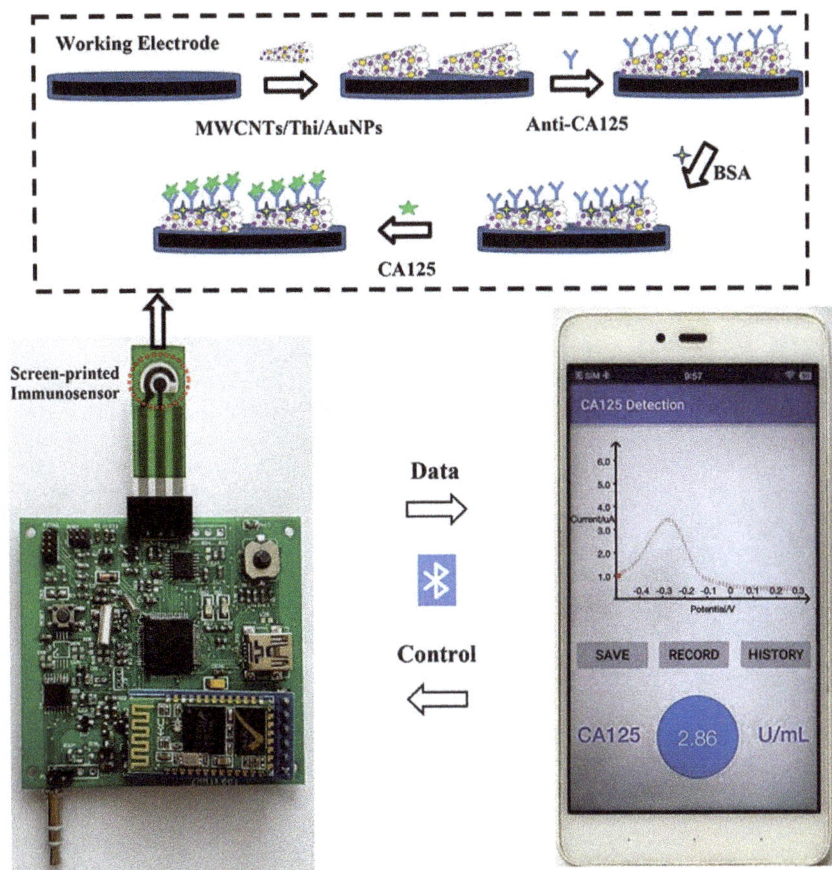

Electrochemical Detector   **Smartphone**

Figure 6. Sensor for detection of cancer antigen (CA125). The screen-printed sensor used electrochemical deterrence for signal measurement and a smartphone with a custom application for readout (reproduced with permission from [132] Copyright 2022, Elsevier).

7. Limitations of Current Paper-Based Methods

Despite the promising future, there are currently several limitations that hinder the large-scale acceptance of paper-based sensors. Ranging from fabrication methods, a requirement of measurement equipment, to regulatory requirements, many limitations need to be addressed before paper sensors are actually put into service. Sensitivity and accuracy are also concerns in paper-based devices.

However, paper-based sensors are excellent for qualitative and semi-quantitative screening. One can further improve the accuracy of paper-based testing through implementing paper-based testing with artificial-intelligence-based analysis. Machine learning and deep learning can even predict the sequence of DNA and RNA that can point to cancer mutations. They could also potentially detect cancer cells in blood droplets. The ability to differentiate between cancer versus normal cells in blood based on AI can be a very powerful approach in making paper-based testing a reality for mass screening [152].

With the exception of laser printing and screen printing, different fabrication methods used for the fabrication of paper-based microfluidic devices such as photo-lithography, e-beam lithography, reactive ion etching, and metal or oxide deposition require extensive capital investment and training, making them non-feasible for use in low- and middle-income countries.

For devices with qualitative measurement of signals, despite the low cost of single paper-based sensing device, expensive equipment such as electrochemical workstations, photo detectors, or electrochemiluminscence detectors are required for a results readout. This not only affects the portability of the device but also increases the overall cost of the screening, and it requires a trained technician to carry out the readout.

Nevertheless, paper-based sensors can make a significant impact in terms of testing blood for infectious diseases and even in the fight against cancer. The qualitative assessment of whether a person has cancer based on biomarkers in blood can reduce the cancer clinical burden in low- and middle-income countries. The paper-based testing method will definitely have an advantage here to reduce the clinical burden through the large-scale screening of populations. People with cancer undergoing chemotherapy may be prone to infectious diseases. Life-threatening infectious diseases can kill people in a few days. Here, POC systems based on paper and colorimetric detection methods for parasites, viruses, and other agents would be highly valuable. Before vaccination, people, including doctors, were dying of COVID-19 within a few days to two weeks. Low-cost qualitative paper-based sensors with instant results through color change are an absolute necessity to differentiate between population who has the virus from those who do not. This may be cheaper than the current PCR test that is used and is done in the laboratory. In the era of the pandemic, low-cost sensors for personal safety are very important. Qualitative paper-based sensors with instant read out will be highly useful in such pandemics as one cannot run tests in the laboratory frequently.

8. Conclusions

Paper-based sensors have immense potential to to act as low-cost tools in the mass screening of cancer. They can be used in cancer-screening camps, especially in low- and middle-income countries. Due to the health care resources being highly stressed in these countries, screening using paper-based sensors will act as a filter. The samples of the patients who test positive in the screening stages can then be treated as a priority. This will in a way help in resource allocation and management in these countries. It must be mentioned that the paper-based sensors with their current state of the art are insufficient for providing data to healthcare professionals in making important decisions. Further, testing will be required to ascertain the stage of the disease before any treatment plan is decided. Many times, the same biomarker is produced for multiple types of cancers. CEA is a common marker for multiple types of cancer such as lung, breast, ovaries, stomach, and intestine, to name a few. Thus, it is necessary to determine the type of cancer and the stage of disease progression before starting the treatment. Normal CEA levels are 2.5 ng/mL. A CEA level of 10 ng/mL would indicate the presence of cancer, and anything above 20 ng/mL would indicate the spread of cancer.

Patients with continuously decreasing levels of CEA do better after treatment than patients with increasing CEA levels. A regular paper-based CEA test with an instant read out can tell the doctors the potential for cancer progression. Their use will reduce the number of expensive tests (CT scans, PET-scans, etc.) a patient undergoes during follow-up visits and also help reduce the cost of treatment. In the era of telemedicine, it is becoming increasingly convenient to deliver healthcare at home. Cancer-screening tests can be conducted at home, and the results of the test can be emailed or texted to a doctor automatically using smart phones. So, paper-based testing is important here for home-based low-cost sensors.

Although paper-based sensors may not potentially be the knight in shining armor against humanity's fight against cancer, they can potentially be the important foot soldier in the fight. Future integrated paper-based sensors where all the sensing and electronic circuitry are printed in paper can make an impact on low-cost testing. Paper-based cantilevers with optical sensors and electronics integrated in a hand-held chip could enable the detection of cancer biomarkers such as prostate-specific antigen (PSA) from blood samples. Thus, there are exciting opportunities for paper-based sensors in the fight against cancer. The places where paper-based sensors along with mobile phones and AI-based techniques could make an impact in the fight against cancer are (1) the low-cost qualitative screening of large populations; (2) reducing the clinical burden through proper resource allocation; (3) estimating cancer prognosis; (4) monitoring cancer treatment; and (5) detecting cancer recurrence qualitatively or semi-quantitatively. With the ability to miniaturize anything, from detectors to spectrometers, one can implement miniaturized low-cost detectors along with paper-based sensors for cancer detection. Such miniaturized sensors are already made by many companies and could be bought off the shelf and integrated with a paper-based sensor. Finally, paper-based methods with artificial intelligence techniques can enable low cost and further improve the sensitivity and accuracy of paper-based sensors.

Author Contributions: Conceptualization, A.D. and A.J.N.; writing—original draft preparation, A.D.; writing—review and editing, A.D., A.J.N. and B.P.; and supervision, A.J.N., B.P., and T.-H.O. All authors have read and agreed to the published version of the manuscript.

Funding: B.P. acknowledges funding from the United States-India Educational Foundation through the Fulbright-Nehru Academic and Research Professional Excellence Award, Grant Number: 2021/APE-R/63. A.J.N. acknowledges the financial support from the Department of Biotechnology, Government of India, through Ramalingaswami Re-entry fellowship (D.O. No. BT/HRD/35/02/2006) and VIT SEED GRANT 2021-2022 (SG20210234).

Institutional Review Board Statement: Not applicable.

Informed Consent Statement: Not applicable.

Data Availability Statement: Not applicable.

Acknowledgments: A.J.N. would like to acknowledge the financialsupport from the Department of Biotechnology, Government of India, through Ramalingaswami Re-entry fellowship (D.O. No. BT/HRD/35/02/2006) and VIT SEED GRANT 2021-2022 (SG20210234). B.P. acknowledges funding from the United States-India Educational Foundation through the Fulbright-Nehru Academic and Research Professional Excellence Award, Grant Number: 2021/APE-R/63.

Conflicts of Interest: The authors declare no conflict of interest.

Abbreviations

The following abbreviations are used in this manuscript:

PoCT	Point of Care Testing
CEA	Carcino Embryonic Antigen
MCF-7	Michigan Cancer Foundation
AFP	Alpha-fetoprotein
MWCNT	Multi-walled carbon nano tubes
CA125	Cancer antigen 125
CA153	Carbohydrate antigen 153
PSA	Prostate-specific antigen
HRP	Horse radish peroxidase
MOF	Metal–Organic Framework
miR-141	microRNA-141
miR-21	microRNA-21
NSE	Neuron-specific enolase
PEAK1	Pseudopodium-enriched atypical kinase one
EGFR	Epidermal Growth Factor Receptor
Cyt C	Cytochrome c
PCF	Photonic crystal fiber
DSN	Duplex-specific nuclease
CPE	Conjugated polyelectrolyte

VEGF-C	Vascular endothelial growth factor C
NMB	New methylene blue
NH2-SWCNTs	Amino-functional single-walled carbon nanotubes
AuNPs	Gold nanoparticles

References

1. Ferlay, J.; Soerjomataram, I.; Dikshit, R.; Eser, S.; Mathers, C.; Rebelo, M.; Parkin, D.M.; Forman, D.; Bray, F. Cancer incidence and mortality worldwide: Sources, methods and major patterns in GLOBOCAN 2012. *Int. J. Cancer* **2015**, *136*, E359–E386. [CrossRef] [PubMed]
2. Bray, F.; Laversanne, M.; Weiderpass, E.; Soerjomataram, I. The ever-increasing importance of cancer as a leading cause of premature death worldwide. *Cancer* **2021**, *127*, 3029–3030. [CrossRef] [PubMed]
3. Danaei, G.; Vander Hoorn, S.; Lopez, A.D.; Murray, C.J.; Ezzati, M. Comparative Risk Assessment Collaborating Group. Causes of cancer in the world: Comparative risk assessment of nine behavioural and environmental risk factors. *Lancet* **2005**, *366*, 1784–1793. [CrossRef]
4. Ferlay, J.; Partensky, C.; Bray, F. More deaths from pancreatic cancer than breast cancer in the EU by 2017. *Acta Oncol.* **2016**, *55*, 1158–1160. [CrossRef] [PubMed]
5. Cancer. Available online: https://www.who.int/health-topics/cancer#tab=tab_1 (accessed on 10 July 2022).
6. Sung, H.; Ferlay, J.; Siegel, R.L.; Laversanne, M.; Soerjomataram, I.; Jemal, A.; Bray, F. Global Cancer Statistics 2020: GLOBOCAN Estimates of Incidence and Mortality Worldwide for 36 Cancers in 185 Countries. *CA Cancer J. Clin.* **2021**, *71*, 209–249. doi: 10.3322/caac.21660. [CrossRef] [PubMed]
7. Damgacioglu, H.; Sonawane, K.; Zhu, Y.; Li, R.; Balasubramanian, B.A.; Lairson, D.R.; Giuliano, A.R.; Deshmukh, A.A. Oropharyngeal cancer incidence and mortality trends in all 50 states in the US, 2001–2017. *JAMA Otolaryngol. Head Neck Surg.* **2022**, *148*, 155–165. [CrossRef]
8. Sekeroglu, B.; Tuncal, K. Prediction of cancer incidence rates for the European continent using machine learning models. *Health Inform. J.* **2021**, *27*, 1460458220983878. [CrossRef]
9. Soerjomataram, I.; Bray, F. Planning for tomorrow: Global cancer incidence and the role of prevention 2020–2070. *Nat. Rev. Clin. Oncol.* **2021**, *18*, 663–672. [CrossRef]
10. Pramesh, C.; Badwe, R.A.; Bhoo-Pathy, N.; Booth, C.M.; Chinnaswamy, G.; Dare, A.J.; de Andrade, V.P.; Hunter, D.J.; Gopal, S.; Gospodarowicz, M.; et al. Priorities for cancer research in low-and middle-income countries: A global perspective. *Nat. Med.* **2022**, *28*, 649–657. [CrossRef]
11. Kretz, A.L.; Trauzold, A.; Hillenbrand, A.; Knippschild, U.; Henne-Bruns, D.; von Karstedt, S.; Lemke, J. TRAILblazing strategies for cancer treatment. *Cancers* **2019**, *11*, 456. [CrossRef]
12. Teleanu, R.I.; Chircov, C.; Grumezescu, A.M.; Teleanu, D.M. Tumor angiogenesis and anti-angiogenic strategies for cancer treatment. *J. Clin. Med.* **2019**, *9*, 84. [CrossRef] [PubMed]
13. L Arias, J. Drug targeting strategies in cancer treatment: An overview. *Mini Rev. Med. Chem.* **2011**, *11*, 1–17. [CrossRef]
14. Isaeva, O.; Osipov, V. Different strategies for cancer treatment: Mathematical modelling. *Comput. Math. Methods Med.* **2009**, *10*, 253–272. [CrossRef]
15. Tunali, I.; Gillies, R.J.; Schabath, M.B. Application of radiomics and artificial intelligence for lung cancer precision medicine. *Cold Spring Harb. Perspect. Med.* **2021**, *11*, a039537. [CrossRef] [PubMed]
16. Abbasi Kajani, A.; Haghjooy Javanmard, S.; Asadnia, M.; Razmjou, A. Recent advances in nanomaterials development for nanomedicine and cancer. *ACS Appl. Bio Mater.* **2021**, *4*, 5908–5925. [CrossRef]
17. Kenner, B.; Chari, S.T.; Kelsen, D.; Klimstra, D.S.; Pandol, S.J.; Rosenthal, M.; Rustgi, A.K.; Taylor, J.A.; Yala, A.; Abul-Husn, N.; et al. Artificial intelligence and early detection of pancreatic cancer: 2020 summative review. *Pancreas* **2021**, *50*, 251. [CrossRef] [PubMed]
18. Hawkes, N. Cancer survival data emphasise importance of early diagnosis. *BMJ* **2019**, *364*, l408. [CrossRef] [PubMed]
19. Why Is Early Diagnosis Important? Cancer Research UK. Available online: https://www.cancerresearchuk.org/about-cancer/cancer-symptoms/why-is-early-diagnosis-important (accessed on 22 June 2021).
20. Brown, M.L.; Yabroff, K.R. 12, Economic impact of cancer in the United States. *Cancer Epidemiol. Prev.* **2006**, *202*, 202–214.
21. IJzerman, M.J.; Berghuis, A.S.; de Bono, J.S.; Terstappen, L.W. Health economic impact of liquid biopsies in cancer management. *Expert Rev. Pharmacoecon. Outcomes Res.* **2018**, *18*, 593–599. [CrossRef]
22. Janovsky, C.C.P.S.; Bittencourt, M.S.; Novais, M.A.P.D.; Maciel, R.; Biscolla, R.P.M.; Zucchi, P. Thyroid cancer burden and economic impact on the Brazilian public health system. *Arch. Endocrinol. Metab.* **2018**, *62*, 537–544. [CrossRef]
23. Shah, S.C.; Kayamba, V.; Peek, R.M., Jr.; Heimburger, D. Cancer control in low-and middle-income countries: Is it time to consider screening? *J. Glob. Oncol.* **2019**, *5*, 1–8. [CrossRef]
24. Rivera-Franco, M.M.; Leon-Rodriguez, E. Delays in breast cancer detection and treatment in developing countries. *Breast Cancer Basic Clin. Res.* **2018**, *12*, 1178223417752677. [CrossRef] [PubMed]
25. Vu, M.; Yu, J.; Awolude, O.A.; Chuang, L. Cervical cancer worldwide. *Curr. Probl. Cancer* **2018**, *42*, 457–465. [CrossRef] [PubMed]

26. Ginsburg, O.; Yip, C.H.; Brooks, A.; Cabanes, A.; Caleffi, M.; Dunstan Yataco, J.A.; Gyawali, B.; McCormack, V.; McLaughlin de Anderson, M.; Mehrotra, R.; et al. Breast cancer early detection: A phased approach to implementation. *Cancer* **2020**, *126*, 2379–2393. [CrossRef]

27. Cortes, J.; Perez-García, J.M.; Llombart-Cussac, A.; Curigliano, G.; El Saghir, N.S.; Cardoso, F.; Barrios, C.H.; Wagle, S.; Roman, J.; Harbeck, N.; et al. Enhancing global access to cancer medicines. *CA Cancer J. Clin.* **2020**, *70*, 105–124. [CrossRef] [PubMed]

28. Hull, R.; Mbele, M.; Makhafola, T.; Hicks, C.; Wang, S.M.; Reis, R.M.; Mehrotra, R.; Mkhize-Kwitshana, Z.; Kibiki, G.; Bates, D.O.; et al. Cervical cancer in low and middle-income countries. *Oncol. Lett.* **2020**, *20*, 2058–2074. [CrossRef]

29. Moodley, J.; Cairncross, L.; Naiker, T.; Constant, D. From symptom discovery to treatment-women's pathways to breast cancer care: A cross-sectional study. *BMC Cancer* **2018**, *18*, 312. [CrossRef]

30. Darj, E.; Chalise, P.; Shakya, S. Barriers and facilitators to cervical cancer screening in Nepal: A qualitative study. *Sex. Reprod. Healthc.* **2019**, *20*, 20–26. [CrossRef]

31. Carrilho, E.; Martinez, A.W.; Whitesides, G.M. Understanding wax printing: A simple micropatterning process for paper-based microfluidics. *Anal. Chem.* **2009**, *81*, 7091–7095. doi: 10.1021/ac901071p. [CrossRef] [PubMed]

32. Derda, R.; Tang, S.K.Y.; Laromaine, A.; Mosadegh, B.; Hong, E.; Mwangi, M.; Mammoto, A.; Ingber, D.E.; Whitesides, G.M. Multizone paper platform for 3D cell cultures. *PLoS ONE* **2011**, *6*, e18940. doi: 10.1371/journal.pone.0018940. [CrossRef] [PubMed]

33. Ellerbee, A.K.; Phillips, S.T.; Siegel, A.C.; Mirica, K.A.; Martinez, A.W.; Striehl, P.; Jain, N.; Prentiss, M.; Whitesides, G.M. Quantifying colorimetric assays in paper-based microfluidic devices by measuring the transmission of light through paper. *Anal. Chem.* **2009**, *81*, 8447–8452. doi: 10.1021/ac901307q. [CrossRef] [PubMed]

34. Glavan, A.C.; Martinez, R.V.; Maxwell, E.J.; Subramaniam, A.B.; Nunes, R.M.D.; Soh, S.; Whitesides, G.M. Rapid fabrication of pressure-driven open-channel microfluidic devices in omniphobic R(F) paper. *Lab Chip* **2013**, *13*, 2922–2930. doi: 10.1039/c3lc50371b. [CrossRef] [PubMed]

35. Glavan, A.C.; Christodouleas, D.C.; Mosadegh, B.; Yu, H.D.; Smith, B.S.; Lessing, J.; Fernández-Abedul, M.T.; Whitesides, G.M. Folding analytical devices for electrochemical ELISA in hydrophobic R(H) paper. *Anal. Chem.* **2014**, *86*, 11999–12007. doi: 10.1021/ac5020782. [CrossRef] [PubMed]

36. Martinez, A.W.; Phillips, S.T.; Wiley, B.J.; Gupta, M.; Whitesides, G.M. FLASH: A rapid method for prototyping paper-based microfluidic devices. *Lab Chip* **2008**, *8*, 2146. doi: 10.1039/b811135a. [CrossRef]

37. Martinez, A.W.; Phillips, S.T.; Whitesides, G.M. Three-dimensional microfluidic devices fabricated in layered paper and tape. *Proc. Natl. Acad. Sci. USA* **2008**, *105*, 19606–19611. doi: 10.1073/pnas.0810903105. [CrossRef]

38. Martinez, A.W.; Phillips, S.T.; Nie, Z.; Cheng, C.M.; Carrilho, E.; Wiley, B.J.; Whitesides, G.M. Programmable diagnostic devices made from paper and tape. *Lab Chip* **2010**, *10*, 2499–2504. doi: 10.1039/c0lc00021c. [CrossRef]

39. Martinez, R.V.; Fish, C.R.; Chen, X.; Whitesides, G.M. Elastomeric Origami: Programmable Paper-Elastomer Composites as Pneumatic Actuators. *Adv. Funct. Mater.* **2012**, *22*, 1376–1384. doi: 10.1002/adfm.201102978. [CrossRef]

40. Nie, Z.; Deiss, F.; Liu, X.; Akbulut, O.; Whitesides, G.M. Integration of paper-based microfluidic devices with commercial electrochemical readers. *Lab Chip* **2010**, *10*, 3163–3169. doi: 10.1039/c0lc00237b. [CrossRef]

41. Nemiroski, A.; Christodouleas, D.C.; Hennek, J.W.; Kumar, A.A.; Maxwell, E.J.; Fernández-Abedul, M.T.; Whitesides, G.M. Universal mobile electrochemical detector designed for use in resource-limited applications. *Proc. Natl. Acad. Sci. USA* **2014**, *111*, 11984–11989. doi: 10.1073/pnas.1405679111. [CrossRef]

42. Lan, W.J.; Maxwell, E.J.; Parolo, C.; Bwambok, D.K.; Subramaniam, A.B.; Whitesides, G.M. Paper-based electroanalytical devices with an integrated, stable reference electrode. *Lab Chip* **2013**, *13*, 4103–4108. doi: 10.1039/c3lc50771h. [CrossRef]

43. Lan, W.J.; Zou, X.U.; Hamedi, M.M.; Hu, J.; Parolo, C.; Maxwell, E.J.; Bühlmann, P.; Whitesides, G.M. Paper-based potentiometric ion sensing. *Anal. Chem.* **2014**, *86*, 9548–9553. doi: 10.1021/ac5018088. [CrossRef] [PubMed]

44. Siegel, A.C.; Phillips, S.T.; Wiley, B.J.; Whitesides, G.M. Thin, lightweight, foldable thermochromic displays on paper. *Lab Chip* **2009**, *9*, 2775–2781. doi: 10.1039/b905832j. [CrossRef]

45. Mosadegh, B.; Dabiri, B.E.; Lockett, M.R.; Derda, R.; Campbell, P.; Parker, K.K.; Whitesides, G.M.r Three-Dimensional Paper-Based Model for Cardiac Ischemia. *Adv. Healthc. Mater.* **2014**, *3*, 1036–1043. doi: 10.1002/adhm.201300575. [CrossRef] [PubMed]

46. Lessing, J.; Glavan, A.C.; Walker, S.B.; Keplinger, C.; Lewis, J.A.; Whitesides, G.M. Inkjet printing of conductive inks with high lateral resolution on omniphobic "R(F) paper" for paper-based electronics and MEMS. *Adv. Mater.* **2014**, *26*, 4677–4682. doi: 10.1002/adma.201401053. [CrossRef]

47. Liu, X.; Mwangi, M.; Li, X.; O'Brien, M.; Whitesides, G.M. Paper-based piezoresistive MEMS sensors. *Lab Chip* **2011**, *11*, 2189–2196. doi: 10.1039/c1lc20161a. [CrossRef]

48. Glavan, A.C.; Martinez, R.V.; Subramaniam, A.B.; Yoon, H.J.; Nunes, R.M.; Lange, H.; Thuo, M.M.; Whitesides, G.M. Omniphobic "RF paper" produced by silanization of paper with fluoroalkyltrichlorosilanes. *Adv. Funct. Mater.* **2014**, *24*, 60–70. [CrossRef]

49. Thuo, M.M.; Martinez, R.V.; Lan, W.J.; Liu, X.; Barber, J.; Atkinson, M.B.; Bandarage, D.; Bloch, J.F.; Whitesides, G.M. Fabrication of low-cost paper-based microfluidic devices by embossing or cut-and-stack methods. *Chem. Mater.* **2014**, *26*, 4230–4237. [CrossRef]

50. Mazzeo, A.D.; Kalb, W.B.; Chan, L.; Killian, M.G.; Bloch, J.F.; Mazzeo, B.A.; Whitesides, G.M. Paper-based, capacitive touch pads. *Adv. Mater.* **2012**, *24*, 2850–2856. [CrossRef] [PubMed]

51. Huang, J.Y.; Lin, H.T.; Chen, T.H.; Chen, C.A.; Chang, H.T.; Chen, C.F. Signal amplified gold nanoparticles for cancer diagnosis on paper-based analytical devices. *ACS Sens.* **2018**, *3*, 174–182. [CrossRef]

52. Hu, J.; Wang, S.; Wang, L.; Li, F.; Pingguan-Murphy, B.; Lu, T.J.; Xu, F. Advances in paper-based point-of-care diagnostics. *Biosens. Bioelectron.* **2014**, *54*, 585–597. [CrossRef]

53. Ito, E.; Iha, K.; Yoshimura, T.; Nakaishi, K.; Watabe, S. Early diagnosis with ultrasensitive ELISA. *Adv. Clin. Chem.* **2021**, *101*, 121–133. [PubMed]

54. Arya, S.K.; Estrela, P. Recent advances in enhancement strategies for electrochemical ELISA-based immunoassays for cancer biomarker detection. *Sensors* **2018**, *18*, 2010. [CrossRef] [PubMed]

55. Hosseini, S.; Vázquez-Villegas, P.; Rito-Palomares, M.; Martinez-Chapa, S.O. Advantages, disadvantages and modifications of conventional ELISA. In *Enzyme-Linked Immunosorbent Assay (ELISA)*; Springer: Berlin/Heidelberg, Germany, 2018; pp. 67–115.

56. Qiu, J.; Keyser, B.; Lin, Z.T.; Wu, T. Autoantibodies as potential biomarkers in breast cancer. *Biosensors* **2018**, *8*, 67. [CrossRef] [PubMed]

57. Arya, S.K.; Estrela, P. Electrochemical ELISA-based platform for bladder cancer protein biomarker detection in urine. *Biosens. Bioelectron.* **2018**, *117*, 620–627. [CrossRef]

58. Al Mughairy, B.; Al-Lawati, H.A. Recent analytical advancements in microfluidics using chemiluminescence detection systems for food analysis. *TrAC Trends Anal. Chem.* **2020**, *124*, 115802. [CrossRef]

59. Tonkinson, J.L.; Stillman, B.A. Nitrocellulose: A tried and true polymer finds utility as a post-genomic substrate. *Front. Biosci.-Landmark* **2002**, *7*, 1–12.

60. Jiang, Q.; Han, T.; Ren, H.; Aziz, A.U.R.; Li, N.; Zhang, H.; Zhang, Z.; Liu, B. Bladder cancer hunting: A microfluidic paper-based analytical device. *Electrophoresis* **2020**, *41*, 1509–1516. [CrossRef]

61. Evans, E.; Gabriel, E.F.M.; Coltro, W.K.T.; Garcia, C.D. Rational selection of substrates to improve color intensity and uniformity on microfluidic paper-based analytical devices. *Analyst* **2014**, *139*, 2127–2132. [CrossRef]

62. Swahn, B. A new micromethod for the determination of total lipids in serum. *Scand. J. Clin. Lab. Investig.* **1952**, *4*, 247–248. [CrossRef]

63. Chandler, R.; Gutierrez, C. The filter-paper method of suction measurement. *Geotechnique* **1986**, *36*, 265–268. [CrossRef]

64. Chardon, W.; Menon, R.; Chien, S. Iron oxide impregnated filter paper (Pi test): A review of its development and methodological research. *Nutr. Cycl. Agroecosystems* **1996**, *46*, 41–51. [CrossRef]

65. Fawcett, R.; Collis-George, N. A filter-paper method for determining the moisture characteristics of soil. *Aust. J. Exp. Agric.* **1967**, *7*, 162–167. [CrossRef]

66. Partridge, S. Filter-paper partition chromatography of sugars: I. General description and application to the quantitative analysis of sugars in apple juice, egg white and foetal blood of sheep. with a note by RG Westall. *Biochem. J.* **1948**, *42*, 238–248. [CrossRef] [PubMed]

67. Tang, R.; Li, M.; Yan, X.; Xie, M.; Liu, L.N.; Li, Z.; Xu, F. Comparison of paper-based nucleic acid extraction materials for point-of-care testing applications. *Cellulose* **2022**, *29*, 2479–2495. [CrossRef] [PubMed]

68. Minamide, L.; Bamburg, J. A filter paper dye-binding assay for quantitative determination of protein without interference from reducing agents or detergents. *Anal. Biochem.* **1990**, *190*, 66–70. [CrossRef]

69. Aryal, R.; Terman, P.; Voss, K.J. Comparison of two filter-based reflectance methods to measure the light absorption by atmospheric aerosols. *J. Atmos. Ocean. Technol.* **2014**, *31*, 923–929. [CrossRef]

70. Rauf, S.; Ali, Y.; Hussain, S.; Ullah, F.; Hayat, A. Design of a novel filter paper-based construct for rapid analysis of acetone. *PLoS ONE* **2018**, *13*, e0199978. [CrossRef]

71. Kurdekar, A.; Chunduri, L.; Bulagonda, E.P.; Haleyurgirisetty, M.K.; Kamisetti, V.; Hewlett, I.K. Comparative performance evaluation of carbon dot-based paper immunoassay on Whatman filter paper and nitrocellulose paper in the detection of HIV infection. *Microfluid. Nanofluidics* **2016**, *20*, 99. [CrossRef]

72. Fierro-Mercado, P.M.; Hernández-Rivera, S.P. Highly sensitive filter paper substrate for SERS trace explosives detection. *Int. J. Spectrosc.* **2012**, *2012*, 716527. [CrossRef]

73. Gan, W.; Zhuang, B.; Zhang, P.; Han, J.; Li, C.X.; Liu, P. A filter paper-based microdevice for low-cost, rapid, and automated DNA extraction and amplification from diverse sample types. *Lab Chip* **2014**, *14*, 3719–3728. [CrossRef]

74. Kim, H.; Prezzi, M.; Salgado, R. Calibration of Whatman Grade 42 filter paper for soil suction measurement. *Can. J. Soil Sci.* **2016**, *97*, 93–98. [CrossRef]

75. Mei, J.V.; Zobel, S.D.; Hall, E.M.; De Jesús, V.R.; Adam, B.W.; Hannon, W.H. Performance properties of filter paper devices for whole blood collection. *Bioanalysis* **2010**, *2*, 1397–1403. [CrossRef] [PubMed]

76. Rottinghaus, E.; Bile, E.; Modukanele, M.; Maruping, M.; Mine, M.; Nkengasong, J.; Yang, C. Comparison of Ahlstrom grade 226, Munktell TFN, and Whatman 903 filter papers for dried blood spot specimen collection and subsequent HIV-1 load and drug resistance genotyping analysis. *J. Clin. Microbiol.* **2013**, *51*, 55–60. [CrossRef] [PubMed]

77. Langer, E.K.; Johnson, K.J.; Shafer, M.M.; Gorski, P.; Overdier, J.; Musselman, J.; Ross, J.A. Characterization of the elemental composition of newborn blood spots using sector-field inductively coupled plasma-mass spectrometry. *J. Expo. Sci. Environ. Epidemiol.* **2011**, *21*, 355–364. [CrossRef] [PubMed]

78. Borman, A.M.; Linton, C.J.; Miles, S.J.; Campbell, C.K.; Johnson, E.M. Ultra-rapid preparation of total genomic DNA from isolates of yeast and mould using Whatman FTA filter paper technology–a reusable DNA archiving system. *Med. Mycol.* **2006**, *44*, 389–398. [CrossRef] [PubMed]

79. da Cunha Santos, G.; Liu, N.; Tsao, M.S.; Kamel-Reid, S.; Chin, K.; Geddie, W.R. Detection of EGFR and KRAS mutations in fine-needle aspirates stored on Whatman FTA cards: Is this the tool for biobanking cytological samples in the molecular era? *Cancer Cytopathol.* **2010**, *118*, 450–456. [CrossRef] [PubMed]

80. Cards, F. FTA Cards for Preservation of Nucleic Acids for Molecular Assays. *Arch. Pathol. Lab. Med.* **2018**, *142*, 308–312.

81. Davis, E.H.; Velez, J.O.; Russell, B.J.; Basile, A.J.; Brault, A.C.; Hughes, H.R. Evaluation of Whatman FTA cards for the preservation of yellow fever virus RNA for use in molecular diagnostics. *PLoS Neglected Trop. Dis.* **2022**, *16*, e0010487. [CrossRef]

82. Kurien, B.T.; Scofield, R.H. Western blotting. *Methods* **2006**, *38*, 283–293. [CrossRef]

83. Mansfield, M.A., The Use of Nitrocellulose Membranes in Lateral-Flow Assays. In *Drugs of Abuse: Body Fluid Testing*; Wong, R.C., Tse, H.Y., Eds.; Humana Press: Totowa, NJ, USA, 2005; pp. 71–85. doi: 10.1007/978-1-59259-951-6_4. [CrossRef]

84. Koga, H.; Nagashima, K.; Suematsu, K.; Takahashi, T.; Zhu, L.; Fukushima, D.; Huang, Y.; Nakagawa, R.; Liu, J.; Uetani, K.; et al. Nanocellulose Paper Semiconductor with a 3D Network Structure and Its Nano–Micro–Macro Trans-Scale Design. *ACS Nano* **2022**, *16*, 8630–8640. [CrossRef]

85. Tang, R.; Liu, L.; Li, M.; Yao, X.; Yang, Y.; Zhang, S.; Li, F. Transparent microcrystalline cellulose/polyvinyl alcohol paper as a new platform for three-dimensional cell culture. *Anal. Chem.* **2020**, *92*, 14219–14227. [CrossRef] [PubMed]

86. Lee, M.J.; Soum, V.; Lee, S.N.; Choi, J.H.; Shin, J.H.; Shin, K.; Oh, B.K. Pumpless three-dimensional photo paper–based microfluidic analytical device for automatic detection of thioredoxin-1 using enzyme-linked immunosorbent assay. *Anal. Bioanal. Chem.* **2022**, *414*, 3219–3230. [CrossRef]

87. Wang, Y.; Sun, S.; Luo, J.; Xiong, Y.; Ming, T.; Liu, J.; Ma, Y.; Yan, S.; Yang, Y.; Yang, Z.; et al. Low sample volume origami-paper-based graphene-modified aptasensors for label-free electrochemical detection of cancer biomarker-EGFR. *Microsyst. Nanoeng.* **2020**, *6*, 32. [CrossRef] [PubMed]

88. Henry, N.L.; Hayes, D.F. Cancer biomarkers. *Mol. Oncol.* **2012**, *6*, 140–146. [CrossRef] [PubMed]

89. Diamandis, E.P. Cancer biomarkers: Can we turn recent failures into success? *J. Natl. Cancer Inst.* **2010**, *102*, 1462–1467. [CrossRef] [PubMed]

90. Dincer, C.; Bruch, R.; Costa-Rama, E.; Fernández-Abedul, M.T.; Merkoçi, A.; Manz, A.; Urban, G.A.; Güder, F. Disposable Sensors in Diagnostics, Food, and Environmental Monitoring. *Adv. Mater.* **2019**, *31*, 1806739. doi: doi: 10.1002/adma. 201806739. [CrossRef] [PubMed]

91. Weber, W.; Fussenegger, M. Emerging biomedical applications of synthetic biology. *Nat. Rev. Genet.* **2012**, *13*, 21–35. [CrossRef]

92. Zourob, M.; Elwary, S.; Khademhosseini, A. *Recognition Receptors in Biosensors*; Springer: Berlin/Heidelberg, Germany, 2010.

93. Justino, C.I.; Freitas, A.C.; Pereira, R.; Duarte, A.C.; Santos, T.A.R. Recent developments in recognition elements for chemical sensors and biosensors. *TrAC Trends Anal. Chem.* **2015**, *68*, 2–17. [CrossRef]

94. Piletsky, S.A.; Whitcombe, M.J. *Designing Receptors for The Next Generation of Biosensors*; Springer Science & Business Media: Berlin/Heidelberg, Germany, 2012; Volume 12.

95. Zhang, X.; Soori, G.; Dobleman, T.J.; Xiao, G.G. The application of monoclonal antibodies in cancer diagnosis. *Expert Rev. Mol. Diagn.* **2014**, *14*, 97–106. [CrossRef]

96. Modjtahedi, H.; Ali, S.; Essapen, S. Therapeutic application of monoclonal antibodies in cancer: Advances and challenges. *Br. Med. Bull.* **2012**, *104*, 41–59. [CrossRef] [PubMed]

97. Peng, Y.; Van Gelder, V.; Amaladoss, A.; Patel, K.H. Covalent binding of antibodies to cellulose paper discs and their applications in naked-eye colorimetric immunoassays. *J. Vis. Exp.* **2016**, e54111. [CrossRef] [PubMed]

98. Henderson, C.A.; McLiesh, H.; Then, W.L.; Garnier, G. Activity and longevity of antibody in paper-based blood typing diagnostics. *Front. Chem.* **2018**, *6*, 193. [CrossRef] [PubMed]

99. Tang, R.H.; Liu, L.N.; Zhang, S.F.; He, X.C.; Li, X.J.; Xu, F.; Ni, Y.H.; Li, F. A review on advances in methods for modification of paper supports for use in point-of-care testing. *Microchim. Acta* **2019**, *186*, 521. [CrossRef] [PubMed]

100. Berry, J.D.; Gaudet, R.G. Antibodies in infectious diseases: Polyclonals, monoclonals and niche biotechnology. *New Biotechnol.* **2011**, *28*, 489–501. [CrossRef] [PubMed]

101. Ambrosi, A.; Airo, F.; Merkoçi, A. Enhanced gold nanoparticle based ELISA for a breast cancer biomarker. *Anal. Chem.* **2010**, *82*, 1151–1156. [CrossRef]

102. Ji, S.; Lee, M.; Kim, D. Detection of early stage prostate cancer by using a simple carbon nanotube@ paper biosensor. *Biosens. Bioelectron.* **2018**, *102*, 345–350. [CrossRef]

103. Shao, N.; Wickstrom, E.; Panchapakesan, B. Nanotube–antibody biosensor arrays for the detection of circulating breast cancer cells. *Nanotechnology* **2008**, *19*, 465101. [CrossRef]

104. Loeian, M.S.; Aghaei, S.M.; Farhadi, F.; Rai, V.; Yang, H.W.; Johnson, M.D.; Aqil, F.; Mandadi, M.; Rai, S.N.; Panchapakesan, B. Liquid biopsy using the nanotube-CTC-chip: Capture of invasive CTCs with high purity using preferential adherence in breast cancer patients. *Lab Chip* **2019**, *19*, 1899–1915. [CrossRef]

105. Khoshroo, A.; Fattahi, A.; Hosseinzadeh, L. Development of paper-based aptasensor for circulating tumor cells detection in the breast cancer. *J. Electroanal. Chem.* **2022**, *910*, 116182. [CrossRef]

106. Prasad, K.S.; Abugalyon, Y.; Li, C.; Xu, F.; Li, X. A new method to amplify colorimetric signals of paper-based nanobiosensors for simple and sensitive pancreatic cancer biomarker detection. *Analyst* **2020**, *145*, 5113–5117. [CrossRef]

107. Zhou, J.; Zheng, Y.; Liu, J.; Bing, X.; Hua, J.; Zhang, H. A paper-based detection method of cancer cells using the photo-thermal effect of nanocomposite. *J. Pharm. Biomed. Anal.* **2016**, *117*, 333–337. [CrossRef] [PubMed]

108. Keefe, A.D.; Pai, S.; Ellington, A. Aptamers as therapeutics. *Nat. Rev. Drug Discov.* **2010**, *9*, 537–550. [CrossRef] [PubMed]

109. Zhou, W.; Huang, P.J.J.; Ding, J.; Liu, J. Aptamer-based biosensors for biomedical diagnostics. *Analyst* **2014**, *139*, 2627–2640. [CrossRef] [PubMed]

110. Song, S.; Wang, L.; Li, J.; Fan, C.; Zhao, J. Aptamer-based biosensors. *TrAC Trends Anal. Chem.* **2008**, *27*, 108–117. [CrossRef]

111. Han, K.; Liang, Z.; Zhou, N. Design strategies for aptamer-based biosensors. *Sensors* **2010**, *10*, 4541–4557. [CrossRef]

112. Khosravi, F.; Loeian, S.M.; Panchapakesan, B. Ultrasensitive Label-Free Sensing of IL-6 Based on PASE Functionalized Carbon Nanotube Micro-Arrays with RNA-Aptamers as Molecular Recognition Elements. *Biosensors* **2017**, *7*, 17. doi: 10.3390/bios7020017. [CrossRef]

113. Liang, L.; Su, M.; Li, L.; Lan, F.; Yang, G.; Ge, S.; Yu, J.; Song, X. Aptamer-based fluorescent and visual biosensor for multiplexed monitoring of cancer cells in microfluidic paper-based analytical devices. *Sens. Actuators B Chem.* **2016**, *229*, 347–354. [CrossRef]

114. Wang, Y.; Xu, H.; Luo, J.; Liu, J.; Wang, L.; Fan, Y.; Yan, S.; Yang, Y.; Cai, X. A novel label-free microfluidic paper-based immunosensor for highly sensitive electrochemical detection of carcinoembryonic antigen. *Biosens. Bioelectron.* **2016**, *83*, 319–326. [CrossRef]

115. Fan, Y.; Liu, J.; Wang, Y.; Luo, J.; Xu, H.; Xu, S.; Cai, X. A wireless point-of-care testing system for the detection of neuron-specific enolase with microfluidic paper-based analytical devices. *Biosens. Bioelectron.* **2017**, *95*, 60–66. [CrossRef]

116. Wang, Y.; Luo, J.; Liu, J.; Sun, S.; Xiong, Y.; Ma, Y.; Yan, S.; Yang, Y.; Yin, H.; Cai, X. Label-free microfluidic paper-based electrochemical aptasensor for ultrasensitive and simultaneous multiplexed detection of cancer biomarkers. *Biosens. Bioelectron.* **2019**, *136*, 84–90. [CrossRef]

117. Cass, A.E.; Davis, G.; Francis, G.D.; Hill, H.A.O.; Aston, W.J.; Higgins, I.J.; Plotkin, E.V.; Scott, L.D.; Turner, A.P. Ferrocene-mediated enzyme electrode for amperometric determination of glucose. *Anal. Chem.* **1984**, *56*, 667–671. [CrossRef] [PubMed]

118. Fan, Y.; Shi, S.; Ma, J.; Guo, Y. A paper-based electrochemical immunosensor with reduced graphene oxide/thionine/gold nanoparticles nanocomposites modification for the detection of cancer antigen 125. *Biosens. Bioelectron.* **2019**, *135*, 1–7. [CrossRef] [PubMed]

119. Wu, Y.; Xue, P.; Kang, Y.; Hui, K.M. based microfluidic electrochemical immunodevice integrated with nanobioprobes onto graphene film for ultrasensitive multiplexed detection of cancer biomarkers. *Anal. Chem.* **2013**, *85*, 8661–8668. [CrossRef] [PubMed]

120. Gao, X.; Boryczka, J.; Kasani, S.; Wu, N. Enabling direct protein detection in a drop of whole blood with an "on-strip" plasma separation unit in a paper-based lateral flow strip. *Anal. Chem.* **2020**, *93*, 1326–1332. [CrossRef] [PubMed]

121. Yildiz, U.H.; Alagappan, P.; Liedberg, B. Naked eye detection of lung cancer associated miRNA by paper-based biosensing platform. *Anal. Chem.* **2013**, *85*, 820–824. [CrossRef]

122. Källander, K.; Tibenderana, J.K.; Akpogheneta, O.J.; Strachan, D.L. Zelee Hill, Augustinus HA ten Asbroek, Lesong Conteh, Betty R Kirkwood, and Sylvia R Meek. 2013. Mobile health (mHealth) approaches and lessons for increased performance and retention of community health workers in low-and middle-income countries: A review. *J. Med. Internet Res.* **2013**, *15*, e17. [CrossRef]

123. Aydindogan, E.; Guler Celik, E.; Timur, S. Paper-Based analytical methods for smartphone sensing with functional nanoparticles: Bridges from smart surfaces to global health. *Anal. Chem.* **2018**, *90*, 12325–12333.

124. Kassal, P.; Horak, E.; Sigurnjak, M.; Steinberg, M.D.; Steinberg, I.M. Wireless and mobile optical chemical sensors and biosensors. *Rev. Anal. Chem.* **2018**, *37*, 20170024. [CrossRef]

125. Chandra Kishore, S.; Samikannu, K.; Atchudan, R.; Perumal, S.; Edison, T.N.J.I.; Alagan, M.; Sundramoorthy, A.K.; Lee, Y.R. Smartphone-Operated Wireless Chemical Sensors: A Review. *Chemosensors* **2022**, *10*, 55. [CrossRef]

126. Xu, J.; Chen, X.; Khan, H.; Yang, L. A dual-readout paper-based sensor for on-site detection of penicillinase with a smartphone. *Sens. Actuators B Chem.* **2021**, *335*, 129707. [CrossRef]

127. Shrivas, K.; Patel, S.; Thakur, S.S.; Shankar, R.; et al. Food safety monitoring of the pesticide phenthoate using a smartphone-assisted paper-based sensor with bimetallic Cu@ Ag core–shell nanoparticles. *Lab Chip* **2020**, *20*, 3996–4006. [CrossRef] [PubMed]

128. Zhang, W.; Niu, X.; Li, X.; He, Y.; Song, H.; Peng, Y.; Pan, J.; Qiu, F.; Zhao, H.; Lan, M. A smartphone-integrated ready-to-use paper-based sensor with mesoporous carbon-dispersed Pd nanoparticles as a highly active peroxidase mimic for H_2O_2 detection. *Sens. Actuators B Chem.* **2018**, *265*, 412–420. [CrossRef]

129. Shrivas, K.; Kant, T.; Karbhal, I.; Kurrey, R.; Sahu, B.; Sinha, D.; Patra, G.K.; Deb, M.K.; Pervez, S.; et al. Smartphone coupled with paper-based chemical sensor for on-site determination of iron (III) in environmental and biological samples. *Anal. Bioanal. Chem.* **2020**, *412*, 1573–1583. [CrossRef] [PubMed]

130. Biswas, S.K.; Chatterjee, S.; Bandyopadhyay, S.; Kar, S.; Som, N.K.; Saha, S.; Chakraborty, S. Smartphone-enabled paper-based hemoglobin sensor for extreme point-of-care diagnostics. *ACS Sens.* **2021**, *6*, 1077–1085. [CrossRef]

131. Wang, K.; Yang, J.; Xu, H.; Cao, B.; Qin, Q.; Liao, X.; Wo, Y.; Jin, Q.; Cui, D. Smartphone-imaged multilayered paper-based analytical device for colorimetric analysis of carcinoembryonic antigen. *Anal. Bioanal. Chem.* **2020**, *412*, 2517–2528. [CrossRef] [PubMed]

132. Fan, Y.; Shi, S.; Ma, J.; Guo, Y. Smartphone-based electrochemical system with multi-walled carbon nanotubes/thionine/gold nanoparticles modified screen-printed immunosensor for cancer antigen 125 detection. *Microchem. J.* **2022**, *174*, 107044. [CrossRef]

133. Feng, Q.M.; Pan, J.B.; Zhang, H.R.; Xu, J.J.; Chen, H.Y. Disposable paper-based bipolar electrode for sensitive electrochemiluminescence detection of a cancer biomarker. *Chem. Commun.* **2014**, *50*, 10949–10951. [CrossRef]

134. Tian, R.; Li, Y.; Bai, J. Hierarchical assembled nanomaterial paper-based analytical devices for simultaneously electrochemical detection of microRNAs. *Anal. Chim. Acta* **2019**, *1058*, 89–96. [CrossRef]

135. Fakhri, N.; Abarghoei, S.; Dadmehr, M.; Hosseini, M.; Sabahi, H.; Ganjali, M.R. Paper-based colorimetric detection of miRNA-21 using Ag/Pt nanoclusters. *Spectrochim. Acta Part Mol. Biomol. Spectrosc.* **2020**, *227*, 117529. [CrossRef]

136. Cai, X.; Zhang, H.; Yu, X.; Wang, W. A microfluidic paper-based laser-induced fluorescence sensor based on duplex-specific nuclease amplification for selective and sensitive detection of miRNAs in cancer cells. *Talanta* **2020**, *216*, 120996. [CrossRef]

137. Eid, M.; Rashed, A.N.Z.; Bulbul, A.A.M.; Podder, E. Mono-rectangular core photonic crystal fiber (MRC-PCF) for skin and blood cancer detection. *Plasmonics* **2021**, *16*, 717–727. [CrossRef]

138. Lee, J.; Lee, Y.J.; Ahn, Y.J.; Choi, S.; Lee, G.J. A simple and facile paper-based colorimetric assay for detection of free hydrogen sulfide in prostate cancer cells. *Sens. Actuators B Chem.* **2018**, *256*, 828–834. [CrossRef]

139. Mazzu-Nascimento, T.; Morbioli, G.G.; Milan, L.A.; Donofrio, F.C.; Mestriner, C.A.; Carrilho, E. Development and statistical assessment of a paper-based immunoassay for detection of tumor markers. *Anal. Chim. Acta* **2017**, *950*, 156–161. [CrossRef] [PubMed]

140. Prasad, K.S.; Cao, X.; Gao, N.; Jin, Q.; Sanjay, S.T.; Henao-Pabon, G.; Li, X. A low-cost nanomaterial-based electrochemical immunosensor on paper for high-sensitivity early detection of pancreatic cancer. *Sens. Actuators B Chem.* **2020**, *305*, 127516. [CrossRef] [PubMed]

141. Sun, X.; Li, B.; Tian, C.; Yu, F.; Zhou, N.; Zhan, Y.; Chen, L. Rotational paper-based electrochemiluminescence immunodevices for sensitive and multiplexed detection of cancer biomarkers. *Anal. Chim. Acta* **2018**, *1007*, 33–39. [CrossRef]

142. Sun, Y.; Ge, S.; Xue, J.; Zhou, X.; Lu, W.; Li, G.; Cao, X. Highly sensitive detection of cytochrome c in the NSCLC serum using a hydrophobic paper-based–gold nanourchin substrate. *Biomed. Opt. Express* **2020**, *11*, 7062–7078. [CrossRef]

143. Wu, L.; Zhang, Y.; Wang, Y.; Ge, S.; Liu, H.; Yan, M.; Yu, J. A paper-based electrochemiluminescence electrode as an aptamer-based cytosensor using PtNi@ carbon dots as nanolabels for detection of cancer cells and for in-situ screening of anticancer drugs. *Microchim. Acta* **2016**, *183*, 1873–1880. [CrossRef]

144. Sun, S.; Wang, Y.; Ming, T.; Luo, J.; Xing, Y.; Liu, J.; Xiong, Y.; Ma, Y.; Yan, S.; Yang, Y.; et al. An origami paper-based nanoformulated immunosensor detects picograms of VEGF-C per milliliter of blood. *Commun. Biol.* **2021**, *4*, 121. [CrossRef]

145. Sharma, B.; Parajuli, P.; Podila, R. Rapid detection of urokinase plasminogen activator using flexible paper-based graphene-gold platform. *Biointerphases* **2020**, *15*, 011004. [CrossRef]

146. Zhou, T.; Huang, R.; Huang, M.; Shen, J.; Shan, Y.; Xing, D. CRISPR/Cas13a powered portable electrochemiluminescence chip for ultrasensitive and specific MiRNA detection. *Adv. Sci.* **2020**, *7*, 1903661. [CrossRef]

147. Mukama, O.; Wu, W.; Wu, J.; Lu, X.; Liu, Y.; Liu, Y.; Liu, J.; Zeng, L. A highly sensitive and specific lateral flow aptasensor for the detection of human osteopontin. *Talanta* **2020**, *210*, 120624. [CrossRef] [PubMed]

148. Ge, S.; Ge, L.; Yan, M.; Song, X.; Yu, J.; Liu, S. A disposable immunosensor device for point-of-care test of tumor marker based on copper-mediated amplification. *Biosens. Bioelectron.* **2013**, *43*, 425–431. [CrossRef] [PubMed]

149. Aydindogan, E.; Ceylan, A.E.; Timur, S. based colorimetric spot test utilizing smartphone sensing for detection of biomarkers. *Talanta* **2020**, *208*, 120446. [CrossRef] [PubMed]

150. Bahavarnia, F.; Saadati, A.; Hassanpour, S.; Hasanzadeh, M.; Shadjou, N.; Hassanzadeh, A. Paper-based immunosensing of ovarian cancer tumor protein CA 125 using novel nano-ink: A new platform for efficient diagnosis of cancer and biomedical analysis using microfluidic paper-based analytical devices (μPAD). *Int. J. Biol. Macromol.* **2019**, *138*, 744–754. [CrossRef]

151. Yen, Y.K.; Chao, C.H.; Yeh, Y.S. A graphene-PEDOT: PSS modified paper-based aptasensor for electrochemical impedance spectroscopy detection of tumor marker. *Sensors* **2020**, *20*, 1372. [CrossRef]

152. Patel, D.; Shah, Y.; Thakkar, N.; Shah, K.; Shah, M. Implementation of artificial intelligence techniques for cancer detection. *Augment. Hum. Res.* **2020**, *5*, 6. [CrossRef]

Review

Recent Advances in Microfluidic Platform for Physical and Immunological Detection and Capture of Circulating Tumor Cells

Mahesh Padmalaya Bhat [1,2], Venkatachalam Thendral [1], Uluvangada Thammaiah Uthappa [3], Kyeong-Hwan Lee [2,4], Madhuprasad Kigga [1], Tariq Altalhi [5], Mahaveer D. Kurkuri [1,*] and Krishna Kant [6,*]

[1]   Centre for Research in Functional Materials (CRFM), Jain Global Campus, Jain University, Bengaluru 562112, Karnataka, India; maheshbhat1306@gmail.com (M.P.B.); thendralsasivenkat@gmail.com (V.T.); madhuprasad@jainuniversity.ac.in (M.K.)

[2]   Agricultural Automation Research Center, Chonnam National University, Gwangju 61186, Korea; khlee@chonnam.ac.kr

[3]   School of Chemical Engineering, Yeungnam University, 280 Daehak-ro, Gyeongsan 38541, Korea; sanjuuthappa@gmail.com

[4]   Department of Convergence Biosystems Engineering, Chonnam National University, Gwangju 61186, Korea

[5]   Department of Chemistry, Faculty of Science, Taif University, Taif 21944, Saudi Arabia; tmmba@windowslive.com

[6]   Departamento de Química Física, Campus Universitario, CINBIO Universidade de Vigo, 36310 Vigo, Spain

*   Correspondence: mahaveer.kurkuri@jainuniversity.ac.in (M.D.K.); krishna.kant@uvigo.es (K.K.)

Abstract: CTCs (circulating tumor cells) are well-known for their use in clinical trials for tumor diagnosis. Capturing and isolating these CTCs from whole blood samples has enormous benefits in cancer diagnosis and treatment. In general, various approaches are being used to separate malignant cells, including immunomagnets, macroscale filters, centrifuges, dielectrophoresis, and immunological approaches. These procedures, on the other hand, are time-consuming and necessitate multiple high-level operational protocols. In addition, considering their low efficiency and throughput, the processes of capturing and isolating CTCs face tremendous challenges. Meanwhile, recent advances in microfluidic devices promise unprecedented advantages for capturing and isolating CTCs with greater efficiency, sensitivity, selectivity and accuracy. In this regard, this review article focuses primarily on the various fabrication methodologies involved in microfluidic devices and techniques specifically used to capture and isolate CTCs using various physical and biological methods as well as their conceptual ideas, advantages and disadvantages.

Keywords: circulating tumor cells (CTCs); microfluidic device; physical method; biological method; cancer diagnostics

Citation: Bhat, M.P.; Thendral, V.; Uthappa, U.T.; Lee, K.-H.; Kigga, M.; Altalhi, T.; Kurkuri, M.D.; Kant, K. Recent Advances in Microfluidic Platform for Physical and Immunological Detection and Capture of Circulating Tumor Cells. *Biosensors* **2022**, *12*, 220. https://doi.org/10.3390/bios12040220

Received: 11 March 2022
Accepted: 4 April 2022
Published: 7 April 2022

1. Introduction

Cancer is defined as the uncontrolled proliferation of aberrant cells in the human body, and it is classified into two types: benign and malignant cancers. A benign tumor that grows slowly and has no negative effects on the human body. Malignant tumors, on the other hand, are aggressive, grow quickly, spread rapidly and eventually kill the patient. During metastasis, some tumor cells at the primary tumor's borders undergo a process known as epithelial-mesenchymal transition (EMT), in which the cells lose their epithelial traits and gain migratory mesenchyme properties [1]. These migratory tumor cells enter adjacent arteries and start travelling along with red and white blood cells throughout the body. CTCs enter the bloodstream through the vasculature and circulate alongside healthy hematological cells before metastasis [2,3]. However, these can only be diagnosed if the patient has progressed to the metastatic stage [4]. These CTCs stop internally at some organs and trigger secondary tumors; from this stage onwards, the cancer enters its deadliest form, and the patient could face fatal consequences [5,6]. Hence, the early detection of these cells or the monitoring of their presence in the bloodstream is required and important for the accurate diagnosis and prognosis of cancer [7]. A survey has shown that malignant tumors will be the major cause of death worldwide by 2030, expected to grow to 20.3 million new cancer cases and 13.2 million deaths [8].

However, CTCs are extremely rare among hematological cells. There are only a few CTCs in a 1.0 mL blood sample, where nearly 5 billion red blood cells (RBCs) and 10 million white blood cells (WBCs) are present. In addition, the CTCs may exist in a single-cell or cluster form, with varied phenotypic properties. Based on the changes in protein expression on CTCs, they can be classified into epithelial-mesenchymal, epithelial, and mesenchymal types [9]. Therefore, collecting and isolating them from other components in the bloodstream is quite difficult and challenging [10]. Detection of these rare cells using sensors would be beneficial. Sensors have previously been used for environmental applications [11–17]. On the other hand, sensors would be ideal for the detection of these rare CTCs. Currently, several techniques such as flow cytometry, enzyme-linked immunosorbent assay (ELISA), Western blotting, quantitative polymerase chain reaction (Q-PCR), magnetic-activated cell sorting (MACS), fluorescence-activated cell sorting (FACS) and centrifugation techniques, and laser-based technology are widely used for the biomolecular or cellular analysis of cancer [18–26]. Although these techniques have several limitations, such as substantial sample consumption, low throughput, lack of real-time monitoring, and high overall operational expenses, there are no other alternative simple techniques available for CTC isolation. As a result, there is a great scientific desire to improve cancer diagnosis using low-cost procedures [27].

In our opinion, microfluidic devices are one of the most intriguing methods for capturing and isolating CTCs from blood samples. Microfluidic devices have many advantages, including their high throughput, low cost, miniaturization, quick analysis, high sensitivity, precise operation, high efficiency, portability, low sample consumption, and accuracy [28–34]. As the name implies, microfluidics is concerned with accurate fluid flow management in microliters (10–6) to picoliters (10–12) within micro-volume channels [35]. Various techniques like 3D printing [36], molding, laminating, and high-resolution nanofabrication are used to create these devices. S.C. Terry reported the first lab-on-a-chip (LOC) analysis system in 1979, which was investigated for gas chromatography applications [37]. Since then, microfluidic devices have been investigated for a variety of applications, including biosensors [38], separation [39], analysis [40], drug delivery [41,42], optoelectronics [43], cell manipulation [44], and chemical synthesis [45,46]. There has been much advancement in surface chemistry, which has enabled the development of smart surfaces and devices for various applications [47,48]. In comparison to other approaches, microfluidic channels have a high surface-to-volume ratio [49]. Microfluidic devices are usually made from polymers such polyethylene glycol diacrylate (PEGDA) [50], parylene [51], and polydimethylsiloxane (PDMS) [52,53]. There are two types of microfluidic technologies for capturing and isolating CTCs: physical and biological methods [54]. The different intrinsic features of cell populations, including their density [55], size [56], compressibility [57], deformability [58], dielectric properties [59], and viscosity [60], are used to physically separate CTCs [61,62]. Deterministic lateral displacement, inertial microfluidics, micropores, micropillar arrays, vortex-mediated deformability cytometry (VDC), inertial focusing dielectrophoresis, acoustic waves, and optical approaches have all been reported for the detection and separation of CTCs [63–69]. Biological approaches, on the other hand, rely on specific surface proteins produced on tumor cells to act as molecular recognizers such as transferrin, peptides, sialic acid, and antibodies to trap and isolate CTCs [70–72]. The two primary kinds of biological techniques are positive and negative sorting. The epithelial cell adhesion molecule (EpCAM) is a one-of-a-kind biomarker for positive sorting, which uses CTCs as target cells. Negative sorting, on the other hand, uses CD1513, anti-CD6647, and anti-CD45 biomarkers to identify leukocytes as target cells. Physical approaches are simple to use and do not require expensive biomarkers or a long incubation period. Nonetheless, they lack specificity and isolation purity. Biological approaches, on the other hand, need a more involved, time-consuming, and costly procedure. Yet, they have a high level of specificity, purity, and efficiency [73]. As a result, a physical-based approach is one of the most effective and straightforward methods for capturing and isolating CTCs. In this review, we have highlighted and critically examined the recent relevant literature on the fabrication of microfluidic devices for CTC isolation and the most promising elements of CTC capture and isolation, employing innovative microfluidic devices such as physical and biological techniques. The general technologies involved in the physical and biological separation of CTCs are depicted in Figure 1.

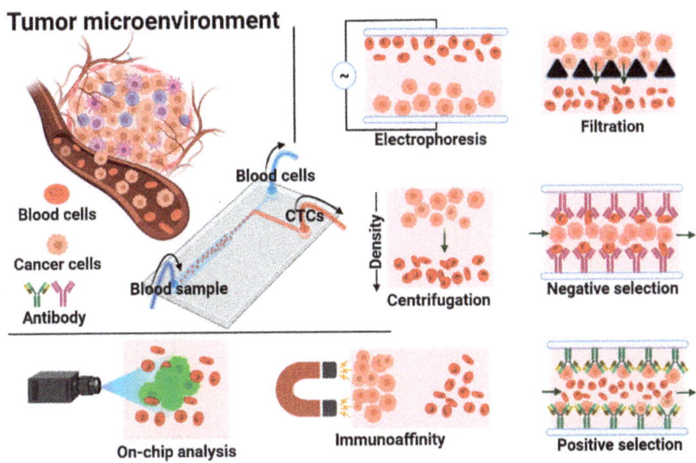

Figure 1. Range of methods involved in physical and biological approaches for early-stage detection and isolation of CTCs.

2. Fabrication of Microfluidic Devices for the Isolation of CTCs

In a very short span of time, microfluidics has emerged in several technological advancements. There are a variety of materials for microfluidic device fabrication, each with different properties according to the requirements. Based on the required specific characteristics of the fabrication material and product requirements, different techniques are used for the development of the device. Another major aspect is the cost of the involved material. In most cases, used devices are disposed of. Thus, the method involved should be economically feasible. Herein, we have classified the most recent techniques adapted in the fabrication of microfluidic devices for the isolation of CTCs.

2.1. Additive Manufacturing

Molding techniques involving PDMS and other thermoplastics are the most common approaches to fabricating microfluidic devices [74]. The disadvantages of traditional fabrication approaches are that they require a cleanroom, are expensive, utilize time-consuming wafer processes, and require the labor-intensive manual assembly of multiple layers. These factors have limited their wide application [75,76]. Furthermore, it is difficult to efficiently fabricate true 3D structures with large surface areas to increase CTC capture efficiency [77,78]. In recent years, 3D printing, which can create 3D objects layer by layer, has received a lot of attention as a potential replacement for the PDMS-based conventional molding process. In the additive manufacturing (AM) approach, the device is fabricated using a 3D printer and computer-aided design (CAD) software to design the desired shape in a short amount of time. Chu et al. created monolithic microfluidic devices to separate CTCs from whole blood samples [79] (Figure 2a). The fabricated device has a 100 mm channel length, 20.5 mm breadth and 19.2 mm width. The microfluidic device is comprised of two inlets for a sample, a buffer, and an outlet for collecting the waste. The main advantages of this device are that during the filtration process, potential cell damage due to handling the sample was eliminated, and the desired pore size could be attained with high resolution in commercially available membrane filters. Further, Gong et al. developed controlled-compression integrated microgaskets (CCIMs) and simple integrated microgaskets (SIMs), which are bound with small chips to form a wider connection of chips accomplished by a microelectromechanical system (MEMS) and nanoelectromechanical system (NEMS) [80]. SIMs or CCIMs are 3D printed as part of the device's fabrication. Thus, no additional materials or components are needed to connect to the larger 3D-printed interface chip. Later, Chen et al. developed a microfluidic device with 3D-printed internal structures to facilitate high fluid flow and surface area [81]. The printed structure was functionalised with EpCAM antibodies to capture CTCs.

2.2. Etching Technique

Etching is the process of protecting the desired area of a substrate while treating the other in order to remove a particular depth of material. The parts that we do not want to etch are usually protected. Liu et al. used wet etching and thermal bonding to create a pyramid-shaped microfluidic device with one inlet and six outlets [82]. The microchamber is a critical functional component of microfluidic devices for CTC separation. A layer of chemical-corrosion-resistant adhesive tape was pasted on a standard glass slide, and a laser ablation system was used to transfer the desired prototype onto the adhesive tape. The first round of tape was then peeled off, and the glass slide with patterned tape was immersed in the etching solution for 25 min at an etch rate of 1 μm/min. The second and third annular tapes were peeled off, and the glass slide with patterned tapes was dipped in the etching solution for 7 and 8 min, respectively. After the device was completed, a laser was used to punch one inlet and six outlets to allow the blood samples to flow. Each outlet was located on a different layer at different heights of the microfluidic device. The first, second, and third steps were respectively 40, 15, and 4~8 μm high. The device showed a throughput of ~99%. The device has the advantages of being simple to set up, having high isolation efficiency, demonstrating improved throughput and not requiring an expensive capture reagent. Further, Yang et al. reported a wet etching and thermal bonding process to create a unique, low-cost, wedge-shaped microfluidic device made of two glass pieces with appropriate specificity and sensitivity [83]. The device is comprised of two inlets, a linear reservoir, and an outlet. After coating a standard glass slide with a chemical-corrosion-resistant adhesive tape, the laser ablation system was used to transfer the microchannel design onto the adhesive tape. The glass slide coated with patterned tape was immersed in a glass etching solution to create a microchannel with a continuously decreasing height (from 60 to 5 μm). Then, two inlets and an outlet (0.5 mm in diameter) were drilled on the glass slide to obtain the final chip. After a dynamic heating and annealing process in a programmable muffle furnace, the two glass slides were bonded together.

2.3. Mold Punching Technique

The fabrication of microstructures via conventional techniques can be costly due to the need for expensive equipment set up and maintenance and the time-consuming nature of the process. If micro- or nano-scale processes can be replicated, manufacturing costs can be drastically lowered. In this technique, micro/nanostructure molds are fabricated once, and products can be duplicated from them. The inverted or negative aspects of the device construction are present in the masters [84]. Liao et al. created an optically induced dielectrophoresis (ODEP) microfluidic device with a T-shaped microchannel made up of four layers: layer A (PDMS), layer B (indium-tin-oxide glass substrate), layer C (double-sided adhesive), and layer D (indium-tin-oxide glass substrate coated with photoconductive material) for the isolation of CTCs using EpCAM/CD45 markers [85]. To facilitate cell suspension transfer, the main channel and side channel's dimensions (L $\times$ W $\times$ H) were set to 2500 $\times$ 1000 $\times$ 60 μm and 2500 $\times$ 400 $\times$ 60 μm, respectively. The junction area in the T-shaped microchannel that was specified for CTC separation was 1400 $\times$ 1000 $\times$ 60 μm. The device consisted of three punch holes for tubing connections, with each hole used for loading the sample, harvesting the fresh, waste cell suspension samples, and collecting the separated cells. The advantages of this device included the fact that the cell manipulation process was simpler and easy to operate.

2.4. Photolithography Technique

Photolithography has been widely used in the fabrication of microfluidic devices. It entails exposing a photoresist-coated substrate to light so that the selectively developed regions can be protected from/subjected to subsequent fabrication processes like etching or deposition [86,87]. This process, however, necessitates the use of costly photolithographic facilities with specialized lighting for working with ultraviolet (UV)-sensitive materials [88] and uses light-sensitive photoresist to transfer a geometric design from a photomask to a smooth surface. On a glass slide, Kwak et al. reported a spiral-shaped channel microfluidic device [89]. Each circular channel measures 250 μm in width and has a gap between them, with a channel depth of 130 μm. The distance between the spiral channel and the magnet (i.e., radius) was reduced from 3500 μm to 500 μm. High throughput and selectivity are two advantages of this design (Figure 2b). Further, Fan et al., on the other hand, devised a novel size-based separation approach for the rapid identification and isolation of CTCs [90]. The authors created a microfluidic device based on a polydimethylsiloxane (PDMS) membrane filter. The device had a thickness of 60 μm, a diameter of 6.9–10.8 μm, and a gap of 25 μm between two holes. The microfilter produced using lithography has several advantages, including precise, uniformly dispersed pores, high porosity, low cost, and quick processing. However, this method is not suitable for mass production. Later, Yan et al. fabricated an electrochemical microchip for high-efficiency CTC isolation to address the limitations of prior efforts [91]. The PDMS micropillar-array-based electrochemical microchip had hierarchical structures spanning from μm to nm, which were created using a traditional soft lithography approach and then gold layer plating for the electrochemical capture and lysing of captured cells. Similarly, Zhou et al. created a PDMS-based multi-flow microfluidic system using dry film resist instead of SU-8, followed by soft photolithography [62]. The developed straight channel had a length, width, and height of 20 mm, 150 μm, and 50 μm on PDMS, which was bound to a glass slide followed by plasma treatment.

Kulasinghe et al. designed a multi-flow straight microchannel of 50 μm height and 150 μm width, with two inputs and two outputs for inertial cell migration [92]. The device uses size-dependent separation from the inertial movement of a mixture of WBCs and CTCs, allowing for the isolation of larger CTC clusters as the channel length increases. The sample was injected through the outer inlet, while the phosphate-buffered saline was injected through the inner inlet (PBS). Cells migrated transversely from the sample zone into the clean buffer flow channel as a result of inertial force. Yoon et al. designed a 4.5 × 4.5 cm² microfluidic device on a surface-oxidized silicon chip [93]. The device has two inlets for the sample and the buffer, which are followed by two outlets for waste and isolated cells. The main channel measured 500 μm in length. The slanted weir runs from the upper side of the main channel wall to the branch point. The height of the slanted weir was 7 μm lower than the height of the main channel. A double-layer photolithographic technique was used to pattern the slanted weir-integrated microfluidic channel. Initially, the first layer was spin-coated with a thickness of 23 μm using negative photoresist SU-8 2050, and the second layer with a weir gap was spin-coated with a thickness of 7 μm using negative photoresist SU-8 2007 to get the expected slanted weir-designed device. In other work, Chen et al. fabricated a PDMS-based microfluidic design consisting of gallium nitride (GaN) and aluminium gallium nitride (AlGaN) layers integrated with a field-effect-transistor (FET) chip of 1.2 × 0.8 mm by the plasma etching and metal deposition method followed by a molecular beam epitaxy process [94]. Photoresist SU-8 was spin-coated on a silicon wafer with a thickness of 30 μm; the length, width and height of the channel were set to 60, 20, and 30 μm, respectively. The upper layer of the device was composed of two inlets for cells and buffered saline with four trapping microchambers for cell capturing. The bottom layer was embedded with an FET sensor array on the epoxy substrate. Raillon et al. printed a circuit board coated with a positive photoresist to fabricate a label-free PDMS microfluidic device for the isolation and enumeration of CTCs from human blood samples [95]. At first, a glass wafer was coated with lift-off resist and positive photoresist, followed by printing electrodes using a laser writer to achieve a glass chip with electrodes. Secondly, an SU-8 mold was used to develop a PDMS chip using standard photolithography. The glass chip and PDMS chip were combined to form a single PDMS impedance chip. The channel dimensions were 70, 16, and 40 μm in depth, length, and width, respectively. The device consisted of a plastic vortex chip and PDMS impedance chip, which were connected for fluidic flow. Syringe pumps were used for the injection of the sample into the channel. The flow rate was optimized for the vortex chip at 7 mL/min and reduced to 100 μL/min for the impedance chip. Further, captured CTCs were flushed out with an increased buffer flow rate of 8 mL/min. An excitation voltage was applied between two electrodes at 500 mV and 460 kHz frequency with a flow rate of 100 μL/min, 10 kHz bandwidth, and 100 kHz sampling frequency to detect cancer cells. The advantages of the specific electrode design chip included its high-frequency measurements, ease of fabrication, and fast particle counting (Figure 2c).

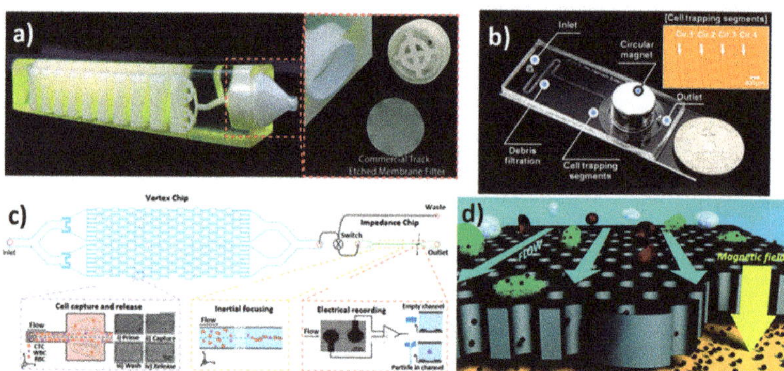

Figure 2. Microfluidic device designs fabricated using various techniques. (**a**) 3D-printed device showing the microchannels with layers and the filter holder; reprinted with permission from ref. [79], 2019, Royal Society of Chemistry. (**b**) Microfluidic device fabricated by photolithography showing spiral channel and cell trapping segments; reprinted with permission from ref. [89], 2018, Elsevier. (**c**) Representation of label-free enumeration of CTCs using a vortex chip connected to an impedance cytometry chip; reprinted with permission from ref. [95], 2019, John Wiley and Sons. (**d**) Schematic of the detection strategy of the micro-aperture chip system for CTC detection; reprinted with permission from ref. [96], 2015, Royal Society of Chemistry.

Similarly, Chen et al. fabricated a PDMS-based hybrid magnet-deformability CTC chip patterned through a photolithographic technique [97]. The thickness of the silicon wafer post-spin coat was 7 μm, where the CTCs were isolated using a magnetic force. The 12 rows of micro-elliptical pillars were designed within the channel. The distance between adjacent micropillars was gradually reduced from 18 to 5 μm for effective CTC removal, while the width between adjacent arrays remained constant at 1500 μm. The presence of a magnet beneath the device aided in increased the capturing efficiency. The micro-ellipse was comprised of three parts, which include a half-ellipse with a semi-long axis of 30 μm, a half-circle with a radius of 15 μm, a rectangle with a length of 30 μm and a device with a depth of 55 μm. Furthermore, Varillas et al. developed a PDMS-based geometrically enhanced mixing (GEM) microfluidic chip with two layers of SU-8 coating (main channel layer and herringbone mixer layer) for the isolation of CTCs using EpCAM antibodies [98]. For the main channel, the thickness of the SU-8 2035 photoresist was 50 μm. The herringbone mixer layer was formed by adding a second layer of SU-8 after UV light exposure and post-soft baking. A precise arrangement between the main channel and the mixer was maintained to create the herringbone mixer pattern. The inlet and outlet wells were created by punching the holes in PDMS after a second exposure was performed. Shamloo et al. fabricated a new integrated Y-shaped microfluidic device consisting of two subunits, a functional unit and a mixing unit, through SU-8 photoresist patterning and a wet etching process for the immunomagnetic separation of CTCs [99]. The blood samples spiked with CTCs were passed through a 500 μm wide inlet channel. The functional unit and mixing unit had dimensions (L × W) of 12 × 4 mm and 9 × 1 mm, respectively, in which the channel was subjected to an alternative voltage by 10 electrodes arranged in a zigzag pattern. The sample flowed for 3 mm before reaching the diverging region, where it extended for 7 mm towards the outlets. Non-tagged cells were collected through the upper outlet, while magnetic-particle-tagged cells were isolated through the lower outlets, which had a magnet beneath them. The important features of the device were its simple geometry, high efficiency, and high feasibility. However, it was lacking in high performance.

Chang et al. used a silicon fabrication process to create a PDMS-based microfluidic chip to capture CTCs [96] (Figure 2d). The device was made up of 8 microchips with dimensions (L × W) of 40 × 20 mm. Each chip had a 9 mm by 3 mm porous area in the center, with a pore area thickness of 50 μm. These microchips were covered with a 1 mm thick glass slide. A PDMS layer of ~2 mm thickness was used as a spacer between the glass slide and microchips to form the fluidic chamber. The dimension of the fluidic chamber was defined by a laser cutter with a 30 mm by 3.8 mm grove. The entire setup was placed on an acrylic stand where a magnet was placed. The inlet and outlet were connected to the sample source and peristaltic pump, respectively. This parallel flow micro-aperture chip system has several advantages, including compatibility, ease of use, and the ability to reuse the chip for cell analysis. Later, Chen et al. used soft lithography to create a microfluidic device with a microwell-structured array for the analysis and isolation of targeted tumor cells [100]. The length and depth of the channel were 10 mm and 60 μm, respectively and the width of the chamber was 2.3 mm. The depth and diameter of four various-sized microwell structures were 5.0 μm/18 μm, 5.0 μm/20 μm, 5.0 μm/22 μm, and 8.0 μm/20 μm, respectively, with excellent selectivity for CTCs. Hoshino et al. designed a PDMS-based immunomagnetic microchip for the capture of CTCs from spiked cultured cancer cell lines by magnetic nanoparticles (MNPs) functionalized with EpCAM antibody [101]. UV-patterned SU8-photoresist coated on the silicon wafer was used as a master. The developed microchannel on PDMS was bonded on a glass substrate with a thickness of 150 μm. The developed microchannel measured 30 mm in length, 20 mm in width, and 500 μm in height. Fallahi et al. used photolithography to create a stretchable, flexible microfluidic device for the size-based separation of CTCs [102]. The channel dimensions of the developed device were 100 mm, 100 μm, and 45 μm in length, width, and height, respectively. There were sample and buffer flow inlets as well as waste, large-cell outlets, and small-cell outlets. The entire chip was placed on a specially designed stretching platform (Figure 3a). However, when compared to other size-based microfluidic separation techniques, the device setup was complicated. Further, Jiang et al. demonstrated the use of microbubbles to extract CTCs in a label-free, high-throughput acoustic microstreaming technique [103]. SU-8 2075 photoresist and soft lithography were used to construct the device on a 4-inch silicon wafer. The system was made up of 101 pairs of lateral cavity acoustic transducers (LCATs), each with one inlet and two outlets. The device had a width of 750 μm and was mounted on a piezoelectric transducer with ultrasonic gel between them. The isolation of CTCs by LCATs depended on the oscillation of trapped microbubbles in lateral slanted dead-end side channels to generate a first-order oscillatory flow at the air–liquid interface followed by a second-order streaming flow that consisted of an open microstreaming flow and a closed-looped microstreaming vortex. The dead-end of the channel was tilted at 15° to allow bulk flow through the microstreaming. The narrow gap in the flow area between the looped microstreaming vortex and the air–liquid interface was controlled by the voltage, which regulated the particle size that flowed through. Cells that were smaller than the gap moved forward along the flow by trapping large CTCs. This method allows rapid isolation with the potential to isolate multiple types of CTCs.

Furthermore, Jou et al. used photolithography to create a silicon-based V-BioChip with nano-pillar arrays with a chip dimension (L × W × H) of 32 × 34 × 0.7 mm [104]. A metal-assisted chemical etching technique was used to create nano-pillars within the microchamber. The chip surface was coated with a layer of polyethylene glycol-biotin (PEG-biotin) using a vapor deposition method. Streptavidin was attached to the biotin end using a liquid deposition method to improve capture efficiency. The surface-modified chip with nano-pillars promotes antigen-antibody interaction between the surface and CTCs, resulting in cell capture. Furthermore, Zhang et al. created a label-free microfluidic device for isolating CTCs from breast cancer patients' blood samples [105]. The photolithography technique was used to develop microchannels to fabricate the microfluidic device. An inlet, a cell intercept area and an outlet were present on the chip. The impurities were filtered through two layers of hexagonal columns in the microchannels. The first and second narrow channels were 50 µm and 20 µm long, respectively, with cell filtration occurring in 30 main channels and 31 side channels. The channels consist of 40 µm cylindrical wells separated by a 100 µm separation distance. Reinholt et al. created a PDMS microfluidic device using photolithography to isolate CTCs using aptamer and extract and to amplify DNA for gene mutation analysis [106]. The device consists of two orthogonal microchannels with two micropillar arrays for CTC isolation at the intersection of the two microchannels and the genomic DNA isolation array downstream of the cell capture array. The cell channel was 1 mm wide, whereas the DNA channel was 500 µm to 1 mm wide and 25 µm deep. Micropillars with a diameter of 50 µm made up the cell capture array. The DNA micropillar array was spaced in a gradient starting at 10 µm and ending at 7 µm. Nasiri et al. developed a hybrid PDMS microfluidic device for CTC isolation via inertial and magnetic separation [107]. For the isolation of CTCs from blood samples, the device consists of an asymmetric serpentine inertial channel, an inertial focusing channel and magnetic cell separation zones. The dimension of the inertial channel was set to 400 µm in width and 80 µm in height, followed by a magnetic separator channel width of 650 µm (Figure 3b).

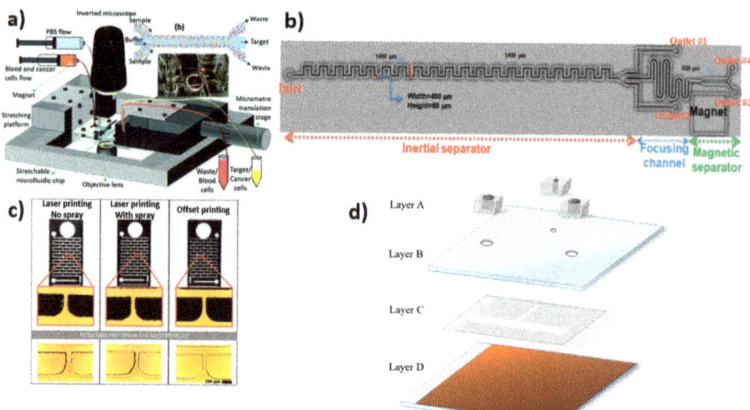

Figure 3. Microfluidic device designs fabricated using various techniques. (**a**) Schematic of the working setup of the stretchable microfluidic device, with an inset showing the multi-flow microchannel with a real stretchable microfluidic device; reprinted with permission from ref. [102], 2021, Royal Society of Chemistry. (**b**) Schematic of hybrid microfluidic cell separation device showing CTC sorting by an inertial focusing microchannel followed by magnetic separation [107]. (**c**) Photomasks fabricated using offset printing showed better resolution and smooth surface over other laser printing techniques [108]. (**d**) Schematic illustration of the optically induced dielectrophoresis (ODEP) microfluidic system assembly where Layer A was composed of fabricated polydimethylsiloxane (PDMS) components; Layer B was composed of indium-tin-oxide (ITO) glass; Layer C was composed of double-sided adhesive tape with microfabricated microchannels; and Layer D was composed of ITO glass substrate coated with a layer of photoconductive material [85].

2.5. Printing Technique

Despite the fact that 3D printing is a cheap, robust and scalable method for producing master molds [109,110], there are still challenges that have prevented microfluidic developers from adopting 3D printing, including resolution, throughput and resin biocompatibility [111]. Attempts to reduce the cost of the technique have focused on UV lighting, laser/offset printing, etc. [112–114] Laser printing and offset printing could be cost-effective alternatives to expensive photolithography technology. Nguyen et al. investigated methacrylate (MA) gel, a type of nail polish that has been shown to work as a photoresist material instead of SU-8, to develop a master mold with additional benefits such as low cost, rapid production, high resolution (100 µm thickness, 100 µm feature size), high accuracy, and reproducibility [108] (Figure 3c). They used laser and offset printing techniques for photomask generation. The fabricated microfluidic device had a diameter of 100 µm and a height of up to 1 mm. They devised a cost-effective method for fabricating microfluidic devices. To save money, standard procedures like spin coating, plasma etching, and aligners were kept out of the device fabrication. Xu et al. created a microfluidic device out of polymethyl methacrylate (PMMA) by using a laser engraving machine to create microchannels on the surface for CTC isolation [115]. The device was divided into two major components: a filtration system on top and a magnetic microfluidic chip at the bottom. The filtration system used a micropore array membrane to isolate CTCs before filtering out the waste cells. The filter membrane measured 20 × 20 mm and had a pore size of 10 µm. The CTCs with trace WBCs were rinsed off the membrane after filtration for further purification. The magnetic microfluidic chip with a magnetic base of 70 mm with a diameter of 50 mm was used for the negative sorting of CTCs. This device showed low capture efficiency and needed two steps for the isolation of CTCs. Recently, Gurudatt et al. fabricated an electrochemical microfluidic channel modified with conducting polymers by a screen printing approach using carbon ink on a glass slide [116]. The developed microchannel exhibited a width and height of 95 ± 2.5 µm and 15 µm, respectively. The screen-printed channel was dried at 60 °C for two days. Further, the channels were covered using a glass slide. Later, for the amplification of separation, the channel wall was modified with a DAT monomer to covalently attach lipids. Further, Nieto et al. fabricated microchips with pillars on a soda-lime glass substrate using a laser-direct writing technique followed by thermal treatment [25]. An aluminum film was placed on the rear side of the soda-lime glass to increase the ablation. A cylindrical array of micro-posts with 420 µm diameter with a pitch and depth of 245 µm was formed. Further, the pillars were functionalised with EpCAM to facilitate CTC isolation.

2.6. Overall Summary of the Fabrication Process

Several fabrication methods have been discussed, each with its own set of characteristics. One must know the minimum feature sizes that the above approaches can produce, as well as a variety of other criteria such as surface roughness, aspect ratio and normal working size, in order to get benefit from the available techniques. Factors such as fluidic outcomes, pressure drop, microchannel, and process time play a major role in the development of the device. Though there are several techniques available for the fabrication of microfluidic devices, photolithography-based devices are determined to be promising in terms of channel dimension precision based on the aforesaid results. The technology, however, can only be used for two-dimensional devices. Additive manufacturing, on the other hand, is cutting-edge, with the potential to create three-dimensional monolithic devices. The main disadvantage is that they are not precise enough for micrometer channels. Therefore, in the near future, 3D printing technology could overcome the challenges and replace the traditional photolithography process for fabricating microfluidic devices. In general, microfluidic devices are fabricated from a variety of materials, including silicon, glass, metals, ceramics, and hard plastics, and they require several fabrication processes, including thermal bonding, chemical etching, and reactive ion etching, which require more time and effort. PDMS-based microfluidic devices, on the other hand, have advantages due to their low cost, optical transparency and biocompatibility.

3. Isolation of Circulating Tumor Cells (CTCs) by Microfluidic Devices

During metastasis, cancer cells detach from the primary tumor and intrude apart into tissues in the bloodstream. To detect and isolate CTCs, various techniques including centrifugation, magnetic separation, microchips, filtration, micro/nano substrates and biomarkers have been used. With the widespread adoption of microfluidic techniques, a large number of researchers have worked hard to develop more efficient and reliable CTC separation technologies ranging from immunomagnetic beads to size-based microfluidic devices. Currently, the major commercialised products for CTC separation techniques include the CellSearch system, which uses immunomagnetic beads, and the CelarCellFX1 system, which uses size-dependent isolation [117]. The methods for isolating CTCs are mostly based on biological qualities of the tumor cells, such as specific antigen expression and receptor, or physical properties of the tumor cells, such as size and deformability. Inertial focusing, acoustics, microfluidic filters, optics and dielectrophoresis are some of the size and deformability-based approaches.

3.1. Size-Based Isolation

Due to its visibility and ease of management, CTC separation based on size and flexibility is one of the oldest approaches. The principle of separating cells from the main flow channel through filtration is, in fact, rather simple. Membrane devices are designed to act as a filter, allowing blood to flow while separating CTCs based on their size and deformability. When diluted blood travels through the main channel, cells greater than a certain size are captured by this membrane filtration set-up within the device, while smaller cells continue on their course and are separated. The risk of clogging, the requirement for frequent maintenance, cleaning, and the incapacity of cells to recover after filtration are all prevalent issues with these devices. This method's most serious flaw is that it can't separate more than one particle type in a single stage. Microfluidic systems capture tumor cells more efficiently via filtering because pore sizes and geometries are carefully controlled by microfabrication. Filters are divided into four types based on their structures: weir-type, pillars, crossflow, and membranes [118]. Size-based filtration using polymer membranes or microsieve membrane filter devices has been shown to extract CTCs from whole blood samples based on the morphological size differences between cancer cells (~15–40 μm in diameter) and leukocytes (~10 μm in diameter) [119]. The size, geometry, and density of the pores in the microfilters can be controlled uniformly and precisely. In addition, this technology can provide maximum sample processing capability via parallel arrays of multiple flow cells, which reduces processing time, cost, and filter clogging while facilitating mass production and high-throughput screening for large-scale clinical studies.

Yoon et al. developed and reported a slanted weir microfluidic channel to reduce haemocyte contamination during CTC isolation [93] (Figure 4a). With a flow rate of 2.5 mL/h and 3.8 mL/h for a breast cancer cell line (LM2 MDA-MB-231) at 0.8° weir to 0.5° weir, a high separation efficiency of ~97% was achieved. The viability of the collected tumor cells was also determined using the trypan blue assay, and it was found to be 97.1% for the 0.8° weir and 95.8% for the 0.5° weir. The viability of the 0.5° weir was slightly lower depending on the high flow rate and shear rate. This chip showed high separation efficiency with minimal contamination. However, the major drawback was its low throughput. Furthermore, Liu et al. developed a simple pyramid-shaped microchamber that is feasible, cost-effective, and highly efficient for CTC separation from breast carcinoma patients [82]. With an optimised flow rate of 200 μL/min, the capture efficiency of the device was assessed with a fresh blood sample in five sequence concentrations of 25–200 cells/mL using four different cancer cell lines (BGC823, H1975, PC-3, and SKBR3) spiked into DMEM medium. As a proof of concept, polystyrene beads with diameters of 10 μm (red beads) and 20 μm (blue beads) were allowed to pass through the pyramid-shaped channel at a flow rate of 200 μL/min. When the flow rate was increased to 300 μL/min, the capture efficiency increased to 92% and 89%, respectively, at different outlet heights of 6 μm and 8 μm. This method has advantages, including lower sample consumption, a simple experimental procedure, high capture efficiency, and ease of observation. Finally, from the DMEM medium, the SKBR3 cell line had a capture efficiency of 93%, while the healthy blood sample had a capture efficiency of 89%. Further, Fan et al. designed and developed a PDMS membrane filter-based technique for the isolation of CTCs [90] (Figure 4b). At a flow rate of 10 mL/h, >90% cancer cell recovery was achieved from a blood sample spiked with lung cancer cells. Later, Zhang et al. created a label-free microfluidic device for isolating CTCs from breast cancer patients [105]. At a flow rate of 10 mL/h, the device demonstrated 73.6% capture efficiency and an 82% recovery rate. The main and side microchannels were 80 μm and 50 μm and 50 μm and 50 μm in width and height, respectively; the filter microchannel was 40 μm in width 10 μm in height. The device was used to isolate CTC cell strains such as SKBR3, MCF-7, and MDAMB231. Immunofluorescence staining was used to identify the cultured cells.

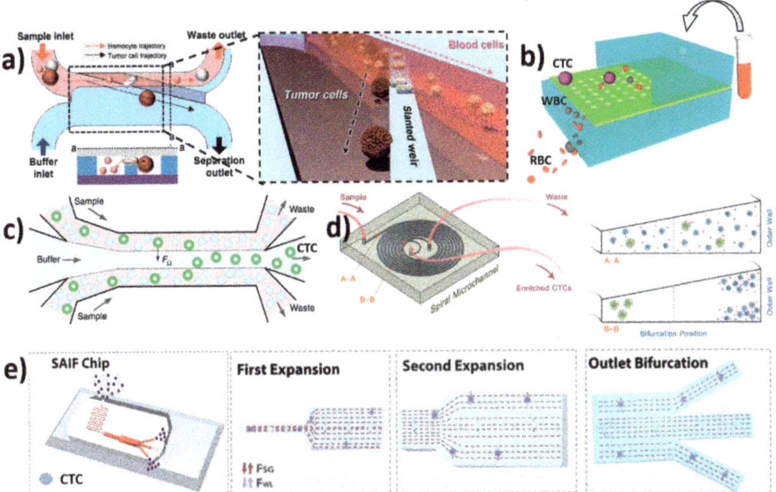

Figure 4. Schematic illustration of microfluidic devices for the isolation of CTCs using various techniques. (**a**) An overview of a slanted weir device; separation of CTCs over a slanted weir based on distinct size and deformability [93]. (**b**) Schematic of the microfluidic device integrated with a PDMS microfiltration membrane for CTC capture; reprinted with permission from ref. [90], 2015, Elsevier. (**c**) Top view of the multi-flow effect of a size-dependent inertial migration microfluidic system representing rotation-induced lift force (FΩ) for the isolation of CTCs [62]. (**d**) Enrichment of CTCs using spiral microfluidic technology utilizing inertial lift force [120]. (**e**) Illustration chip, self-amplified inertial-focused cell bifurcation of CTCs in the microfluidic channel; reprinted with permission from ref. [121], 2020, American Chemical Society.

3.2. Inertial Focusing Microchannel-Based Isolation

Inertial focusing is a phenomenon that occurs when suspended particles in a fluid stream migrate across flow lines and arrange themselves in equilibrium positions at specific cross-sectional positions. This behavior is caused by inertial forces within the channel and is controlled by channel geometry and flow conditions [122,123]. This phenomenon occurs in straight channels due to a balance of two dominating forces such as shear gradient inertial lift force (FSL), caused by the curvature of the fluid velocity profile and wall induced inertial lift force (FWL), caused by the particle's interaction with the nearby wall. The particles are pushed toward the channel walls by FSL, while they are moved away from the walls and toward the channel center by FWL [124,125]. As a result, the particles tend to attain a state of equilibrium where these forces are equal.

Zhou et al. designed a new multi-flow effect of a size-dependent inertial migration microfluidic (MFM) system for the precise detection and isolation of CTCs from spiked blood samples (H460 and HCC827) [62] (Figure 4c). The separation efficiency and purity of CTCs were obtained to be >99% and >87%, respectively, from CTC-spiked blood samples. At a concentration of 10 cells per 5 mL, the device had an efficiency of >83%. The study showed that the average size of WBCs measured around 9 μm, and the average size of the detected CTCs was 30 μm. Additionally, the channel was examined for isolating CTCs from patient blood samples (stage IV lung cancer). The device has the advantage of having a high recovery rate even at very low concentrations, throughput and sensitivity; it had a disadvantage in terms of its performance and recovery rate due to the significant size overlap between target and non-target cells. Later, Gao et al. designed a label-free CTC isolation microfluidic device utilising the advantage of hydrodynamic forces [126]. The chip has a fishbone-shaped channel, rectangular reservoir and inertial focusing microchannel for CTC isolation. RBCs spiked with U87 cells were injected at a flow rate of 9 μL/min, showing 90% separation efficiency with 84.96% purity. Kulasinghe et al. designed a spiral microfluidic chip for the isolation of head and neck cancer cells (HNCs) [120] (Figure 4d). The chip was tested with patients' blood samples at a flow rate of 1.7 mL/min. The chip utilises inherent Dean vortex flow along with inertial lift force, which drives smaller hematologic cells towards the outer wall by facilitating the efficient separation of CTCs. The chip showed 54% detection efficiency. Furthermore, Warkiani et al. reported the label-free spiral microfluidic chip for the size-based separation of CTCs from the sample using hydrodynamic forces [127]. At a flow rate of 100 μL/min, the chip achieved ≥85% isolation efficiency. The chip could isolate CTCs from a 7.5 mL sample in less than 40 min. However, stacking three chips together yielded better results by isolating CTCs from a 7.5 mL samples in less than 10 min. Thus, the chip showed high throughput. Later, Ozbry et al. developed a microfluidic chip with a symmetrically curved channel for continuous and high-throughput isolation of cancer cells [128]. The cancer cell lines MDA-MB-231, Jurkat, K562, and HeLa were injected into the curvilinear channel at a curvature angle of 280°. The flow rate was increased constantly from 400 μL/min to 2700 μL/min at an interval of 90 s for each 100 μL increase in the injection volume. The study revealed cell size based on flow velocity. The chip exhibits high viability of >94%.

Nam et al. fabricated a capillary inserted microfluidic device for the isolation tumor cells via viscoelastic flow [129]. The capillary tube facilitates 3D particle pre-alignment prior to separation. The presence of two outlets facilities the isolation of migrated particles with 5 and 10 μm diameter exhibiting ~99% isolation efficiency. At a flow rate of 200 μL/min, 94% of MCF-7 cells were isolated from leukocytes with 97% purity. Further, Abdulla et al. developed a self-amplified inertial focused (SAIF) microfluidic device for the size-based, high throughput isolation of CTCs [121] (Figure 4e). The device demonstrated a narrow zigzag microchannel connected to expansion sites to enable size-based separation. The tested cancer cells such as lung cancer cells (A549), breast cancer cells (MCF-7), and cervical cancer cells (HeLa) isolation efficiency of ~80%. Che et al. developed label-free, size-based isolation of CTCs using vertex microfluidic chip [130]. At a flow rate of 8 mL/min (for diluted blood) and 800 μL/min (for whole blood); 83% capture efficiency was recorded. Thanormsridetchai et al. developed a spiral microfluidic device for capturing of CTCs [131]. The device with five spiral microchannels (500 μm height, 130 μm width, 5.5 mm length) was injected with samples at a flow rate of 1.0 mL/min. The device showed 90% capture efficiency.

3.3. Dielectrophoresis-Based Isolation

Dielectrophoresis with external electric field sources is a quick, simple and well-known technique for manipulating a variety of biological particles within a microchannel [132]. It is also used to separate the movement of distinct cancer cells [133,134]. Cancerous cells could be separated from normal blood cells or the cell sample solution using the dielectrophoresis method based on cell properties such as size, morphology, deformability, mechanical, electrical and magnetic properties [122].

Chiu et al. investigated the size-dependent separation of cancer cell clusters using an optically induced dielectrophoresis (ODEP)-based microfluidic system [135]. The device was tested with a human prostate cancer cell line (PC-3) and leukocytes to evaluate its performance. The device could isolate as low as 15 cells/mL with a recovery rate of 41.5%. Overall, the proposed method could isolate CTCs with purity as high as 100% at a sample flow rate of 2.5 μL/min. Thus, the method was found to be promising in the isolation of CTCs with high sensitivity without interference from leukocytes. In another study, Li et al. demonstrated the dielectrophoresis technique using an array of wireless bipolar electrodes for the high-throughput isolations of CTCs [136]. The 32 parallel microchannels with 2950 μm, 200 μm, and 25 μm length, width, and height, respectively, were fabricated using the photolithography technique. The device could throughput 100 μL/h samples with a 39.6 mm^2 device footprint. Further, Kim et al. developed a dielectrophoresis cell-trapping method for the trapping of cancer cells using a microfluidic device [137]. At a flow rate of 100 μL/min, 92 ± 9% of cells were isolated at the designated location. The technique enables the isolation of very low concentrations of cancer cells from large volumes of samples with high recovery. Liao et al. developed an optically induced dielectrophoresis (ODEP)-based microfluidic device for the isolation of high-purity CD45neg/EpCAMneg cells from the blood samples of cancer patients [85]. To recognize the EpCAM, surface marker-positive CTCs and CD45 surface marker-positive leucocytes were stained using fluorescent dyes. The diameters of PC-3 and SW620 cancerous cells were found to be 20.1 ± 1.5 and 1 μm, respectively. The device demonstrated 100% CTC capture purity in capturing live CD45neg/EpCAMneg cells. The device takes around 4 h for the analysis of 4 mL of sample suspension. The recovery rate of the microfluidic device was found to be 81.0 ± 0.7%.

3.4. Magnetic Field-Based Isolation

Magnetic field-derived microfluidic chips are broadly classified as labelled methods and label-free methods of isolation. Positive and negative selection are the two most common methods of labelled magnetic isolation. CTCs can be actively isolated using functionalized magnetic nanoparticles (MNPs) when a magnetic field is applied. Specific antigen-coupled MNPs can bind to specific surface proteins expressions on CTCs, resulting in positive CTC selection [138]. Due to the diversity of cancer cells, CTCs shed from primitive tumors are highly heterogeneous, including epithelial cancer cells such as gastric cancer, mesenchymal cancer cells such as osteosarcoma and other cancer cells such as leukemia. This enables a wide range of antigens to be used to label different CTCs with antiepithelial cell adhesion molecule (EpCAM), which is the most commonly used antigen. On the other hand, negative enrichment of CTCs based on WBC depletion was achieved using anti-CD45 surface antigens because the antigens are particularly expressed on the surface of WBCs [139]. Due to inter-patient and intra-patient heterogeneity in tumor biology, particularly in the case of epithelial-mesenchymal transition (EMT), identifying CTC-specific markers becomes difficult. Meanwhile, label-free magnetic isolation isolates CTCs based on their size difference from hematological cells using magnetic fluids such as paramagnetic salt solutions or ferrofluids as media.

3.4.1. Immunomagnetic (Label)-Based Isolation

Chang et al. developed a novel parallel flow micro-aperture chip system for CTC isolation in the spiked MCF-7 cell line at a flow rate of 2 mL/min [96]. CTCs with sizes ranging from 10 to 30 μm were found in the sample solution after it had been coated with antibody-mediated magnetic beads. The chip detected approximately 89% of the spiked MCF-7 breast cancer cell lines. The device has several advantages, including its ease of use, robustness, compatibility and versatility. The device was integrated with a PDMS microfiltration membrane for CTC capture and a parallel flow micro-aperture chip system for capturing CTCs. Furthermore, clinical samples revealed the possibility of isolating cancer cells (non-small-cell lung cancer cell line and pancreatic cancer cell line) that were bound on beads and captured on the chip's surface. Furthermore, Kwak et al. investigated the selectivity and capture efficiency of the developed spiral-shaped channel device for two types of tumor cell lines, MDA-MB-231 and MCF-7, based on the level of EpCAM antigen expression [89]. The results showed that the capture efficiency of MDA-MB-231 and MCF-7 cells were 81.2 ± 3.5% and 96.3 ± 1.5%, respectively, at a flow rate of 150 µL/min. MDA-MB-231 cells had an average purity of 82.8%, while MCF-7 cells had an average purity of 85.9%. However, because of the low EpCAM expression in this reported device, several heterogeneous CTCs could not be detected and quantified. Recently, Kang et al. developed a positive and negative method for the isolation of CTCs (MDA-MB-231, PC-3, SKBR3, and MCF-7) by lateral magnetophoresis using magnetic nanobead-functionalized EpCAM and CD45/CD66b antibodies [140]. The lateral magnetophoresis technique was used to design a disposable chip with a microchannel on a multipurpose substrate fixed to ferromagnetic wires. The device works both on positive and negative methods for the isolation of CTCs using anti-EpCAM and anti-CD45/CD66b nanobeads. The ferromagnetic wires were inlaid at 5.7° towards the flow direction on the substrate. As the blood flowed through the lateral magnetophoretic microchannel, the residual magnetic nanobeads were bound to the ferromagnetic wires. The silicon-coated polymer film with a thickness of 12 µm was bonded to a microstructure PDMS replica to form a disposable microchannel substrate. The flow rate and suction rate for the sample and buffer were optimized in the positive method to 2 mL/h and 3.2 mL/h, respectively, resulting in the release of CTCs in the outlet at a flow rate of 0.8 mL/h. This device was evaluated for the isolation of the SKBR3 and MCF-7 cell lines, and the recovery rates were 93.9 ± 1.0% and 98.4 ± 1.5%, respectively. However, this method resulted in low EpCAM expression in MDA-MB-231 and PC-3 cells. Further, the flow rate for the sample and buffer was optimized to 2.8 mL/h for the negative method. The method yielded recovery rates of 85.2 ± 4.2 and 80.7 ± 7.6% for SKBR3 and MCF-7 cell lines, respectively, and 91.0 ± 2.0% and 75.7 ± 9.3% for MDA-MB-231 and PC-3 cells, respectively. A fluorescence microscope was then used to enumerate WBCs and CTCs from the outlet. The positive method produced more pure isolated CTCs than the negative method. Following this, Chen et al. developed a size-based microfluidic device with high capture efficiency for CTC isolation [97]. A few strong permanent magnets were fixed beneath the glass substrate to capture the magnetized CTCs. Three different cancerous cell lines (HCT116, SW480, and MCF-7) were tested with different EpCAM antibody expression levels to evaluate the device. Capture efficiency for MCF-7, HTC116, and SW480 was found to be up to 97.2 ± 6%, 85.7 ± 14.3%, and 91.5 ± 8.9%, respectively. Due to cell line accumulation, capture efficiency was decreased. The flow rate was optimised to 1.5 mL/h for the system operated without a magnet, which showed a capture efficiency of around 90%. The magnetic bead at a high processing rate of 3 mL/h showed a capture efficiency above 90% within 20 min. The live/dead assay revealed 96% cell viability. The reverse flushing process removed the majority of the CTCs from the channel. Despite the device's high processing rate, there was a lack of capture efficiency.

Furthermore, Shamloo et al. created a PDMS-based integrated microfluidic platform for CTC capture using an immunomagnetic technique [99]. The separation and mixing units, as mentioned in the fabrication section, use electric and magnetic forces for high throughput to increase the purity and capture efficiency in the microfluidic system. To evaluate the device's capture efficiency, anti-EpCAM functionalized iron nanoparticles were tagged to different types of blood samples spiked with 100,000 cancerous cells, such as SKBR3 (human breast cancer cell line), PC-3 (prostate cancer cell line) and Colo205 (colon cancer cell line). The capture rate for SKBR3 and Colo205 cell lines was up to 97%, while the PC-3 cell line was 107%. As a result, this integrated microfluidic device has high compatibility and feasibility in cancer research. Later, Poudineh et al. developed magnetic raking cytometry to generate a phenotypic expression of captured CTCs [141]. The device consisted of circular nickel micromagnets with an array of X-shaped structures. The size of the micromagnets was increased along the channel to enhance the CTC capture efficiency. CTCs coated with anti-EpCAM-functionalised immunomagnetic beads were retained at the capture zone of the device. In addition, Poudineh et al. reported a microfluidic approach for profiling functional and biochemical phenotypes of CTCs [142]. The device consisted of four capture zones with an X-shaped morphology and a single-cell isolation area. The aptamer-coated CTCs functionalised with MNPs were captured at four capture zones by EpCAM expression. This was followed by releasing them to a single-cell isolation area using antisense DNA. The device showed 79 ± 4% recovery efficiency. Recently, Yin et al. constructed a dual-antibody (PSMA and EpCAM)-functionalised microfluidic device for the isolation of CTCs [143]. The dual-antibody-functionalised strategy showed a significant increase in the capture efficiency for LnCAP and LnCAP-EMP cancer cell lines. The device consists of antibody- and Fe3O4@microbead-functionalised Ni (nickel) micropillars under external magnetic conditions and a chaotic herringbone platform (Figure 5a). The device could successfully identify CTCs from 20 out of 24 blood samples.

3.4.2. Label-Free-Based Magnetic Isolation

Zhao et al. demonstrated size-based ferrohydrodynamic HeLa cell isolation using a microfluidic device [144]. Cell mixtures (HeLa cells, blood cells) and ferrofluids were mixed, then injected at a flow rate of 8 μL/min. The magnetic buoyancy force caused deflections of cells from their laminar flow patterns when the magnet was placed close to the channel. The force operating on cells inside ferrofluids is a body force proportional to cell volume, resulting in the spatial separation of cells of various sizes at the microchannel's end. As a result, larger HeLa cells and smaller blood cells emerge through distinct pathways (Figure 5b). The device exhibited >99% capture efficiency. The method was found to be cost-effective with high throughput. Furthermore, Zhao et al. used label-free size-based ferrohydrodynamic CTC isolation using a microfluidic device [145]. The device showed a high throughput of 6 mL/h with a recovery rate of 92.9%. The device could isolate CTCs as low as ~100 cells/mL. In addition, the device demonstrated recovery rates for cancer cells line such as H1299 (92.3%), A549 (88.3%), H3122 (93.7%), PC-3 (95.3%), MCF-7 (94.7%), and HCC1806 (12.2%). The device showed short-term cell viability, normal proliferation, and unaffected key biomarker expression. Later, Zhao et al. developed a label-free isolation method using ferrofluids to separate low-concentration cancer cells from cell culture lines in microfluidics [146]. The isolation depended on the variation in size of CTCs with WBCs in biocompatible ferrofluids. At a throughput of 1.2 mL/h, the device showed isolation efficiencies of 80 ± 3%, 81 ± 5%, 82 ± 5%, 82 ± 4%, and 86 ± 6% for A549 lung cancer, H1299 lung cancer, MCF-7 breast cancer, MDA-MB-231 breast cancer, and PC-3 prostate cancer cell lines, respectively.

3.5. Acoustic-Based Isolation

An acoustic wave is a form of a mechanical wave that propagates across a longitudinal wave and is generated by mechanical stress from a piezoelectric transducer. Surface acoustic waves (SAWs) and bulk acoustic waves (BAWs) are the two forms of acoustic waves. Both have been widely employed in the field of microfluidics to manipulate micro-objects [147]. Travelling SAWs (TSAWs) and standing SAWs (SSAWs) are the two types of SAW-driven microfluidics. SAWs that propagate in one direction and radiate away from acoustic sources are known as travelling surface acoustic waves (TSAWs). Two opposing travelling SAWs interfering or a reflecting travelling SAW create stationary nodes and antinodes in an open or limited domain, resulting in standing surface acoustic waves (SSAWs). Alternatively, bulk acoustic waves (BAWs) are standing waves that propagate within the microchannel's resonant chamber. To generate BAWs, a piezoelectric transducer is bonded to the microchannels and actuated by an AC power supply in BAW-based microfluidic devices. Unlike SAWs, which propagate along the material's surface, bulk acoustic waves propagate within the material's core. As a result, BAW-based microfluidic devices require more energy to create identical acoustic effects to SAW-based microfluidic devices [148,149].

Jiang et al. used the LCATs technique for the isolations of CTCs from breast cancer patients with different stages of cancer [103] (Figure 5c). The advantage of LCATs was their ability to pump samples and trap CTCs without the use of a syringe pump. The device captured 230,000 cells with 200 pairs of dead-end side channels at 6 VP-P (peak-to-peak voltage) and 5.2 kHz, with an average of 1150 cells per pair of dead-end side channels. In less than 8 min, the device could process 7.5 mL of blood samples. However, the real CTC-spiked blood samples showed a capture efficiency of 92.8% with 90% viability. As a result, the technique must be improved in order to achieve higher capture efficiency in real-time applications. Wu et al. examined the acoustic separation of CTCs from leukocytes [150]. A piezoelectric substrate bound to a pair of interdigital transducers (IDTs) in a microfluidic channel generated two Rayleigh waves in opposite directions, resulting in periodic wave nodes and antinodes. In order to facilitate high throughput, a PDMS-glass hybrid channel was used to produce acoustic waves. At a throughput of 7.5 mL/h, 86% CTCs were recovered from the sample. Furthermore, Wang et al. developed a multi-stage device consisting of a pair of interdigital transducers (IDTs) and focused interdigital transducers (FIDTs) using microelectromechanical systems (MEMS) for the separation of CTCs by SAWs [151]. The acoustic waves generated by IDTs enabled the cells to be placed at pressure nodes (Figure 5d), whereas acoustic waves generated by FIDTs push the RBCs from CTCs, resulting in isolation. At a flow rate of 0.3 µL/min, the device showed ~90% isolation efficiency for U87 glioma cells. Karthick et al. developed the acoustic impedance size-independent isolation of CTCs using a microfluidic device [152]. At an optimized flow rate, the device could recover 86% of HeLa cells and 88% of MDA-MA-231 CTCs. Later, Xue et al. presented an acoustic multifunctional micromanipulation (AMM) microstreaming device capable of patterning, tapping, isolating, and rotating microparticles with respect to size and shape [153]. A microcavity array with an inner micro vortex and outer micro vortex was generated by acoustic waves to achieve cell manipulation. The device showed ~90% isolation efficiency. Recently, Cushing et al. reported continuous-flow acoustophoretic negative selections of WBCs from CTCs with the help of negative acoustic contrast elastomeric particles (EPs) functionalised with CD45 antibodies [154]. EP-bound WBC aligned at the channel wall, enabling unbound CTCs to flow through the channel centre. The device facilitated the isolation of label-free CTCs from WBCs with a recovery rate of ~85–90%.

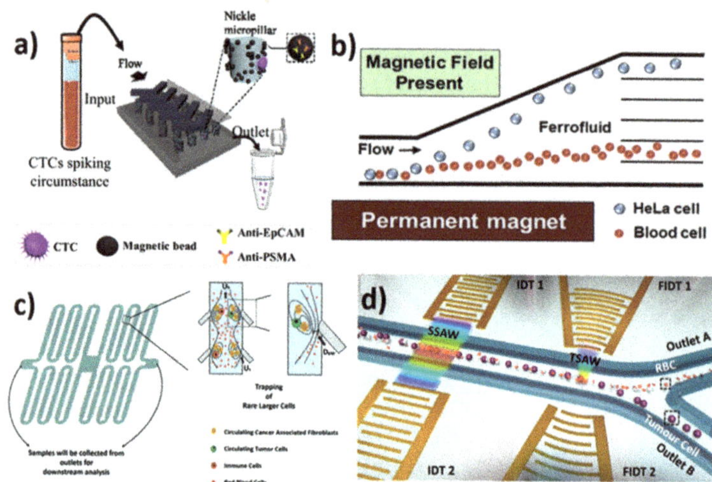

Figure 5. Schematic illustration of microfluidic devices for the isolation of CTCs using various techniques. (**a**) Schematic of the working mechanism of a dual-antibody-functionalised microfluidic device for the isolation of CTCs using magnetic beads; reprinted with permission from ref. [143], 2018, American Chemical Society. (**b**) Schematics of label-free isolation of HeLa cells in ferrofluids under magnetic fields by magnetic buoyancy forces; reprinted with permission from ref. [144], 2015, Wlsevier. (**c**) Schematic of CTC isolation in bubble-based acoustic microstreaming, which releases smaller cells by trapping larger CTCs; reprinted with permission from ref. [103], 2020, Royal Society of Chemistry. (**d**) Schematic illustration of a multi-stage device consisting of a pair of IDTs and FIDTs to generate SSAWs and TSAWs for the isolation of CTCs; reprinted with permission from ref. [151], 2018, Elsevier.

3.6. Combined Method-Based Isolation

The combination of two or more modes of isolation techniques in a microfluidic device facilitates the highly efficient isolation of CTCs. Nasiri et al. fabricated a hybrid microfluidic system that uses inertial flow and magnetophoresis to isolate CTCs [107]. The MCF-7 cells were conjugated with EpCAM antibodies and MNPs to improve magnetic susceptibility. These surface-modified cells were mixed with blood cells and were injected into the hybrid device at a flow rate of 1000 µL/min. The device exhibited a separation efficiency of ~95% with a purity of ~93%. Furthermore, Raillon et al. combined a vortex chip and an impedance chip to create microfluidic devices for label-free, high-throughput CTC isolation and enumeration [95]. Firstly, a vortex chip was used to purify the cancer cells. Later, an impedance chip with a pair of electrodes measured the fluctuation of an applied electric field in the presence of CTCs. This device was subjected to beads and tumor cells as proof of concept. PEEK/Tefzel tubings were used to form connections along with the vortex chip, impedance chip and syringe-containing samples. In the vortex chip, the flow rate to capture CTCs was optimized to 100 µL/min. The channel was validated with 8, 15, and 20 µm fluorescent beads through which the vortex chip enriched beads with an amplitude ranging from 250 nA to 100–250 nA. By using an impedance chip, 1477 beads were detected, and 1294 beads were enumerated from the device. Finally, MCF-7 cells were assessed in the channel at an optimized flow rate of 100 µL/min. RBCs and PBMCs (peripheral blood mononuclear cells) were separated using Ficoll. Thus, it was observed that at 60 nA, 95% of MCF-7 cells were separated from RBCs and PBMCs by leaving 5% of MCF-7 as a false negative. Later, Shamloo et al. employed a passive and a hybrid centrifugal device design to isolate tumor cells with the help of MNPs [155]. In the passive design, a contraction–expansion array (CEA) microchannel with a bifurcation region was used to isolate tumor cells through inertial effects and bifurcation law. In the hybrid design, a CEA microchannel with stacks of magnets was used to isolate magnetically labelled tumor cells. The devices were utilised to isolate human breast cancer cells (MCF-7). The devices were performed with various centrifugal speeds, demonstrating a recovery rate of 76% at 2100 rpm for the passive design. On the other hand, the hybrid design showed an 85% recovery rate at 1200 rpm. Though the hybrid design showed a high recovery percentage, the passive design was less space-, cost-, and time-consuming.

Chen developed a triplet microchannel spiral microfluidic chip that interconnected with many tilted slits based on inertial and deformability principles for the continuous isolation of CTCs [156]. Using inertial and viscous drag forces, cells of various sizes were made to achieve different equilibrium throughout the microchannel. The bigger CTCs were gathered at the central streamline. The chip showed a high isolation capacity of 90% at a flow rate of 80 mL/h. Later, Antfolk et al. fabricated a microfluidic device with two inlets and three outlets for the label-free, on-chip separation and enumeration of target tumor cells [157]. They bound together acoustic and dielectrophoresis chips through plasma treatment. The outlet of the acoustic chip was aligned to an inlet of the dielectrophoresis chip for the efficient isolation of target cells. Prostate tumor cells (DU145) were effectively isolated from peripheral blood mononuclear cells at a recovery rate of 76.2 ± 5.9%. Furthermore, Liu et al. designed a label-free inertial-ferrohydrodynamic CTC-capturing microfluidic device [158] (Figure 6a). The technique enabled the high-throughput, high-resolution isolation of CTCs. The method could differentiate the ~1–2 µm diameter difference in cells for efficient separation. The developed method showed a recovery rate of 94% with high purity. In addition, Xu et al. created an integrated microfluidic device for CTC isolation [115]. The prefiltered CTCs were subjected to magnet-assisted isolation on a microfluidic chip comprised of anti-CD45 antibody-functionalized magnetic beads (Figure 6b). For PC-9-spiked blood samples, the device demonstrated a capture efficiency of ~85% and a purity of 60.4%. Despite the fact that the method involved two steps of isolation with high throughput and minimal cell damage, the device lacked capture efficacy. Later, Garg et al. presented a multi-functional microfluidic microstreaming LCAT-based device for the size-based isolation, enrichment, and in situ biomarker-based sorting of cells from blood [159]. At a flow rate of 25 μL/min, targeted MCF-7 cells were trapped in microstreaming vortices at ~100% efficiency.

3.7. Electrochemical-Based Isolation

Electrochemical detection relies on the transfer of electrons at the analyte-electrode interface, which is frequently accompanied by the process of analyte-receptor recognition. Electrochemical procedures have a fast response time, cheap cost, simplicity, clinically appropriate sensitivity, specificity and the potential to miniaturize when compared to other analytical methods [160]. Meanwhile, they are frequently used in conjunction with other technologies to achieve multimode detection with increased accuracy and sensitivity.

Yan et al. fabricated a micropillar array electrochemical microchip for the isolation and analysis of CTCs [91]. The device surface was coated with a gold layer, followed by oligonucleotide modification via gold-thiol (Figure 6c). Further, avidin and EpCAM antibodies were functionalised. In order to lyse the cells, the device was modified with two slices of gold to use as the working electrodes. By applying a voltage, the captured cells were lysed. The –OH ions generated during electrochemical lysis broke down the lipid bilayer of the captured cells. The device showed a capture efficiency of 85–100%. Furthermore, Gurudatt et al. developed an electrochemical microfluidic system that combines CTC separation, enrichment, and detection [116]. Whole blood cells flowing through a microchannel were initially functionalized with electroactive daunomycin (DM, an anticancer drug that can selectively interact with CTCs). The target species in the microfluidic channel exhibited a wave-like motion when an alternating current perpendicular to the hydrodynamic flow was applied and was segregated and enriched in a size-dependent manner. The CTCs were subsequently examined using a direct DM oxidation method with an electrochemical sensor at the channel end. With a separation efficiency of 92.0 ± 0.5% and a detection rate of 90.9%, this device is capable of successfully discriminating various cancer cells in patients' blood samples.

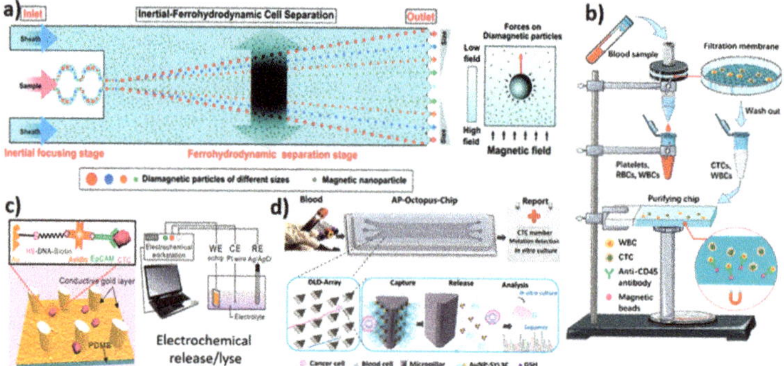

Figure 6. Schematic illustration of microfluidic devices for the isolation of CTCs using various techniques. (**a**) Schematic illustration of working principle of an inertial-ferrohydrodynamic cell separation chip in ferrofluids under a magnetic field; reprinted with permission from ref. [158], 2021, Royal Society of Chemistry. (**b**) Schematic of isolation of CTCs through filtration, followed by anti-CD45 antibody functionalized magnetic beads [115]. (**c**) Schematic illustration of e-chip exhibiting a conductive gold layer functionalised with EpCAM antibodies responsible for the capture and electrochemical release/lyse of CTCs; reprinted with permission from ref. [91], 2017, American Chemical Society. (**d**) Schematic of DLD working principle of AP-Octopus-Chip, where CTCs interact with micropillar-functionalised AuBO-SYL3C to get captured and released by Au-S bond disruption; reprinted with permission from ref. [161], 2019, John Wiley and Sons.

3.8. Biological Interaction-Based Isolation

Though CTCs are found in the bloodstream, they retain the characteristics of their original tumor cell from the metastatic sites. The expression of EpCAM is a pervasive biological property of CTCs. As a result, EpCAM was used as a specific biomarker for CTC isolation in positive selection. However, the EpCAM protein is present on CTCs but not on blood cells. Thus, other markers such as CD1513, CD6647, and CD45 are used as specific biomarkers for blood cells for negative selection. Stott et al. developed a herringbone microfluidic device by photolithography [162]. The microchannels were functionalised with EpCAM antibodies to facilitate CTC isolation. The presence of a herringbone pattern generates micro vortices, which results in thorough mixing of blood samples, facilitating the high interaction between the functionalised channel surface and CTCs. The device could isolate CTCs from patients' blood with advanced prostate and lung cancer with a success rate of 93%. The device showed high throughput and promising results. Later, Song et al. developed an aptamer-tailed octopus chip (AP-Octopus-Chip) for capturing CTCs [161]. To improve capture efficiency, a deterministic lateral displacement (DLD)-patterned microfluidic chip was altered with multivalent aptamer-functionalized nanospheres (Figure 6d). CTCs were forced to transverse streamlines and interact with AuNP-SYL3C modified micropillars. Blood cells that are smaller than CTCs stayed inside the initial flow streamline, and bigger CTCs interacted with them. The enriched CTCs were released after capture when the -AuS bond was broken by excess glutathione. Sheng et al. developed a geometrically enhanced mixing (GEM) chip for the capture and isolation of CTCs from pancreatic cancer cell lines [163] (Figure 7a). Initially, anti-EpCAM was biotinylated and loaded to the surface of a microfluidic channel containing L3.6pl, BxPC-3, and MIAPaCa-2 cells in order to capture CTCs. Flow cytometry results show that L3.6pl cells bind strongly to anti-EpCAM, whereas MIAPaCa-2 cells do not. For capturing CTCs, the flow rate and velocity were optimised to 1 µL/s and 0.75 mm/s, respectively. The GEM chip detected ~23 CTCs from 7.5 mL of blood, with the capture efficiency of $90 \pm 2\%$ for the L3.6pl cells line and $92 \pm 4\%$ for the BxPC-3 cells. The device has the advantage of being able to isolate CTCs with sufficient throughput in 17 min. Overall, the device achieved >90% capture efficiency, >84% purity and a throughput of 3.6 mL of blood in 1 h. However, the device falls short in terms of CTC capture purity. Furthermore, Nieto et al. developed a soda-lime glass-based microfluidic device by using the laser-ablation direct writing method and laser-assisted thermal treatment for the isolation of CTCs [25]. With this treatment, the roughness, optical transparency and reshaping of the microstructures were improved. The surface-modified microchannel with EpCAM antibodies developed by this approach trapped the CTCs. The results showed a capture efficiency of ~76% for HEC-1A tumor cells.

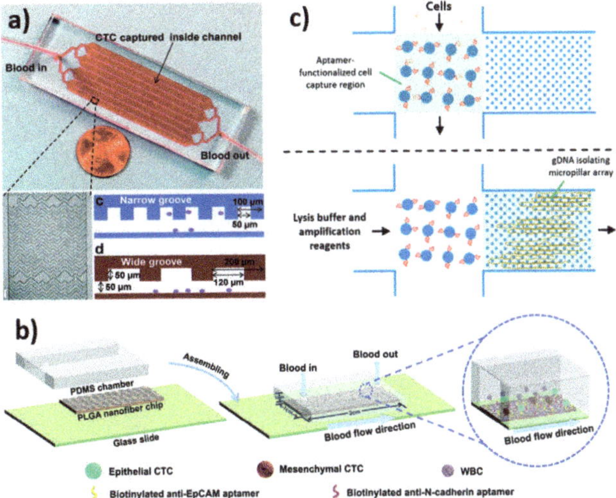

Figure 7. Schematic illustration of microfluidic devices for the isolation of CTCs using various techniques. (**a**) GEM chip with eight parallel channels with an inlet and an outlet showing asymmetric

herringbone grooves inside the channel; reprinted with permission from ref. [163], 2013, Royal Society of Chemistry. (**b**) Schematic of dual aptamer-functionalised PLGA nanofiber-based microfluidic chip for the isolation of various phenotypic CTCs; reprinted with permission from ref. [164], 2021, Royal Society of Chemistry. (**c**) Schematic of microchannel design with aptamer-modified micropillar array for capturing cancer cells and isolating their gDNA; reprinted with permission from ref. [106], 2018, American Chemical Society.

In addition, Jou et al. created the V-BioChip for isolating SKOV3 ovarian tumor cells from epithelial ovarian cancer patients' blood samples [104]. Using anti-EpCAM antibody interactions on the device's surface at a flow rate of 0.6 mL/h, the device demonstrated a capture efficiency of 48.3%. The combination of anti-EpCAM antibody and anti-N-cadherin antibody on the device surface resulted in a capture efficacy of 89.6%. Despite the functionalised surface, the obtained results showed a lower capture efficiency. Further, Wu et al. created a PLGA nanofiber, aptamer-functionalized microfluidic device for isolating ovarian cancer cells such as A2780 and OVCAR-3 cells [164] (Figure 7b). The EpCAM-functionalised chip demonstrated a good capture efficiency of 89% for OVCAR-3 cells, while NC3S demonstrated high efficiency of 91% for A2780 with a release efficiency of 88% and 92%, respectively. Later, Reinholt et al. developed a PDMS microfluidic system to isolate HeLa (cervical cancer cell line) and CAOV-3 (ovarian cancer cell line) cancer cells [106] (Figure 7c). For the capture of CTCs via a streptavidin–biotin conjugation, the microchannel surface was functionalized with a DNA aptamer. The capture efficiency was great when the CTCs were suspended in PBS buffer and flushed into the microchannel at a flow rate of 5 µL/min. The collected cells were also lysed using a DNA array channel. The cellular contents were allowed to flow out while the gDNA was isolated on the micropillar. The use of gDNA allows for the extraction of enormous amounts of data from a small number of cells without the need for genome amplification. In another study, Pulikkathodi et al. developed an AlGaN/GaN high-electron-mobility (HEMT) biosensor array for the detection and isolation of CTCs [165]. Furthermore, these chips are mounted on a thermos-curable polymeric substrate. The formed array has several aptamer-immobilized areas, which are sensitive to CTCs. The device showed high sensitivity and selectivity, making it a potential device for CTC isolation. Zhang et al. combined a size-based microfluidic device with surface-enhanced Raman spectroscopy (SERS) for the detection of tumor cells [166] (Figure 8a). Three kinds of SERS aptamer nano vectors were utilised for the detection of breast cancer cell lines in accordance with surface protein expressions. Initially, at a flow rate of 1 µL/min, tumor cells were separated through filtration. Then, SERS receptors were used to analyse the captured CTCs. Recently, Chen et al. developed a 3D-printed microfluidic device for the isolation of CTC from a blood sample [81]. The channel surface was functionalised with EpCAM antibodies to capture EpCAM-positive cancer cell lines, such as MCF-7, SW480, PC-3, and EpCAM-negative 293T cells (Figure 8b). At a flow rate of 1 mL/h with a 2 cm channel length, the device showed a capture efficiency of up to ~92% for MCF-7, ~87.74 for SW480, and ~89.35 for PC-3.

Cheng et al. designed and developed a 3D scaffold microfluidic device with a thermosensitive coating for the isolation and release of CTCs [167]. Gelatin hydrogel was coated on the surface of Ni (nickel) foam. In addition, the surface of the gelatin was functionalised with an anti-EpCAM monoclonal antibody to capture MCF-7 cells (Figure 8c). At an optimised flow rate of 50 µL/min, CTCs were captured. Further, the chip was transferred to an incubator at 37 °C in order to dissolve the gelatin hydrogel to facilitate the release of captured CTCs. The chip showed ~88% capture efficiency. The isolation of platelet-covered CTCs is extremely difficult due to the masking of surface epitopes. Furthermore, Jiang et al. designed a herringbone macromixing microfluidic platform using stealth CTCs as surface markers for the isolation of CTCs [168]. They used epithelial and mesenchymal phenotypes for the platelet-targeted isolation of CTCs. At first, the free platelets were isolated by hydrodynamic size-based isolation. Further, EpCAM/CD41 antibodies were employed for the isolation of platelet-covered CTCs. The device isolated 66% of lung, 60% of breast, and 80% of melanoma cancer cells. Zeinali et al. demonstrated the integrated immunoaffinitybased isolation of CTCs from pancreatic cancer patients [169]. The device could isolate epithelial and epithelial-to-mesenchymal transition CTCs simultaneously by using EpCAM and CD133 antibodies. At a flow rate of 1 mL/h, the device showed ≥97% CTC recovery with >76% purity. Yin et al. designed a micruifluidic device with a silicon filter with a pyramidal microcavity array (MCA) for the isolation of CTCs [170]. In order to improve the capture efficiency, the surface of the MCA filter was modified with an anti-EpCAM antibody (Figure 8d). The device showed a capture efficiency of ~80% for MCF-7, SW620, and HeLa cell lines spiked in whole blood. The device could effectively filter various sizes of CTCs with high capture efficiency. Kermanshah et al. applied magnetic ranking cytometry (MagRC) to a biologically relevant study [171]. Nickel micromagnets of different sizes were developed to create isolation zones to capture magnetized CTCs. The blood samples of mice containing prostate cancer cells were mixed with EpCAM antibody-modified MNPs and were analysed using the MagRC device. Furthermore, Sun et al. developed a size-based separation where the microfluidic device has ~103 pores/mm^2, exhibiting 68,000 effective pores with a pore diameter of 8 μm [172]. The capture efficiency for MCF-7 cells on the device was found to be 72 ± 10.6% when using the traditional ISET (isolation by size of epithelial tumor cell) technique at a flow rate of 1 mL/min, whereas the capture efficiency of M-ISET (microbeads assisting ISET) was found to be 93.3 ± 3%. As a result, the M-ISET method was found to be a powerful tool for improving the efficiency of CTC separation.

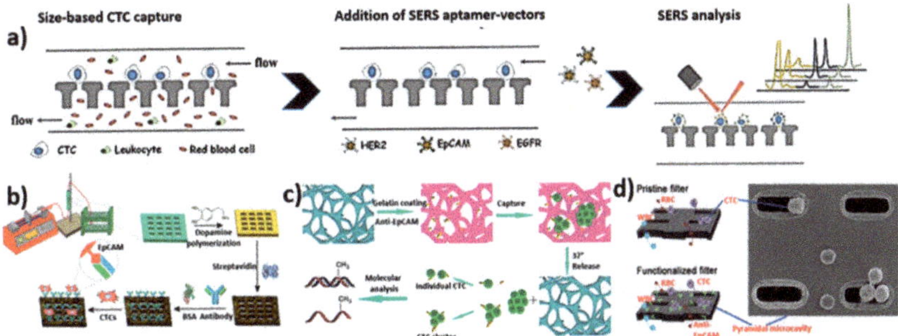

Figure 8. Schematic illustration of microfluidic devices for the isolation of CTCs using various techniques. (**a**) Working strategy of SERS nano vectors for CTC capture, cell phe-notype profiling and multivariate analysis for in situ profiling of CTCs; reprinted with permission from ref. [166], 2018, John Wiley and Sons. (**b**) Schematic of the working setup of the microfluidic platform and surface modification of 3D-printed microfluidic device with an-ti-EpCAM antibody for the isolation of CTCs; reprinted with permission from ref. [81], 2020, Elsevier. (**c**) Schematic surface modification of 3D Ni foam scaffold with gelatin and anti-EpCAM to capture CTCs; these were released at 37 °C for molecular analysis; reprinted with permission from ref. [167], 2017, American Chemical Society. (**d**) Schematic of CTC isolation by a filtration chip functionalised with anti-EpCAM antibody and SEM image of captured cells on pyramidal MCA; reprinted with permission from ref. [170], 2019, Elsevier.

3.9. Overview of Microfluidic Device Performance for the Isolation of Circulating Tumor Cells

Importantly, there are two types of CTC isolation methods: physical and biological. Physical approaches are typically based on physical properties, such as size, volume, deformability, density, dielectric properties, and viscosity, with benefits such as high capturing efficiency, simple sample preparation, and cost-effectiveness. On the other hand, biological approaches are based on antigen-antibody interactions. The main disadvantage, in this case, is that it is an expensive and time-consuming method. In addition, there are some challenges and drawbacks in identifying and separating CTCs. When dealing with microfluidic devices, five different technological criteria are to be considered: the detection limit, capture speed, biocompatibility, purity, and high throughput. There are various devices mentioned, such as spiral-shaped, slanted weir, T-shaped microchannel, and multi-flow microfluidic (MFM) systems, geometrically enhanced mixing (GEM) chips, PDMS-based integrated microfluidic platforms, pyramid-shaped microchambers, ODEP-based microfluidic devices, parallel flow micro-aperture chip systems, a label-free microfluidic device for the detection and separation of CTCs with different capture efficiency. Table 1 summarizes the details of microfluidic devices for CTC isolation.

Table 1. Overview of microfluidic devices with CTC isolation mechanism, chip fabrication and other technical parameters.

Isolation Method	Device Fabrication	Device Dimension	Flow Rate	Efficiency	Cancer Cell Lines	Ref.
Size-based isolation						
Size and deformability	Double-layer photolithography	L = 500 μm T = 23 μm	2.5 mL/h	~97%	LM2 MDA-MB-231	[93]
Size	Wet etching technique and thermal bonding technique	L = 22 mm H = 40 μm	200 μL/min	85%	BGC823, H1975, PC-3, SKBR3	[82]
Size-based PDMS microflitration membrane	Photolithography	T = 60 μm	10 mL/h	>90%	A549, SK-MES-1, H446	[90]
Size	Photolithography	Main channel L = 80 μm; Main channel L = 50 μm H = 50 μm	10 mL/h	82%	SKBR3, MCF-7, MDAMB231	[105]
Inertial focusing microchannel-based isolation						
Label-free, inertial migration of cells	Photolithography	L = 20 mm W = 150 μm H = 50 μm	300 μL/min	>99%	H460, HCC827	[62]
Rotation-induced inertial lift force	photolithography	W = 100, 200, 400 μm D = 30 μm	9 μL/min	90%	U87	[126]
Dean vortex flow, inertial lift force	Photolithography	-	1.7 mL/min	54%	FaDu, CAL27, RPMI2650, UD-SCC9 HNC cells, MDA-MB-468	[120]
Inertial and Dean drag forces	Photolithography	W = 500 μm H = 170 μm	100 μL/min	≥85%	MDA-MB-231, MCF-7, T24	[127]
Inertial microfluidics and Dean flow physics	Photolithography	L = 9.75 mm W = 350 μm	400–2700 μL/min	>94%	MDA-MB-231, Jurkat, K562, HeLa	[128]
Size-dependent lateral migration	Photolithography	Capillary inner and outer diameter = 50 and 360 μm; H = 200 μm L = 5 and 1 cm	200 μL/min	94%	MCF-7	[129]
Self-amplified inertial-focused (SAIF) separation	Photolithography	Zigzag channel W = 40 μm; First expansion region W = 0.84 mm; Second expansion region W = 1.64 mm; H = 50 μm	0.4 mL/min	~80%	A549, MCF-7, HeLa	[121]
Vortex and inertial cell focusing lift force	Photolithography	L = 1000 μm W = 40 μm H = 70 μm; Trapping zone L, W = 720, 230 μm	8 mL/min	83%	MCF-7	[130]
Inertial lift force and Dean drag force	Photolithography	L = 5.5 mm W = 130 μm H = 500 μm	1 mL/min	90%	MCTC	[131]
Dielectrophoresis-based isolation						
Optically induced dielectrophoretic (ODEP) force	Metal mould-punching	Main channel, L = 25 mm, W = 1000 μm, H = 100 μm; Side channel, L = 15 mm, W = 400 μm, H = 100 μm	2.5 μL/min	41.5%	PC-3	[135]

Table 1. *Cont.*

Isolation Method	Device Fabrication	Device Dimension	Flow Rate	Efficiency	Cancer Cell Lines	Ref.
Dielectrophoresis at wireless bipolar electrode (BPE) array	Photolithography	L = 2.95 mm W = 200 µm H = 25 µm	20 µm/s	96%	MDA-MB-231, Jurkat E6-1 T	[136]
Dielectrophoresis (DEP) force	Photolithography and wet etching	L = 7 mm H = 50 µm	100 µL/min	92 ± 9%	NCI-H1975	[137]
Optically induced dielectrophoresis (ODEP)	Metal mould-punching	Main channel, L = 2500 µm, W = 1000 µm, H = 60 µm; Side channel, L = 2500 µm, W = 400 µm, H = 60 µm	-	81.0 ± 0.7%	PC-3, SW620	[85]
Magnetic field-based isolation						
Immunomagnetics and size-based filtration	Photolithography	T = 50 µm	2 mL/min	~89%	MCF-7	[96]
EpCAM-specific conjugation of MNPs	Photolithography	Microchannel W = 250 µm; Trapping site H = 400 µm, W = 100 µm	150 µL/min	~81.2–96.3%	MDA-MB-231, MCF-7	[89]
EpCAM-based positive method and CD45/CD66b-based negative method by lateral magnetophoresis	Photolithography	Free-bead capture microchannel, L = 42.5 mm, W = 1 mm, H = 50 µm; Lateral magnetophoretic microchannel, L = 42.5 mm, W = 2.8 mm, H = 100 µm	2 mL/h and 3.2 mL/h	83.1%	MDA-MB-231, PC-3, SKBR3, MCF-7	[140]
Magnet deformability	Photolithography	L = 49,000 µm W = 10,000 µm	3 mL/h	90%	HCT116, SW480, MCF-7	[97]
Immunomagnetic technique	Photolithography	L = 9 mm W = 1 mm	-	97–107%	SKBR3, PC-3, Colo205	[99]
Magnetic-ranking cytometry and phenotypic profiling of CTCs	Photolithography	L = 8.75 cm H = 50 µm	500 µL/h	>90%	SKBR3, PC-3, MDA-MB-231	[141]
MNP-labeled aptamers	Photolithography	-	25 mL/h	~79%	PC-3, SKBR3	[142]
Magnetic-bead-mediated dual-antibody functionalised microfluidics	Photolithography	-	0.8 mL/h	>85%	LnCAP and LnCAP-EMP	[143]
Cell size difference in ferrofluids under permanent magnetic influence	Photolithography	L = 2.54 mm W, H = 635 µm	8 µL/min	>99%	HeLa	[144]
Ferrodynamic cell separation	Photolithography	L = 4.94 cm W = 900 µm	6 mL/h	~92.9%	H1299, A549, H3122, PC-3, MCF-7, HCC1806	[145]
Acoustic-based isolation						
Cell size difference in ferrofluids	Photolithography	L = 5.81 cm W = 900 µm	20 µL/min	82.2%	A549, H1299, MCF-7, MDA-MB-231	[146]
Lateral cavity acoustic transducers	Photolithography	W = 750 µm H = 100 µm	25 µL/min	94%	Breast, bone, lung cancer cells	[103]
Hydrodynamic and SAW focusing separation	Photolithography	-	7.5 mL/h	>86%	MCF-7, HeLa, PC-3, LNCaP	[150]
Interdigital transducers (IDTs) and focused interdigital transducers (FIDTs) generating standing SAWs and travelling pulsed SAWs	Photolithography	W = 65 µm H = 50 µm	0.3 µL/min	~90%	U87	[151]

Table 1. *Cont.*

Isolation Method	Device Fabrication	Device Dimension	Flow Rate	Efficiency	Cancer Cell Lines	Ref.
Acoustic impedance contrast	Photolithography and deep reactive ion etching (DRIE)	L = 20 mm W = 380 μm H = 200 μm	20–60 μL/min	>86%	HeLa, MDA-MA-231	[152]
Microvortices generated by acoustic vibration	Photolithography	L = 50 mm W = 40 mm H = 200 μm	10 μL/min	>90%	DU145	[153]
Continuous flow acoustophoretic negative selection	Photolithography	Maun channel, L = 20 mm, W = 375 μm, H = 150 μm; Sub channel, L = 10 mm, W = 300 μm, H = 150 μm	100, 400 μL/min	>98%	MCF-7, DU145	[154]
Combined method-based isolation						
Inertial and magnetic method	Photolithography	W = 400 μm H = 80 μm	1000 μL/min	~95%	MCF-7	[107]
Vortex trapping and impedance cytometry	-	L = 1 cm H = 70 μm	100 μL/min	~ 98%	MCF-7, LoVo, HT-29 human colon cells,	[95]
Inertial hydrodynamic forces and bifurcation law	CNC micromachining	W = 0.26 mm H = 0.2 mm	-	85%	MCF-7	[155]
Inertial and deformability-based principle	Photolithography	L = 1–1.5 cm W = 400, 300, 200 μm	80 mL/h	>90%	MCF-7	[156]
Integrated device with acoustofluidic label-free separation and direct dielectrophoretic cell trapping	Photolithography	L = 2.3 cm W = 375 μm H = 150 μm	80, 160 μL/min	~76%	DU145	[157]
Inertial-ferrohydrodynamic cell separation	Photolithography	H = 60 μm	~60 mL/h	94.8%	H1299, MDA-MB-231, MCF-7, H3122	[158]
Micropore-arrayed filtration and magnetic bead-functionalised antibody-mediated detection	Molding technique	Micropore L, W = 20 mm, diameter = 10 μm	-	~85%	PC-9	[115]
Lateral cavity acoustic transducers (LCAT) and biomarker-based immuno-labelling	Photolithography	Main, side channel W = 500, 100 μm H = 100 μm	25 μL/min	~100%	MCF-7, SKBR3	[159]
Electrochemical isolation						
Antibody-mediated electrochemical release and lysis	Photolithography	L = 40 mm W = 20 mm	1 mL/h	85–100%	PC-3, MCF-7, NCl-H1650	[91]
Electrochemical detection and electric-filed influenced hydrodynamic flow	Screen printing	W = 95 ± 2.5 μm H = 15 ± 1.5 μm	5 μL/min	92 ± 0.5%	HEK-293, HeLa	[116]
Anti-EpCAM-coated channel surface with herringbone grooves	Photolithography	L = 50 mm W = 2.1 mm H = 50 μm	1 μL/s	>90%	L3.6pl, BxPC-3, MIAPaCa-2	[163]
Biological interaction-based isolation						
EpCAM-expressing cells using antibody-coated microposts	Photolithography	L = 20 mm H = 50–100 μm	1.5–2.5 mL/h	93%	PC-3	[162]

Table 1. *Cont.*

Isolation Method	Device Fabrication	Device Dimension	Flow Rate	Efficiency	Cancer Cell Lines	Ref.
Aptamer-functionalized micropillars	Photolithography	-	1 mL/h	80%	W480 colorectal, LNCap prostate, KATO III gastric cancer cells, K-562 chronic myelogenous leukemia cells	[161]
EpCAM antibody-functionalised pillars	Laser direct-write technique	Micropost diameter = 420 µm; Pitch = 245 µm	90 µL/min	~76%	HEC-1A	[25]
Combination of anti-EpCAM antibody and anti-N-cadherin antibody	Photolithography	L = 32 mm W = 34 mm H = 0.7 mm	0.6 mL/h	89.6%	SKOV-3 ovarian tumor cells	[104]
Dual aptamer (EpCAM-5-1 and NC3S)-modified poly(lactic-co-glycolic acid) (PLGA) nanofiber	Electrospinning	L = 2 cm W = 1 cm H = 1 mm	300 µL/min	89–91%	A2780, OVCAR-3	[164]
Aptamer-immobilized microchannel	Photolithography	Cell channel W = 1 mm; DNA channel W = 0.5–1 mm H = ~25 µm	5 µL/min	-	HeLa, CAOV-3	[106]
AlGaN/GaN HEMT biosensor array	Photolithography	L = 22 mm W = 13 mm	-	-	HCT-8	[165]
Size-based and multiplex SERS nanovectors	-	Filter gap = 12 µm, H = 40 µm	1 µL/min	~87–93%	SKBR3, MCF7, and MDA-MB-231	[166]
Microchannel functionalised with anti-EpCAM	3D printing	L = 2 cm	1 mL/h	~87–92%	MCF-7, SW480, PC-3, 293T	[81]
Gelatin-coated Ni foam functionalised with anti-WpCAM	Ni foam surface modification	L = 20 mm W = 4 mm H = 1 mm	50 µL/min	~88%	MCF-7	[167]
Lateral displacement (DLD) and herringbone CTC chip functionalised with EpCAM and CD41 antibodies	Deep reactive ion etching	H = 150 µm	1.14 ± 0.24 mL/h	60–83%	Lung, breast, melanoma cancer cells	[168]
EpCAM and CD133 antibodies functionalised hexagonal array of posts	Photolithography	L = 44.6 mm W = 16.9 mm H = 100 µm	1 mL/h	13.6–97.5%	HT-29, Panc-1, PC-3, Hs-578T, Capan-1	[169]
Microcavity array functionalised with anti-EpCAM	Photolithography	H = 200 ± 10 µm Microcavity L, W = 30, 8 µm	0.1 mL/min	~76–83%	MCF-7, SW620	[170]
Magnetic ranking cytometry and CTC surface marker expression	Photolithography	L = 5.4 cm W = 4.3 cm H = 50 µm, Radii of Ni magnet = 145–235 µm	400 µL/h	>90%	LNCaP, PC-3, PC-3M	[171]
Isolation by size of epithelial tumor cell (ISET) and microbeads assisting ISET	-	L = 4 mm W = 17 mm H = 300 µm	1 mL/min	~72–93%	MCF-7, KATO III, PC-3	[172]

4. Conclusions and Prospects

CTCs play an important role in cancer metastasis and are studied clinically for cancer prognosis and diagnosis, known as liquid biopsy. Though there are several commercial technologies available such as CellSearch, CytoQuest, GILUPI CellCollector, ApoStream, Screencell, and ISET for CTC isolation, these technologies still have some drawbacks to their application in clinics. These devices have tedious fabrication and operational protocols, resulting in limited batch fabrication for large-scale production. In addition, they lack the sensitivity to satisfy clinical demands due to the presence of various kinds of tumor cell types. Hence, there is a need for greater effort to develop CTC isolation techniques. The developed CTC isolation technology should be easy to fabricate and operate. The detection strategy should be simple, fast, and accurate. In this regard, microfluidic technology is a multidisciplinary research field that can be used for capturing and isolating CTCs due to their numerous advantages over traditional separation techniques. When compared to traditional macro-scale devices, the microfluidic technique has numerous benefits, including portability, improved sensitivity, lower operating costs, and higher throughput.

We summarized perspectives on the various strategical microfluidic devices regarding both label-free isolation and label-based detection of CTCs using various methods such as dielectrophoresis, inertial migration, the electrochemical method, the M-ISET method, the hydrodynamic method, the sandwich moulding method, deformability-based separation, the label-free immune separation method, and the label-based method in this review. Microfluidic devices for physical approaches can be easily fabricated due to their reusability and the absence of expensive antibodies. Thus, physical-based devices allow researchers to detect unidentified biological markers, which could lead to breakthrough results in the near future. On the other hand, there is still no efficient method for capturing, isolating, enumerating, and characterising CTCs. We are expecting improved CTC capture methodologies in several aspects in the future. The microfluidic device should be reliable and stable in its isolation of CTCs. It should be able to detect as many CTC variants as possible in real-time to meet the clinical demand. In order to fully realise the potential of microfluidics, positive isolation, negative enrichment and highly integrated devices need to be developed to analyse the phenotype and genotype properties of CTCs. A standard operating procedure (SOP) is required for efficiently capturing and isolating CTCs. The design and strategy of capture can vary greatly from device to device. Furthermore, we believe that these novel microfluidic technologies for CTC capture and isolation will be approved by regulatory agencies and used as real-time equipment in the near future.

Author Contributions: Conceptualization, M.P.B. and V.T.; methodology, M.P.B. and V.T.; software, U.T.U.; validation, K.-H.L.; resources, T.A.; data curation, U.T.U., M.K. and T.A.; writing—original draft preparation, M.P.B. and V.T.; visualization, M.K. and M.D.K.; supervision, M.D.K. and K.K.; project administration, M.D.K. and K.K.; funding acquisition, T.A., M.D.K. and K.K. All authors have read and agreed to the published version of the manuscript.

Funding: This research was funded by Department of Science and Technology (TDP/BDTD/32/2019 and DST/TDT/DDP-31/2021), Taif University (TURSP-2020/04), European Union's Horizon 2020 research and innovation program under the Marie Sklodowska-Curie (894227).

Institutional Review Board Statement: Not applicable.

Informed Consent Statement: Not applicable.

Data Availability Statement: The data that supports this study are available from corresponding authors upon reasonable request.

Acknowledgments: We acknowledge the financial support from DST (TDP/BDTD/32/2019 and DST/TDT/DDP-31/2021) India for funding. Thanks to Taif University Researches Supporting Project (TURSP-2020/04), Taif University, Taif, Saudi Arabia. Krishna Kant also acknowledges the European Union's Horizon 2020 research and innovation program under the Marie Sklodowska-Curie grant agreement no. (894227). Finally, we acknowledge Jain University, India for providing funds and other facilities.

Conflicts of Interest: The authors declare no conflict of interest.

References

1. Wu, S.; Liu, S.; Liu, Z.; Huang, J.; Pu, X.; Li, J.; Yang, D.; Deng, H.; Yang, N.; Xu, J. Classification of Circulating Tumor Cells by Epithelial-Mesenchymal Transition Markers. *PLoS ONE* **2015**, *10*, e0123976. [CrossRef] [PubMed]
2. Cho, H.; Kim, J.; Song, H.; Sohn, K.Y.; Jeon, M.; Han, K.-H. Microfluidic technologies for circulating tumor cell isolation. *Analyst* **2018**, *143*, 2936–2970. [CrossRef] [PubMed]
3. Jackson, J.M.; Witek, M.A.; Kamande, J.W.; Soper, S.A. Materials and microfluidics: Enabling the efficient isolation and analysis of circulating tumour cells. *Chem. Soc. Rev.* **2017**, *46*, 4254–4280. [CrossRef] [PubMed]
4. Potdar, P.D.; Lotey, N.K. Role of circulating tumor cells in future diagnosis and therapy of cancer. *J. Cancer Metastasis Treat.* **2015**, *1*, 44–56. [CrossRef]
5. Garrido-Navas, C.; de Miguel-Pérez, D.; Exposito-Hernandez, J.; Bayarri, C.; Amezcua, V.; Ortigosa, A.; Valdivia, J.; Guerrero, R.; Garcia Puche, J.L.; Lorente, J.A. Cooperative and escaping mechanisms between circulating tumor cells and blood constituents. *Cells* **2019**, *8*, 1382. [CrossRef]
6. Kurkuri, M.D.; Al-Ejeh, F.; Shi, J.Y.; Palms, D.; Prestidge, C.; Griesser, H.J.; Brown, M.P.; Thierry, B. Plasma functionalized PDMS microfluidic chips: Towards point-of-care capture of circulating tumor cells. *J. Mater. Chem.* **2011**, *21*, 8841–8848. [CrossRef]
7. Liu, H.Y.; Hille, C.; Haller, A.; Kumar, R.; Pantel, K.; Hirtz, M. Highly efficient capture of circulating tumor cells by microarray in a microfluidic device. *FASEB J.* **2019**, *33*, lb230. [CrossRef]
8. Bray, F.; Jemal, A.; Grey, N.; Ferlay, J.; Forman, D. Global cancer transitions according to the Human Development Index (2008–2030): A population-based study. *Lancet Oncol.* **2012**, *13*, 790–801. [CrossRef]
9. Jie, X.-X.; Zhang, X.-Y.; Xu, C.-J. Epithelial-to-mesenchymal transition, circulating tumor cells and cancer metastasis: Mechanisms and clinical applications. *Oncotarget* **2017**, *8*, 81558. [CrossRef]
10. Cheng, S.-B.; Chen, M.-M.; Wang, Y.-K.; Sun, Z.-H.; Xie, M.; Huang, W.-H. Current techniques and future advance of microfluidic devices for circulating tumor cells. *TrAC Trends Anal. Chem.* **2019**, *117*, 116–127. [CrossRef]
11. Madhuprasad; Bhat, M.P.; Jung, H.-Y.; Losic, D.; Kurkuri, M.D. Anion Sensors as Logic Gates: A Close Encounter? *Chem. Eur. J.* **2016**, *22*, 6148–6178. [CrossRef] [PubMed]
12. Bhat, M.P.; Patil, P.; Nataraj, S.K.; Altalhi, T.; Jung, H.-Y.; Losic, D.; Kurkuri, M.D. Turmeric, naturally available colorimetric receptor for quantitative detection of fluoride and iron. *Chem. Eng. J.* **2016**, *303*, 14–21. [CrossRef]
13. Patil, P.; Bhat, M.P.; Gatti, M.G.; Kabiri, S.; Altalhi, T.; Jung, H.-Y.; Losic, D.; Kurkuri, M. Chemodosimeter functionalized diatomaceous earth particles for visual detection and removal of trace mercury ions from water. *Chem. Eng. J* **2017**, *327*, 725–733. [CrossRef]
14. Patil, P.; Ajeya, K.V.; Bhat, M.P.; Sriram, G.; Yu, J.; Jung, H.-Y.; Altalhi, T.; Kigga, M.; Kurkuri, M.D. Real-Time Probe for the Efficient Sensing of Inorganic Fluoride and Copper Ions in Aqueous Media. *ChemistrySelect* **2018**, *3*, 11593–11600. [CrossRef]
15. Bhat, M.P.; Kigga, M.; Govindappa, H.; Patil, P.; Jung, H.-Y.; Yu, J.; Kurkuri, M. A reversible fluoride chemosensor for the development of multi-input molecular logic gates. *New J. Chem.* **2019**, *43*, 12734–12743. [CrossRef]
16. Bhat, M.P.; Vinayak, S.; Yu, J.; Jung, H.-Y.; Kurkuri, M. Colorimetric Receptors for the Detection of Biologically Important Anions and Their Application in Designing Molecular Logic Gate. *ChemistrySelect* **2020**, *5*, 13135–13143. [CrossRef]
17. Pirzada, M.; Altintas, Z. Nanomaterials for healthcare biosensing applications. *Sensors* **2019**, *19*, 5311. [CrossRef]
18. Nolan, J.; Nedosekin, D.A.; Galanzha, E.I.; Zharov, V.P. Detection of apoptotic circulating tumor cells using in vivo fluorescence flow cytometry. *Cytom. Part A* **2019**, *95*, 664–671. [CrossRef]
19. Safaei, T.S.; Mohamadi, R.M.; Sargent, E.H.; Kelley, S.O. In situ electrochemical ELISA for specific identification of captured cancer cells. *ACS Appl. Mater. Interfaces* **2015**, *7*, 14165–14169. [CrossRef]
20. Huaman, J.; Naidoo, M.; Zang, X.; Ogunwobi, O.O. Fibronectin regulation of integrin B1 and SLUG in circulating tumor cells. *Cells* **2019**, *8*, 618. [CrossRef]
21. Andergassen, U.; Zebisch, M.; Kölbl, A.C.; König, A.; Heublein, S.; Schröder, L.; Hutter, S.; Friese, K.; Jeschke, U. Real-time qPCR-based detection of circulating tumor cells from blood samples of adjuvant breast cancer patients: A preliminary study. *Breast Care* **2016**, *11*, 194–198. [CrossRef] [PubMed]
22. Wang, X.; Sun, L.; Zhang, H.; Wei, L.; Qu, W.; Zeng, Z.; Liu, Y.; Zhu, Z. Microfluidic chip combined with magnetic-activated cell sorting technology for tumor antigen-independent sorting of circulating hepatocellular carcinoma cells. *PeerJ* **2019**, *7*, e6681. [CrossRef] [PubMed]
23. Gossett, D.R.; Weaver, W.M.; Mach, A.J.; Hur, S.C.; Tse, H.T.K.; Lee, W.; Amini, H.; Di Carlo, D. Label-free cell separation and sorting in microfluidic systems. *Anal. Bioanal. Chem.* **2010**, *397*, 3249–3267. [CrossRef] [PubMed]
24. Xiao, Y.; Shen, M.; Shi, X. Design of functional electrospun nanofibers for cancer cell capture applications. *J. Mater. Chem. B* **2018**, *6*, 1420–1432. [CrossRef] [PubMed]
25. Nieto, D.; Couceiro, R.; Aymerich, M.; Lopez-Lopez, R.; Abal, M.; Flores-Arias, M.T. A laser-based technology for fabricating a soda-lime glass based microfluidic device for circulating tumour cell capture. *Colloids Surf. B Biointerfaces* **2015**, *134*, 363–369. [CrossRef]
26. Bhat, M.P.; Kurkuri, M.; Losic, D.; Kigga, M.; Altalhi, T. New optofluidic based lab-on-a-chip device for the real-time fluoride analysis. *Anal. Chim. Acta* **2021**, *1159*, 338439. [CrossRef]
27. Leung, C.-H.; Wu, K.-J.; Li, G.; Wu, C.; Ko, C.-N.; Ma, D.-L. Application of label-free techniques in microfluidic for biomolecules detection and circulating tumor cells analysis. *TrAC Trends Anal. Chem.* **2019**, *117*, 78–83. [CrossRef]

28. Edd, J.F.; Mishra, A.; Dubash, T.D.; Herrera, S.; Mohammad, R.; Williams, E.K.; Hong, X.; Mutlu, B.R.; Walsh, J.R.; de Carvalho, F.M. Microfluidic concentration and separation of circulating tumor cell clusters from large blood volumes. *Lab Chip* **2020**, *20*, 558–567. [CrossRef]

29. Gharaghani, F.M.; Akhond, M.; Hemmateenejad, B. A three-dimensional origami microfluidic device for paper chromatography: Application to quantification of Tartrazine and Indigo carmine in food samples. *J. Chromatogr. A* **2020**, *1621*, 461049. [CrossRef]

30. He, G.; Yang, C.; Feng, J.; Wu, J.; Zhou, L.; Wen, R.; Huang, S.; Wu, Q.; Liu, F.; Chen, H.J. Hierarchical spiky microstraws-integrated microfluidic device for efficient capture and in situ manipulation of cancer cells. *Adv. Funct. Mater.* **2019**, *29*, 1806484. [CrossRef]

31. Chen, J.; Chen, D.; Xie, Y.; Yuan, T.; Chen, X. Progress of microfluidics for biology and medicine. *Nano-Micro Lett.* **2013**, *5*, 66–80. [CrossRef]

32. Kim, M.; Mo Jung, S.; Lee, K.H.; Jun Kang, Y.; Yang, S. A microfluidic device for continuous white blood cell separation and lysis from whole blood. *Artif. Organs* **2010**, *34*, 996–1002. [CrossRef] [PubMed]

33. Qin, Y.; Yang, X.; Zhang, J.; Cao, X. Developing a non-fouling hybrid microfluidic device for applications in circulating tumour cell detections. *Colloids Surf. B Biointerfaces* **2017**, *151*, 39–46. [CrossRef] [PubMed]

34. Halldorsson, S.; Lucumi, E.; Gómez-Sjöberg, R.; Fleming, R.M. Advantages and challenges of microfluidic cell culture in polydimethylsiloxane devices. *Biosens. Bioelectron.* **2015**, *63*, 218–231. [CrossRef]

35. Panesar, S.; Neethirajan, S. Microfluidics: Rapid diagnosis for breast cancer. *Nano-Micro Lett.* **2016**, *8*, 204–220. [CrossRef]

36. Liu, G.; Bhat, M.P.; Kim, C.S.; Kim, J.; Lee, K.-H. Improved 3D-Printability of Cellulose Acetate to Mimic Water Absorption in Plant Roots through Nanoporous Networks. *Macromolecules* **2022**, *55*, 1855–1865. [CrossRef]

37. Castillo-León, J. Microfluidics and lab-on-a-chip devices: History and challenges. In *Lab-on-a-Chip Devices and Micro-Total Analysis Systems*; Springer: Cham, Switzerland, 2015; pp. 1–15.

38. Muhsin, S.A.; Al-Amidie, M.; Shen, Z.; Mlaji, Z.; Liu, J.; Abdullah, A.; El-Dweik, M.; Zhang, S.; Almasri, M. A microfluidic biosensor for rapid simultaneous detection of waterborne pathogens. *Biosens. Bioelectron.* **2022**, *203*, 113993. [CrossRef]

39. Leong, S.Y.; Ong, H.B.; Tay, H.M.; Kong, F.; Upadya, M.; Gong, L.; Dao, M.; Dalan, R.; Hou, H.W. Microfluidic Size Exclusion Chromatography (µSEC) for Extracellular Vesicles and Plasma Protein Separation. *Small* **2022**, *18*, 2104470. [CrossRef]

40. Livak-Dahl, E.; Sinn, I.; Burns, M. Microfluidic Chemical Analysis Systems. *Annu. Rev. Chem. Biomol. Eng.* **2011**, *2*, 325–353. [CrossRef]

41. Sanjay, S.T.; Zhou, W.; Dou, M.; Tavakoli, H.; Ma, L.; Xu, F.; Li, X. Recent advances of controlled drug delivery using microfluidic platforms. *Adv. Drug Deliv. Rev.* **2018**, *128*, 3–28. [CrossRef]

42. Riahi, R.; Tamayol, A.; Shaegh, S.A.M.; Ghaemmaghami, A.M.; Dokmeci, M.R.; Khademhosseini, A. Microfluidics for advanced drug delivery systems. *Curr. Opin. Chem. Eng.* **2015**, *7*, 101–112. [CrossRef] [PubMed]

43. Borecki, M.; Korwin-Pawlowski, M.L.; Beblowska, M.; Szmidt, J.; Jakubowski, A. Optoelectronic Capillary Sensors in Microfluidic and Point-of-Care Instrumentation. *Sensors* **2010**, *10*, 3771–3797. [CrossRef] [PubMed]

44. Luo, T.; Fan, L.; Zhu, R.; Sun, D. Microfluidic single-cell manipulation and analysis: Methods and applications. *Micromachines* **2019**, *10*, 104. [CrossRef] [PubMed]

45. Bendre, A.; Bhat, M.P.; Lee, K.-H.; Altalhi, T.; Alruqi, M.A.; Kurkuri, M. Recent developments in microfluidic technology for synthesis and toxicity-efficiency studies of biomedical nanomaterials. *Mater. Today Adv.* **2022**, *13*, 100205. [CrossRef]

46. Elvira, K.S.; Wootton, R.C.; de Mello, A.J. The past, present and potential for microfluidic reactor technology in chemical synthesis. *Nat. Chem.* **2013**, *5*, 905–915. [CrossRef]

47. Zhou, Y.; Dong, Z.; Andarge, H.; Li, W.; Pappas, D. Nanoparticle modification of microfluidic cell separation for cancer cell detection and isolation. *Analyst* **2020**, *145*, 257–267. [CrossRef]

48. Qu, L.; Xu, J.; Tan, X.; Liu, Z.; Xu, L.; Peng, R. Dual-aptamer modification generates a unique interface for highly sensitive and specific electrochemical detection of tumor cells. *ACS Appl. Mater. Interfaces* **2014**, *6*, 7309–7315. [CrossRef]

49. Safdar, M.; Jänis, J.; Sanchez, S. Microfluidic fuel cells for energy generation. *Lab Chip* **2016**, *16*, 2754–2758. [CrossRef]

50. Aykar, S.S.; Reynolds, D.E.; McNamara, M.C.; Hashemi, N.N. Manufacturing of poly(ethylene glycol diacrylate)-based hollow microvessels using microfluidics. *RSC Adv.* **2020**, *10*, 4095–4102. [CrossRef]

51. Jung, B.-J.; Kim, J.; Kim, J.-a.; Jang, H.; Seo, S.; Lee, W. PDMS-parylene hybrid, flexible microfluidics for real-time modulation of 3D helical inertial microfluidics. *Micromachines* **2018**, *9*, 255. [CrossRef]

52. Friend, J.; Yeo, L. Fabrication of microfluidic devices using polydimethylsiloxane. *Biomicrofluidics* **2010**, *4*, 026502. [CrossRef] [PubMed]

53. Eddings, M.A.; Johnson, M.A.; Gale, B.K. Determining the optimal PDMS–PDMS bonding technique for microfluidic devices. *J. Micromech. Microeng.* **2008**, *18*, 067001. [CrossRef]

54. Yin, J.; Deng, J.; Du, C.; Zhang, W.; Jiang, X. Microfluidics-based approaches for separation and analysis of circulating tumor cells. *TrAC Trends Anal. Chem.* **2019**, *117*, 84–100. [CrossRef]

55. Murlidhar, V.; Zeinali, M.; Grabauskiene, S.; Ghannad-Rezaie, M.; Wicha, M.S.; Simeone, D.M.; Ramnath, N.; Reddy, R.M.; Nagrath, S. A Radial Flow Microfluidic Device for Ultra-High-Throughput Affinity-Based Isolation of Circulating Tumor Cells. *Small* **2014**, *10*, 4895–4904. [CrossRef]

56. Lee, A.; Park, J.; Lim, M.; Sunkara, V.; Kim, S.Y.; Kim, G.H.; Kim, M.-H.; Cho, Y.-K. All-in-one centrifugal microfluidic device for size-selective circulating tumor cell isolation with high purity. *Anal. Chem.* **2014**, *86*, 11349–11356. [CrossRef]

57. Sato, K. Microdevice in Cellular Pathology: Microfluidic Platforms for Fluorescence in situ Hybridization and Analysis of Circulating Tumor Cells. *Anal. Sci.* **2015**, *31*, 867–873. [CrossRef]
58. Park, E.S.; Jin, C.; Guo, Q.; Ang, R.R.; Duffy, S.P.; Matthews, K.; Azad, A.; Abdi, H.; Todenhöfer, T.; Bazov, J.; et al. Continuous Flow Deformability-Based Separation of Circulating Tumor Cells Using Microfluidic Ratchets. *Small* **2016**, *12*, 1909–1919. [CrossRef]
59. Zhang, X.; Xu, X.; Ren, Y.; Yan, Y.; Wu, A. Numerical simulation of circulating tumor cell separation in a dielectrophoresis based Y-Y shaped microfluidic device. *Sep. Purif. Technol.* **2021**, *255*, 117343. [CrossRef]
60. Shi, W.; Wang, S.; Maarouf, A.; Uhl, C.G.; He, R.; Yunus, D.; Liu, Y. Magnetic particles assisted capture and release of rare circulating tumor cells using wavy-herringbone structured microfluidic devices. *Lab Chip* **2017**, *17*, 3291–3299. [CrossRef]
61. Rafeie, M.; Zhang, J.; Asadnia, M.; Li, W.; Warkiani, M.E. Multiplexing slanted spiral microchannels for ultra-fast blood plasma separation. *Lab Chip* **2016**, *16*, 2791–2802. [CrossRef]
62. Zhou, J.; Kulasinghe, A.; Bogseth, A.; O'Byrne, K.; Punyadeera, C.; Papautsky, I. Isolation of circulating tumor cells in non-small-cell-lung-cancer patients using a multi-flow microfluidic channel. *Microsyst. Nanoeng.* **2019**, *5*, 8. [CrossRef] [PubMed]
63. Tian, F.; Liu, C.; Lin, L.; Chen, Q.; Sun, J. Microfluidic analysis of circulating tumor cells and tumor-derived extracellular vesicles. *TrAC Trends Anal. Chem.* **2019**, *117*, 128–145. [CrossRef]
64. Zhang, J.; Yan, S.; Yuan, D.; Alici, G.; Nguyen, N.-T.; Ebrahimi Warkiani, M.; Li, W. Fundamentals and applications of inertial microfluidics: A review. *Lab Chip* **2016**, *16*, 10–34. [CrossRef]
65. Wang, Z.; Xu, D.; Wang, X.; Jin, Y.; Huo, B.; Wang, Y.; He, C.; Fu, X.; Lu, N. Size-matching hierarchical micropillar arrays for detecting circulating tumor cells in breast cancer patients' whole blood. *Nanoscale* **2019**, *11*, 6677–6684. [CrossRef] [PubMed]
66. Che, J.; Yu, V.; Garon, E.B.; Goldman, J.W.; Di Carlo, D. Biophysical isolation and identification of circulating tumor cells. *Lab Chip* **2017**, *17*, 1452–1461. [CrossRef]
67. Brimmo, A.T.; Menachery, A.; Qasaimeh, M.A. Microelectrofluidic probe for sequential cell separation and patterning. *Lab Chip* **2019**, *19*, 4052–4063. [CrossRef] [PubMed]
68. Wu, Z.; Jiang, H.; Zhang, L.; Yi, K.; Cui, H.; Wang, F.; Liu, W.; Zhao, X.; Zhou, F.; Guo, S. The acoustofluidic focusing and separation of rare tumor cells using transparent lithium niobate transducers. *Lab Chip* **2019**, *19*, 3922–3930. [CrossRef]
69. Hu, X.; Zhu, D.; Chen, M.; Chen, K.; Liu, H.; Liu, W.; Yang, Y. Precise and non-invasive circulating tumor cell isolation based on optical force using homologous erythrocyte binding. *Lab Chip* **2019**, *19*, 2549–2556. [CrossRef]
70. Zhao, W.; Liu, Y.; Jenkins, B.D.; Cheng, R.; Harris, B.N.; Zhang, W.; Xie, J.; Murrow, J.R.; Hodgson, J.; Egan, M.; et al. Tumor antigen-independent and cell size variation-inclusive enrichment of viable circulating tumor cells. *Lab Chip* **2019**, *19*, 1860–1876. [CrossRef]
71. Ahmed, M.G.; Abate, M.F.; Song, Y.; Zhu, Z.; Yan, F.; Xu, Y.; Wang, X.; Li, Q.; Yang, C. Isolation, Detection, and Antigen-Based Profiling of Circulating Tumor Cells Using a Size-Dictated Immunocapture Chip. *Angew. Chem. Int. Ed.* **2017**, *56*, 10681–10685. [CrossRef]
72. Cheng, J.; Liu, Y.; Zhao, Y.; Zhang, L.; Zhang, L.; Mao, H.; Huang, C. Nanotechnology-assisted isolation and analysis of circulating tumor cells on microfluidic devices. *Micromachines* **2020**, *11*, 774. [CrossRef] [PubMed]
73. Zou, D.; Cui, D. Advances in isolation and detection of circulating tumor cells based on microfluidics. *Cancer Biol. Med.* **2018**, *15*, 335–353. [PubMed]
74. Sackmann, E.K.; Fulton, A.L.; Beebe, D.J. The present and future role of microfluidics in biomedical research. *Nature* **2014**, *507*, 181–189. [CrossRef] [PubMed]
75. Au, A.K.; Lee, W.; Folch, A. Mail-order microfluidics: Evaluation of stereolithography for the production of microfluidic devices. *Lab Chip* **2014**, *14*, 1294–1301. [CrossRef]
76. Tseng, P.; Murray, C.; Kim, D.; Di Carlo, D. Research highlights: Printing the future of microfabrication. *Lab Chip* **2014**, *14*, 1491–1495. [CrossRef] [PubMed]
77. Sochol, R.; Sweet, E.; Glick, C.; Venkatesh, S.; Avetisyan, A.; Ekman, K.; Raulinaitis, A.; Tsai, A.; Wienkers, A.; Korner, K. 3D printed microfluidic circuitry via multijet-based additive manufacturing. *Lab Chip* **2016**, *16*, 668–678. [CrossRef]
78. Bhattacharjee, N.; Urrios, A.; Kang, S.; Folch, A. The upcoming 3D-printing revolution in microfluidics. *Lab Chip* **2016**, *16*, 1720–1742. [CrossRef]
79. Chu, C.-H.; Liu, R.; Ozkaya-Ahmadov, T.; Boya, M.; Swain, B.E.; Owens, J.M.; Burentugs, E.; Bilen, M.A.; McDonald, J.F.; Sarioglu, A.F. Hybrid negative enrichment of circulating tumor cells from whole blood in a 3D-printed monolithic device. *Lab Chip* **2019**, *19*, 3427–3437. [CrossRef]
80. Gong, H.; Woolley, A.T.; Nordin, G.P. 3D printed high density, reversible, chip-to-chip microfluidic interconnects. *Lab Chip* **2018**, *18*, 639–647. [CrossRef]
81. Chen, J.; Liu, C.-Y.; Wang, X.; Sweet, E.; Liu, N.; Gong, X.; Lin, L. 3D printed microfluidic devices for circulating tumor cells (CTCs) isolation. *Biosens. Bioelectron.* **2020**, *150*, 111900. [CrossRef]
82. Liu, F.; Wang, S.; Lu, Z.; Sun, Y.; Yang, C.; Zhou, Q.; Hong, S.; Wang, S.; Xiong, B.; Liu, K.; et al. A simple pyramid-shaped microchamber towards highly efficient isolation of circulating tumor cells from breast cancer patients. *Biomed. Microdevices* **2018**, *20*, 83. [CrossRef] [PubMed]
83. Yang, C.; Zhang, N.; Wang, S.; Shi, D.; Zhang, C.; Liu, K.; Xiong, B. Wedge-shaped microfluidic chip for circulating tumor cells isolation and its clinical significance in gastric cancer. *J. Transl. Med.* **2018**, *16*, 139. [CrossRef] [PubMed]
84. Scott, S.M.; Ali, Z. Fabrication Methods for Microfluidic Devices: An Overview. *Micromachines* **2021**, *12*, 319. [CrossRef] [PubMed]

85. Liao, C.-J.; Hsieh, C.-H.; Chiu, T.-K.; Zhu, Y.-X.; Wang, H.-M.; Hung, F.-C.; Chou, W.-P.; Wu, M.-H. An Optically Induced Dielectrophoresis (ODEP)-Based Microfluidic System for the Isolation of High-Purity CD45(neg)/EpCAM(neg) Cells from the Blood Samples of Cancer Patients-Demonstration and Initial Exploration of the Clinical Significance of These Cells. *Micromachines* **2018**, *9*, 563. [CrossRef] [PubMed]

86. Lim, C.T.; Low, H.Y.; Ng, J.K.; Liu, W.-T.; Zhang, Y. Fabrication of three-dimensional hemispherical structures using photolithography. *Microfluid. Nanofluid.* **2009**, *7*, 721. [CrossRef]

87. Tian, W.-C.; Finehout, E. *Microfluidics for Biological Applications*; Springer Science & Business Media: Berlin, Germany, 2009; Volume 16.

88. Gale, B.K.; Jafek, A.R.; Lambert, C.J.; Goenner, B.L.; Moghimifam, H.; Nze, U.C.; Kamarapu, S.K. A Review of Current Methods in Microfluidic Device Fabrication and Future Commercialization Prospects. *Inventions* **2018**, *3*, 60. [CrossRef]

89. Kwak, B.; Lee, J.; Lee, J.; Kim, H.S.; Kang, S.; Lee, Y. Spiral shape microfluidic channel for selective isolating of heterogenic circulating tumor cells. *Biosens. Bioelectron.* **2018**, *101*, 311–316. [CrossRef]

90. Fan, X.; Jia, C.; Yang, J.; Li, G.; Mao, H.; Jin, Q.; Zhao, J. A microfluidic chip integrated with a high-density PDMS-based microfiltration membrane for rapid isolation and detection of circulating tumor cells. *Biosens. Bioelectron.* **2015**, *71*, 380–386. [CrossRef]

91. Yan, S.; Chen, P.; Zeng, X.; Zhang, X.; Li, Y.; Xia, Y.; Wang, J.; Dai, X.; Feng, X.; Du, W.; et al. Integrated Multifunctional Electrochemistry Microchip for Highly Efficient Capture, Release, Lysis, and Analysis of Circulating Tumor Cells. *Anal. Chem.* **2017**, *89*, 12039–12044. [CrossRef]

92. Kulasinghe, A.; Zhou, J.; Kenny, L.; Papautsky, I.; Punyadeera, C. Capture of Circulating Tumour Cell Clusters Using Straight Microfluidic Chips. *Cancers* **2019**, *11*, 89. [CrossRef]

93. Yoon, Y.; Lee, J.; Ra, M.; Gwon, H.; Lee, S.; Kim, M.Y.; Yoo, K.-C.; Sul, O.; Kim, C.G.; Kim, W.-Y.; et al. Continuous Separation of Circulating Tumor Cells from Whole Blood Using a Slanted Weir Microfluidic Device. *Cancers* **2019**, *11*, 200. [CrossRef] [PubMed]

94. Chen, Y.-H.; Pulikkathodi, A.K.; Ma, Y.-D.; Wang, Y.-L.; Lee, G.-B. A microfluidic platform integrated with field-effect transistors for enumeration of circulating tumor cells. *Lab Chip* **2019**, *19*, 618–625. [CrossRef] [PubMed]

95. Raillon, C.; Che, J.; Thill, S.; Duchamp, M.; Desbiolles, B.X.E.; Millet, A.; Sollier, E.; Renaud, P. Toward Microfluidic Label-Free Isolation and Enumeration of Circulating Tumor Cells from Blood Samples. *Cytom. Part A* **2019**, *95*, 1085–1095. [CrossRef] [PubMed]

96. Chang, C.-L.; Huang, W.; Jalal, S.I.; Chan, B.-D.; Mahmood, A.; Shahda, S.; O'Neil, B.H.; Matei, D.E.; Savran, C.A. Circulating tumor cell detection using a parallel flow micro-aperture chip system. *Lab Chip* **2015**, *15*, 1677–1688. [CrossRef]

97. Chen, H.; Zhang, Z.; Liu, H.; Zhang, Z.; Lin, C.; Wang, B. Hybrid magnetic and deformability based isolation of circulating tumor cells using microfluidics. *AIP Adv.* **2019**, *9*, 025023. [CrossRef]

98. Varillas, J.I.; Zhang, J.; Chen, K.; Barnes, I.I.; Liu, C.; George, T.J.; Fan, Z.H. Microfluidic Isolation of Circulating Tumor Cells and Cancer Stem-Like Cells from Patients with Pancreatic Ductal Adenocarcinoma. *Theranostics* **2019**, *9*, 1417–1425. [CrossRef] [PubMed]

99. Shamloo, A.; Ahmad, S.; Momeni, M. Design and Parameter Study of Integrated Microfluidic Platform for CTC Isolation and Enquiry; A Numerical Approach. *Biosensors* **2018**, *8*, 56. [CrossRef]

100. Chen, Q.; Wu, J.; Zhang, Y.; Lin, Z.; Lin, J.-M. Targeted isolation and analysis of single tumor cells with aptamer-encoded microwell array on microfluidic device. *Lab Chip* **2012**, *12*, 5180–5185. [CrossRef]

101. Hoshino, K.; Huang, Y.-Y.; Lane, N.; Huebschman, M.; Uhr, J.W.; Frenkel, E.P.; Zhang, X. Microchip-based immunomagnetic detection of circulating tumor cells. *Lab Chip* **2011**, *11*, 3449–3457. [CrossRef]

102. Fallahi, H.; Yadav, S.; Phan, H.-P.; Ta, H.; Zhang, J.; Nguyen, N.-T. Size-tuneable isolation of cancer cells using stretchable inertial microfluidics. *Lab Chip* **2021**, *21*, 2008–2018. [CrossRef]

103. Jiang, R.; Agrawal, S.; Aghaamoo, M.; Parajuli, R.; Agrawal, A.; Lee, A.P. Rapid isolation of circulating cancer associated fibroblasts by acoustic microstreaming for assessing metastatic propensity of breast cancer patients. *Lab Chip* **2021**, *21*, 875–887. [CrossRef] [PubMed]

104. Jou, H.-J.; Chou, L.-Y.; Chang, W.-C.; Ho, H.-C.; Zhang, W.-T.; Ling, P.-Y.; Tsai, K.-H.; Chen, S.-H.; Chen, T.-H.; Lo, P.-H.; et al. An Automatic Platform Based on Nanostructured Microfluidic Chip for Isolating and Identification of Circulating Tumor Cells. *Micromachines* **2021**, *12*, 473. [CrossRef] [PubMed]

105. Zhang, X.; Lu, X.; Gao, W.; Wang, Y.; Jia, C.; Cong, H. A label-free microfluidic chip for the highly selective isolation of single and cluster CTCs from breast cancer patients. *Transl. Oncol.* **2021**, *14*, 100959. [CrossRef] [PubMed]

106. Reinholt, S.J.; Craighead, H.G. Microfluidic Device for Aptamer-Based Cancer Cell Capture and Genetic Mutation Detection. *Anal. Chem.* **2018**, *90*, 2601–2608. [CrossRef] [PubMed]

107. Nasiri, R.; Shamloo, A.; Akbari, J. Design of a Hybrid Inertial and Magnetophoretic Microfluidic Device for CTCs Separation from Blood. *Micromachines* **2021**, *12*, 877. [CrossRef] [PubMed]

108. Nguyen, H.-T.; Thach, H.; Roy, E.; Huynh, K.; Perrault, C.M.-T. Low-Cost, Accessible Fabrication Methods for Microfluidics Research in Low-Resource Settings. *Micromachines* **2018**, *9*, 461. [CrossRef] [PubMed]

109. Waldbaur, A.; Rapp, H.; Länge, K.; Rapp, B.E. Let there be chip—Towards rapid prototyping of microfluidic devices: One-step manufacturing processes. *Anal. Methods* **2011**, *3*, 2681–2716. [CrossRef]

110. Kamei, K.-i.; Mashimo, Y.; Koyama, Y.; Fockenberg, C.; Nakashima, M.; Nakajima, M.; Li, J.; Chen, Y. 3D printing of soft lithography mold for rapid production of polydimethylsiloxane-based microfluidic devices for cell stimulation with concentration gradients. *Biomed. Microdevices* **2015**, *17*, 36. [CrossRef] [PubMed]

111. Waheed, S.; Cabot, J.M.; Macdonald, N.P.; Lewis, T.; Guijt, R.M.; Paull, B.; Breadmore, M.C. 3D printed microfluidic devices: Enablers and barriers. *Lab Chip* **2016**, *16*, 1993–2013. [CrossRef]

112. Li, Y.; Wu, P.; Luo, Z.; Ren, Y.; Liao, M.; Feng, L.; Li, Y.; He, L. Rapid fabrication of microfluidic chips based on the simplest LED lithography. *J. Micromech. Microeng.* **2015**, *25*, 055020. [CrossRef]

113. Isiksacan, Z.; Guler, M.T.; Aydogdu, B.; Bilican, I.; Elbuken, C. Rapid fabrication of microfluidic PDMS devices from reusable PDMS molds using laser ablation. *J. Micromech. Microeng.* **2016**, *26*, 035008. [CrossRef]

114. Thaweskulchai, T.; Schulte, A. A Low-Cost 3-in-1 3D Printer as a Tool for the Fabrication of Flow-Through Channels of Microfluidic Systems. *Micromachines* **2021**, *12*, 947. [CrossRef] [PubMed]

115. Xu, M.; Liu, W.; Zou, K.; Wei, S.; Zhang, X.; Li, E.; Wang, Q. Design and Clinical Application of an Integrated Microfluidic Device for Circulating Tumor Cells Isolation and Single-Cell Analysis. *Micromachines* **2021**, *12*, 49. [CrossRef]

116. Gurudatt, N.G.; Chung, S.; Kim, J.-M.; Kim, M.-H.; Jung, D.-K.; Han, J.-Y.; Shim, Y.-B. Separation detection of different circulating tumor cells in the blood using an electrochemical microfluidic channel modified with a lipid-bonded conducting polymer. *Biosens. Bioelectron.* **2019**, *146*, 111746. [CrossRef] [PubMed]

117. Yin, J.; Wang, Z.; Li, G.; Lin, F.; Shao, K.; Cao, B.; Hou, Y. Characterization of circulating tumor cells in breast cancer patients by spiral microfluidics. *Cell Biol. Toxicol.* **2019**, *35*, 59–66. [CrossRef] [PubMed]

118. Autebert, J.; Coudert, B.; Bidard, F.-C.; Pierga, J.-Y.; Descroix, S.; Malaquin, L.; Viovy, J.-L. Microfluidic: An innovative tool for efficient cell sorting. *Methods* **2012**, *57*, 297–307. [CrossRef]

119. Lim, L.S.; Hu, M.; Huang, M.C.; Cheong, W.C.; Gan, A.T.L.; Looi, X.L.; Leong, S.M.; Koay, E.S.-C.; Li, M.-H. Microsieve lab-chip device for rapid enumeration and fluorescence in situ hybridization of circulating tumor cells. *Lab Chip* **2012**, *12*, 4388–4396. [CrossRef]

120. Kulasinghe, A.; Tran, T.H.P.; Blick, T.; O'Byrne, K.; Thompson, E.W.; Warkiani, M.E.; Nelson, C.; Kenny, L.; Punyadeera, C. Enrichment of circulating head and neck tumour cells using spiral microfluidic technology. *Sci. Rep.* **2017**, *7*, 42517. [CrossRef]

121. Abdulla, A.; Zhang, T.; Ahmad, K.Z.; Li, S.; Lou, J.; Ding, X. Label-free Separation of Circulating Tumor Cells Using a Self-Amplified Inertial Focusing (SAIF) Microfluidic Chip. *Anal. Chem.* **2020**, *92*, 16170–16179. [CrossRef]

122. Zhou, Y.; Ma, Z.; Ai, Y. Sheathless inertial cell focusing and sorting with serial reverse wavy channel structures. *Microsyst. Nanoeng.* **2018**, *4*, 5. [CrossRef]

123. Martel, J.M.; Toner, M. Inertial Focusing in Microfluidics. *Annu. Rev. Biomed. Eng.* **2014**, *16*, 371–396. [CrossRef] [PubMed]

124. Wang, C.; Sun, S.; Chen, Y.; Cheng, Z.; Li, Y.; Jia, L.; Lin, P.; Yang, Z.; Shu, R. Inertial particle focusing and spacing control in microfluidic devices. *Microfluid. Nanofluidics* **2018**, *22*, 25. [CrossRef]

125. Ying, Y.; Lin, Y. Inertial Focusing and Separation of Particles in Similar Curved Channels. *Sci. Rep.* **2019**, *9*, 16575. [CrossRef] [PubMed]

126. Gao, R.; Cheng, L.; Wang, S.; Bi, X.; Wang, X.; Wang, R.; Chen, X.; Zha, Z.; Wang, F.; Xu, X.; et al. Efficient separation of tumor cells from untreated whole blood using a novel multistage hydrodynamic focusing microfluidics. *Talanta* **2020**, *207*, 120261. [CrossRef]

127. Warkiani, M.E.; Khoo, B.L.; Wu, L.; Tay, A.K.P.; Bhagat, A.A.S.; Han, J.; Lim, C.T. Ultra-fast, label-free isolation of circulating tumor cells from blood using spiral microfluidics. *Nat. Protoc.* **2016**, *11*, 134–148. [CrossRef]

128. Ozbey, A.; Karimzadehkhouei, M.; Kocaturk, N.M.; Bilir, S.E.; Kutlu, O.; Gozuacik, D.; Kosar, A. Inertial focusing of cancer cell lines in curvilinear microchannels. *Micro Nano Eng.* **2019**, *2*, 53–63. [CrossRef]

129. Nam, J.; Tan, J.K.S.; Khoo, B.L.; Namgung, B.; Leo, H.L.; Lim, C.T.; Kim, S. Hybrid capillary-inserted microfluidic device for sheathless particle focusing and separation in viscoelastic flow. *Biomicrofluidics* **2015**, *9*, 064117. [CrossRef]

130. Che, J.; Yu, V.; Dhar, M.; Renier, C.; Matsumoto, M.; Heirich, K.; Garon, E.B.; Goldman, J.; Rao, J.; Sledge, G.W.; et al. Classification of large circulating tumor cells isolated with ultra-high throughput microfluidic Vortex technology. *Oncotarget* **2016**, *7*, 12748. [CrossRef]

131. Thanormsridetchai, A.; Ketpun, D.; Srituravanich, W.; Piyaviriyakul, P.; Sailasuta, A.; Jeamsaksiri, W.; Sripumkhai, W.; Pimpin, A. Focusing and sorting of multiple-sized beads and cells using low-aspect-ratio spiral microchannels. *J. Mech. Sci. Technol.* **2017**, *31*, 5397–5405. [CrossRef]

132. Khoshmanesh, K.; Nahavandi, S.; Baratchi, S.; Mitchell, A.; Kalantar-zadeh, K. Dielectrophoretic platforms for bio-microfluidic systems. *Biosens. Bioelectron.* **2011**, *26*, 1800–1814. [CrossRef]

133. Abd Rahman, N.; Ibrahim, F.; Yafouz, B. Dielectrophoresis for Biomedical Sciences Applications: A Review. *Sensors* **2017**, *17*, 449. [CrossRef] [PubMed]

134. Chan, J.Y.; Kayani, A.B.A.; Ali, M.A.M.; Kok, C.K.; Majlis, B.Y.; Hoe, S.L.L.; Marzuki, M.; Khoo, A.S.-B.; Ostrikov, K.; Rahman, M.A.; et al. Dielectrophoresis-based microfluidic platforms for cancer diagnostics. *Biomicrofluidics* **2018**, *12*, 011503. [CrossRef] [PubMed]

135. Chiu, T.-K.; Chou, W.-P.; Huang, S.-B.; Wang, H.-M.; Lin, Y.-C.; Hsieh, C.-H.; Wu, M.-H. Application of optically-induced-dielectrophoresis in microfluidic system for purification of circulating tumour cells for gene expression analysis—Cancer cell line model. *Sci. Rep.* **2016**, *6*, 32851. [CrossRef]

136. Li, M.; Anand, R.K. High-Throughput Selective Capture of Single Circulating Tumor Cells by Dielectrophoresis at a Wireless Electrode Array. *J. Am. Chem. Soc.* **2017**, *139*, 8950–8959. [CrossRef] [PubMed]

137. Kim, S.H.; Ito, H.; Kozuka, M.; Hirai, M.; Fujii, T. Localization of low-abundant cancer cells in a sharply expanded microfluidic step-channel using dielectrophoresis. *Biomicrofluidics* **2017**, *11*, 054114. [CrossRef]

138. Chikaishi, Y.; Yoneda, K.; Ohnaga, T.; Tanaka, F. EpCAM-independent capture of circulating tumor cells with a 'universal CTC-chip'. *Oncol. Rep.* **2017**, *37*, 77–82. [CrossRef]

139. Liu, Z.; Fusi, A.; Klopocki, E.; Schmittel, A.; Tinhofer, I.; Nonnenmacher, A.; Keilholz, U. Negative enrichment by immunomagnetic nanobeads for unbiased characterization of circulating tumor cells from peripheral blood of cancer patients. *J. Med.* **2011**, *9*, 70. [CrossRef]

140. Kang, H.; Kim, J.; Cho, H.; Han, K.-H. Evaluation of Positive and Negative Methods for Isolation of Circulating Tumor Cells by Lateral Magnetophoresis. *Micromachines* **2019**, *10*, 386. [CrossRef]

141. Poudineh, M.; Aldridge, P.M.; Ahmed, S.; Green, B.J.; Kermanshah, L.; Nguyen, V.; Tu, C.; Mohamadi, R.M.; Nam, R.K.; Hansen, A.; et al. Tracking the dynamics of circulating tumour cell phenotypes using nanoparticle-mediated magnetic ranking. *Nat. Nanotechnol.* **2017**, *12*, 274–281. [CrossRef]

142. Poudineh, M.; Labib, M.; Ahmed, S.; Nguyen, L.N.M.; Kermanshah, L.; Mohamadi, R.M.; Sargent, E.H.; Kelley, S.O. Profiling Functional and Biochemical Phenotypes of Circulating Tumor Cells Using a Two-Dimensional Sorting Device. *Angew. Chem. Int. Ed.* **2017**, *56*, 163–168. [CrossRef]

143. Yin, C.; Wang, Y.; Ji, J.; Cai, B.; Chen, H.; Yang, Z.; Wang, K.; Luo, C.; Zhang, W.; Yuan, C.; et al. Molecular Profiling of Pooled Circulating Tumor Cells from Prostate Cancer Patients Using a Dual-Antibody-Functionalized Microfluidic Device. *Anal. Chem.* **2018**, *90*, 3744–3751. [CrossRef] [PubMed]

144. Zhao, W.; Zhu, T.; Cheng, R.; Liu, Y.; He, J.; Qiu, H.; Wang, L.; Nagy, T.; Querec, T.D.; Unger, E.R.; et al. Label-Free and Continuous-Flow Ferrohydrodynamic Separation of HeLa Cells and Blood Cells in Biocompatible Ferrofluids. *Adv. Funct. Mater.* **2016**, *26*, 3990–3998. [CrossRef]

145. Zhao, W.; Cheng, R.; Jenkins, B.D.; Zhu, T.; Okonkwo, N.E.; Jones, C.E.; Davis, M.B.; Kavuri, S.K.; Hao, Z.; Schroeder, C.; et al. Label-free ferrohydrodynamic cell separation of circulating tumor cells. *Lab Chip* **2017**, *17*, 3097–3111. [CrossRef] [PubMed]

146. Zhao, W.; Cheng, R.; Lim, S.H.; Miller, J.R.; Zhang, W.; Tang, W.; Xie, J.; Mao, L. Biocompatible and label-free separation of cancer cells from cell culture lines from white blood cells in ferrofluids. *Lab Chip* **2017**, *17*, 2243–2255. [CrossRef]

147. Lenshof, A.; Evander, M.; Laurell, T.; Nilsson, J. Acoustofluidics 5: Building microfluidic acoustic resonators. *Lab Chip* **2012**, *12*, 684–695. [CrossRef] [PubMed]

148. Gao, Y.; Wu, M.; Lin, Y.; Xu, J. Acoustic Microfluidic Separation Techniques and Bioapplications: A Review. *Micromachines* **2020**, *11*, 921. [CrossRef]

149. Ding, X.; Li, P.; Lin, S.-C.S.; Stratton, Z.S.; Nama, N.; Guo, F.; Slotcavage, D.; Mao, X.; Shi, J.; Costanzo, F.; et al. Surface acoustic wave microfluidics. *Lab Chip* **2013**, *13*, 3626–3649. [CrossRef]

150. Wu, M.; Huang, P.-H.; Zhang, R.; Mao, Z.; Chen, C.; Kemeny, G.; Li, P.; Lee, A.V.; Gyanchandani, R.; Armstrong, A.J.; et al. Circulating Tumor Cell Phenotyping via High-Throughput Acoustic Separation. *Small* **2018**, *14*, 1801131. [CrossRef]

151. Wang, K.; Zhou, W.; Lin, Z.; Cai, F.; Li, F.; Wu, J.; Meng, L.; Niu, L.; Zheng, H. Sorting of tumour cells in a microfluidic device by multi-stage surface acoustic waves. *Sens. Actuators B Chem.* **2018**, *258*, 1174–1183. [CrossRef]

152. Karthick, S.; Pradeep, P.N.; Kanchana, P.; Sen, A.K. Acoustic impedance-based size-independent isolation of circulating tumour cells from blood using acoustophoresis. *Lab Chip* **2018**, *18*, 3802–3813. [CrossRef]

153. Bai, X.; Bin, S.; Yuguo, D.; Wei, Z.; Yanmin, F.; Yuanyuan, C.; Deyuan, Z.; Fumihito, A.; Lin, F. Parallel trapping, patterning, separating and rotating of micro-objects with various sizes and shapes using acoustic microstreaming. *Sens. Actuators A Phys.* **2020**, *315*, 112340. [CrossRef]

154. Cushing, K.; Undvall, E.; Ceder, Y.; Lilja, H.; Laurell, T. Reducing WBC background in cancer cell separation products by negative acoustic contrast particle immuno-acoustophoresis. *Anal. Chim. Acta* **2018**, *1000*, 256–264. [CrossRef] [PubMed]

155. Shamloo, A.; Naghdloo, A.; Besanjideh, M. Cancer cell enrichment on a centrifugal microfluidic platform using hydrodynamic and magnetophoretic techniques. *Sci. Rep.* **2021**, *11*, 1939. [CrossRef] [PubMed]

156. Chen, H. A Triplet Parallelizing Spiral Microfluidic Chip for Continuous Separation of Tumor Cells. *Sci. Rep.* **2018**, *8*, 4042. [CrossRef] [PubMed]

157. Antfolk, M.; Kim, S.H.; Koizumi, S.; Fujii, T.; Laurell, T. Label-free single-cell separation and imaging of cancer cells using an integrated microfluidic system. *Sci. Rep.* **2017**, *7*, 46507. [CrossRef]

158. Liu, Y.; Zhao, W.; Cheng, R.; Puig, A.; Hodgson, J.; Egan, M.; Cooper Pope, C.N.; Nikolinakos, P.G.; Mao, L. Label-free inertial-ferrohydrodynamic cell separation with high throughput and resolution. *Lab Chip* **2021**, *21*, 2738–2750. [CrossRef]

159. Garg, N.; Westerhof, T.M.; Liu, V.; Liu, R.; Nelson, E.L.; Lee, A.P. Whole-blood sorting, enrichment and in situ immunolabeling of cellular subsets using acoustic microstreaming. *Microsyst. Nanoeng.* **2018**, *4*, 17085. [CrossRef]

160. Li, X.-R.; Zhou, Y.-G. Electrochemical detection of circulating tumor cells: A mini review. *Electrochem. Commun.* **2021**, *124*, 106949. [CrossRef]

161. Song, Y.; Shi, Y.; Huang, M.; Wang, W.; Wang, Y.; Cheng, J.; Lei, Z.; Zhu, Z.; Yang, C. Bioinspired Engineering of a Multivalent Aptamer-Functionalized Nanointerface to Enhance the Capture and Release of Circulating Tumor Cells. *Angew. Chem. Int. Ed.* **2019**, *58*, 2236–2240. [CrossRef]

162. Stott, S.L.; Hsu, C.-H.; Tsukrov, D.I.; Yu, M.; Miyamoto, D.T.; Waltman, B.A.; Rothenberg, S.M.; Shah, A.M.; Smas, M.E.; Korir, G.K.; et al. Isolation of circulating tumor cells using a microvortex-generating herringbone-chip. *Proc. Natl. Acad. Sci. USA* **2010**, *107*, 18392–18397. [CrossRef]

163. Sheng, W.; Ogunwobi, O.O.; Chen, T.; Zhang, J.; George, T.J.; Liu, C.; Fan, Z.H. Capture, release and culture of circulating tumor cells from pancreatic cancer patients using an enhanced mixing chip. *Lab Chip* **2014**, *14*, 89–98. [CrossRef] [PubMed]

164. Wu, Z.; Pan, Y.; Wang, Z.; Ding, P.; Gao, T.; Li, Q.; Hu, M.; Zhu, W.; Pei, R. A PLGA nanofiber microfluidic device for highly efficient isolation and release of different phenotypic circulating tumor cells based on dual aptamers. *J. Mater. Chem. B* **2021**, *9*, 2212–2220. [CrossRef] [PubMed]

165. Pulikkathodi, A.K.; Sarangadharan, I.; Hsu, C.-P.; Chen, Y.-H.; Hung, L.-Y.; Lee, G.-Y.; Chyi, J.-I.; Lee, G.-B.; Wang, Y.-L. Enumeration of circulating tumor cells and investigation of cellular responses using aptamer-immobilized AlGaN/GaN high electron mobility transistor sensor array. *Sens. Actuators B Chem.* **2018**, *257*, 96–104. [CrossRef]

166. Zhang, Y.; Wang, Z.; Wu, L.; Zong, S.; Yun, B.; Cui, Y. Combining Multiplex SERS Nanovectors and Multivariate Analysis for In Situ Profiling of Circulating Tumor Cell Phenotype Using a Microfluidic Chip. *Small* **2018**, *14*, 1704433. [CrossRef]

167. Cheng, S.-B.; Xie, M.; Chen, Y.; Xiong, J.; Liu, Y.; Chen, Z.; Guo, S.; Shu, Y.; Wang, M.; Yuan, B.-F.; et al. Three-Dimensional Scaffold Chip with Thermosensitive Coating for Capture and Reversible Release of Individual and Cluster of Circulating Tumor Cells. *Anal. Chem.* **2017**, *89*, 7924–7932. [CrossRef]

168. Jiang, X.; Wong, K.H.K.; Khankhel, A.H.; Zeinali, M.; Reategui, E.; Phillips, M.J.; Luo, X.; Aceto, N.; Fachin, F.; Hoang, A.N.; et al. Microfluidic isolation of platelet-covered circulating tumor cells. *Lab Chip* **2017**, *17*, 3498–3503. [CrossRef]

169. Zeinali, M.; Murlidhar, V.; Fouladdel, S.; Shao, S.; Zhao, L.; Cameron, H.; Bankhead, A., III; Shi, J.; Cuneo, K.C.; Sahai, V.; et al. Profiling Heterogeneous Circulating Tumor Cells (CTC) Populations in Pancreatic Cancer Using a Serial Microfluidic CTC Carpet Chip. *Adv. Biosyst.* **2018**, *2*, 1800228. [CrossRef]

170. Yin, J.; Mou, L.; Yang, M.; Zou, W.; Du, C.; Zhang, W.; Jiang, X. Highly efficient capture of circulating tumor cells with low background signals by using pyramidal microcavity array. *Anal. Chim. Acta* **2019**, *1060*, 133–141. [CrossRef]

171. Kermanshah, L.; Poudineh, M.; Ahmed, S.; Nguyen, L.N.M.; Srikant, S.; Makonnen, R.; Pena Cantu, F.; Corrigan, M.; Kelley, S.O. Dynamic CTC phenotypes in metastatic prostate cancer models visualized using magnetic ranking cytometry. *Lab Chip* **2018**, *18*, 2055–2064. [CrossRef]

172. Sun, N.; Li, X.; Wang, Z.; Li, Y.; Pei, R. High-purity capture of CTCs based on micro-beads enhanced isolation by size of epithelial tumor cells (ISET) method. *Biosens. Bioelectron.* **2018**, *102*, 157–163. [CrossRef]

MDPI

St. Alban-Anlage 66

4052 Basel

Switzerland

www.mdpi.com

Biosensors Editorial Office

E-mail: biosensors@mdpi.com

www.mdpi.com/journal/biosensors